Air Pollution XXIX

WIT*PRESS*

WIT Press publishes leading books in Science and Technology.
Visit our website for the current list of titles.
www.witpress.com

WIT*eLibrary*

Home of the Transactions of the Wessex Institute.
Papers contained in this volume are archived in the WIT eLibrary in volume 252 of WIT Transactions on Ecology and the Environment (ISSN 1743-3541).
The WIT eLibrary provides the international scientific community with immediate and permanent access to individual papers presented at WIT conferences.
Visit the WIT eLibrary at www.witpress.com.

TWENTY NINTH INTERNATIONAL CONFERENCE ON MODELLING, MONITORING AND MANAGEMENT OF AIR POLLUTION

Air Pollution 2021

Conference Chairmen

Juan Casares
University of Santiago de Compostela, Spain
Member of WIT Board of Directors

James Longhurst
University of the West of England, UK

Joanna Barnes
University of the West of England, UK

International Scientific Advisory Committee

Organised by

Wessex Institute, UK

University of The West of England, UK

University of Santiago de Compostela, Spain

Sponsored by

WIT Transactions on Ecology and the Environment

International Journal of Environmental Impacts

WIT Transactions

Wessex Institute
Ashurst Lodge, Ashurst
Southampton SO40 7AA, UK

Senior Editors

Editorial Board

Air Pollution XXIX

Editors

Juan Casares
University of Santiago de Compostela, Spain
Member of WIT Board of Directors

James Longhurst
University of the West of England, UK

Joanna Barnes
University of the West of England, UK

WIT*PRESS* Southampton, Boston

Editors:

Juan Casares
University of Santiago de Compostela, Spain
Member of WIT Board of Directors

James Longhurst
University of the West of England, UK

Joanna Barnes
University of the West of England, UK

Published by

WIT Press
Ashurst Lodge, Ashurst, Southampton, SO40 7AA, UK
Tel: 44 (0) 238 029 3223; Fax: 44 (0) 238 029 2853
E-Mail: witpress@witpress.com
http://www.witpress.com

For USA, Canada and Mexico

Computational Mechanics International Inc
25 Bridge Street, Billerica, MA 01821, USA
Tel: 978 667 5841; Fax: 978 667 7582
E-Mail: infousa@witpress.com
http://www.witpress.com

British Library Cataloguing-in-Publication Data

A Catalogue record for this book is available
from the British Library

ISBN: 978-1-78466-439-8
eISBN: 978-1-78466-440-4
ISSN: 1746-448X (print)
ISSN: 1743-3541 (on-line)

The texts of the papers in this volume were set individually by the authors or under their supervision. Only minor corrections to the text may have been carried out by the publisher.

Preface

Air Pollution 2021 is the 29th Annual Meeting in the successful series of international conferences organised by the Wessex Institute of Technology and the University of the West of England, Bristol concerned with advances in Modelling, Monitoring and Management of Air Pollution.

The series started in Monterrey (1993) and continued in Barcelona (1994), Halkidiki (1995), Toulouse (1996), Bologna (1997), Genova (1998), San Francisco (1999), Cambridge (2000), Ancona (2001), Segovia (2002), Catania (2003), Rhodes (2004), Cordoba (2005), New Forest (2006), Algarve (2007), Skiathos (2008), Tallinn (2009), Kos (2010), Malta (2011), A Coruna (2012), Siena (2013), Split (2014), Valencia (2015), Crete (2016), Cadiz (2017), Naples (2019) and for the first time was delivered in an online format in 2020.

The 29th meeting of the International Conference on Modelling, Monitoring and Management of Air Pollution was delivered online from the 26th to the 28th of October 2021. It was hoped that the meeting might take place face to face but the continuing Covid-19 pandemic meant that this was not possible. That the conference was delivered successfully in an online format was due to the many WIT colleagues who worked hard to ensure that the event provided effective virtual delegate interaction.

The conference brought together contributions from scientists from around the world to present recent work on various aspects of the air pollution phenomena. Discussion ranged across the terrain of air quality modelling, monitoring and management with specific sessions covering modelling studies, indoor air pollution, emission studies, air pollution management, policy and legislation and monitoring approaches. In addition, there were a number of novel papers presented on the interrelations between Covid-19 restrictions and air quality.

Each of the meetings in this series have provided opportunities for delegate discussion about the many and varied challenges of air pollution. This discussion has been enriched and informed by the perspectives of the many international delegates at the meetings. The conference findings and conclusions continue to serve the needs of the air pollution science and policy communities for a rapid, authoritative and accessible volume.

Air pollution sources and impacts remain one of the most challenging problems facing the international community. The series has demonstrated the widespread nature of the air pollution phenomena and has contributed to the evolving understanding of the science and policy contexts of air pollution management. The importance of the conference themes was further highlighted in 2021 by the publication of the World Health Organisation's Air Quality Guidelines (WHO, 2021a). These revised and updated guidelines cover particulate matter, nitrogen dioxide, sulphur dioxide, ozone and carbon monoxide and relate to both ambient (outdoor) and indoor (but not occupational) exposure. In most cases the guidelines have become much more stringent and challenging for those concerned with managing air pollution at a national or local level. What was once thought to be an appropriate concentration to provide public health protection from a named air pollutant is no longer the case. Whist these guidelines do not consider the technical, political or economic barriers to implementation that air quality managers grapple with, they do provide the reference point for public health protection.

It will be interesting to see how nation states consider these new guidelines in the setting of future air quality regulations. n publishing the new guidelines the WHO warned that exceeding the new air quality guideline levels is associated with significant risks to health whilst adhering to them could save millions of lives (WHO, 2021b). The WHO also note that action on air pollution will have co-benefits for action to mitigate emissions of greenhouse gases.

Future conferences will no doubt have plenty to discuss when considering the appropriate level of protection that national air quality standards provide. A recent European Environment Agency study has illustrated the scale of health benefits accruing from the implementation of these new guidelines (EEA, 2021a). If the new air quality guidelines for fine particulate matter had been met across Europe in 2019, then this would have delivered a 72% reduction in premature deaths across the EU compared with 2005 levels. The EEA continues to describe air pollution is the biggest environmental health risk in Europe (EEA, 2021b). The WHO also describe air pollution in this way and note that 'almost 80% of deaths related to $PM_{2.5}$ could be avoided in the world if the current air pollution levels were reduced to those proposed in the updated guideline'(WHO, 2021b).

The WHO estimates that across the world ambient air pollution causes some 4.2 million premature deaths per year, principally due to fine particulates and nitrogen dioxide, although more recent studies suggest that this may be a conservative estimate (Lelieveld *et al.*, 2020; Vohra *et al.*, 2021). This burden of disease is concentrated in Low and Middle Income (LMI) countries and in urban environments. With recent medical research pointing to a range of health impacts arising from exposure to air pollution across the human life course it is clear that much needs to be done to limit the impacts on health and the environment and to create effective policy and regulatory instruments.

In high income countries but not necessarily in LMIs many of the easy air quality policy wins have been secured and the causes of the remaining challenges are deeply entrenched and resistant to change. Policy development and its effective implementation is becoming tougher as the direct costs, financial and political, of improving air quality increase. Yet the public's concern is increasing. This challenge is one the air quality community must recognise, confront and overcome. It will require recognition that air quality is a multifaceted, emergent problem, necessitating the integration of social science and natural science perspective to inform policy and to avoid unintended consequences. A systems approach will be needed to derive appropriate solutions in the long and short term. Key to success will be the integration of effective air quality management into land use, transport, industrial and economic planning processes. Air quality can be seen as an emergent property arising from the system of anthropogenic activities and resulting from policy decisions from differing governmental spheres, often disconnected and asynchronous. True management of air pollution can be only be achieved through a systematic understanding of the often multiple causes, feedback loops and impacts. Ultimately, air pollution is a choice society makes through its collective and individual behaviours and collective practices. However, the consequences of those behaviours and choices will play out in many different ways with those who are least able to exercise choice having air pollution concentrations imposed upon them. However, history shows that concerted, collective and sustained action can lead to dramatic improvements in air quality.

There remains much to do and the conference series will continue to play an important role in providing an opportunity for an international audience to discuss both long-standing and emergent issues in air pollution science and policy.

The papers selected for presentation and published in this volume are part of the Transactions on Ecology and the Environment. They have been archived online in the WIT eLibrary (www.witpress.com/elibrary) where they are easily and permanently available to the international scientific community. These collected papers provide an invaluable record of the development of science and policy pertaining to air pollution.

The Editors wish to thank the authors for their contributions to the conference and to acknowledge the assistance of the eminent members of the International Scientific Advisory Committee for their support for the conference and in particular for their peer reviewing of the manuscripts.

The 29th Conference will be the last devoted solely to air pollution. The 30th conference will be held jointly with WIT's sister conference on Water Pollution. The Editors look forward to welcoming delegates to Milan in July 2022 for this important meeting jointly organised by the Wessex Institute of Technology, the Polytechnic of Milan and the University of the West of England, UK

The Editors, December 2021

References

European Environment Agency. (2021a). *Air quality in Europe 2021*. 15/2021. EEA. https://www.eea.europa.eu/publications/air-quality-in-europe-2021/

European Environment Agency. (2021b). *Air pollution*. https://www.eea.europa.eu/themes/air

Lelieveld, J., Pozzer, A., Pöschl, U., Fnais, M., Haines, A., Münzel, T., 2020. Loss of life expectancy from air pollution compared to other risk factors: a worldwide perspective. *Cardiovascular Research* **116**, pp. 1910–1917. https://doi.org/10.1093/cvr/cvaa025

Vohra, K., Vodonos, A., Schwartz, J., Marais, E.A., Sulprizio, M.P., Mickley, L.J., 2021. Global mortality from outdoor fine particle pollution generated by fossil fuel combustion: Results from GEOS-Chem. *Environmental Research* **195**, 110754. https://doi.org/10.1016/j.envres.2021.110754

World Health Organization. (2021a). *WHO global air quality guidelines: particulate matter (PM2.5 and PM10), ozone, nitrogen dioxide, sulfur dioxide and carbon monoxide*. World Health Organization. https://apps.who.int/iris/handle/10665/345329

World Health Organization. (2021b) *News*. https://www.who.int/news/item/22-09-2021-new-who-global-air-quality-guidelines-aim-to-save-millions-of-lives-from-air-pollution

Contents

Section 6: Global, regional and local studies

SECTION 1
AIR POLLUTION MODELLING

HOW EFFECTIVE ARE NATURE-BASED SOLUTIONS IN DIFFERENT ENVIRONMENTS?

ANA ASCENSO[1], CARLA GAMA[1], PETER ROEBELING[1,2] & ANA ISABEL MIRANDA[1]
[1]CESAM – Centre for Environmental and Marine Studies, Department of Environment and Planning, University of Aveiro, Portugal
[2]Wageningen Economic Research, Wageningen University and Research, The Netherlands

ABSTRACT

Air pollution is a major global concern in urban areas and it is considered the greatest environmental risk to health. Nature-based solutions (NBS) can help improve air quality and reduce the urban heat island effect. The impacts of urban vegetation on air quality and ambient temperature depend on many factors, such as vegetation type, location, pollutants, climate conditions and topography. Therefore, the implementation of NBS needs to be tailored for each city. Within the context of the H2020 UNaLab project, the main objective of this work is to assess the potential of NBS to improve air quality across three European cities with different climates: (i) Eindhoven, The Netherlands; (ii) Tampere, Finland; and (iii) Genova, Italy. The WRF-CHEM model was applied for the hottest week in a present climate reference year. The baseline case (without NBS) and two NBS scenarios were simulated for each city. These scenarios (green roofs and green parks) were implemented in the model by modifying the land-use type and the emissions of the model grid cells. According to the model results, the city that least benefited from NBS was Tampere with an average reduction of 5% in surface temperature, and 1% in nitrogen dioxide (NO_2) and ozone (O_3) concentrations. Temperature-wise Genova and Eindhoven had similar results, approximately 6% reduction, while Genova showed the largest improvement in NO_2 (12%). These results indicate that NBS are more effective in high temperature and high air pollution cities, such as Genova. Moreover, this study reinforces the importance of studying case-specific solutions, considering environmental characteristics and challenges.

Keywords: air quality, temperature, green roofs, green parks, modelling, urban areas.

1 INTRODUCTION

Urban areas need to achieve a balance between sustainable development and population growth. Climate change is making this balance even more difficult to attain. The increase in duration, intensity and frequency of extreme weather events will lead to more environmental, social and economic problems [1]. Among these environmental problems are air pollution and urban heating, which are causing high morbidity and mortality [2]. Air pollution is considered by the World Health Organization (WHO) to be the greatest environmental risk to health. Outdoor air pollution is responsible for 4.2 million worldwide deaths every year [3]. The most recent numbers show that 95%–98% of the European Union population was exposed to ozone (O_3) concentrations above the WHO guideline values set for the protection of human health [4].

Cities are more vulnerable to climate change and are also the most polluted areas. Urban areas need new and innovative approaches, such as nature-based solutions (NBS), to increase city resilience to climate change and promote better air quality [5]. The European Commission defines nature-based solutions as "solutions that are inspired and supported by nature, which are cost-effective, simultaneously provide environmental, social and economic benefits and help build resilience". NBS, such as street vegetation, green roofs

WIT Transactions on Ecology and the Environment, Vol 252, © 2021 WIT Press
www.witpress.com, ISSN 1743-3541 (on-line)
doi:10.2495/AIR210011

and walls, can provide insulation and shade to buildings and therefore have the potential to reduce energy demands, by cutting back on heating and cooling needs, and indirectly reduce air pollutant emissions [6].

It is common to assess the impact of NBS on air quality by estimating the amount of pollutant removed through deposition [7]–[9]. However, there are also various studies assessing the impact of NBS on the urban air quality based on modelling approaches [6], [10]–[16], which allows the calculation of pollutants concentrations in the air integrating the effects of NBS on temperature and dispersion conditions. Most of the available air quality modelling studies concern microscale (in the range of metres) [10], [17], [18], but there are also several mesoscale (in the range of kilometres) modelling studies [6], [12], [13]. Therefore, the estimation of the NBS potential to improve air quality can change depending on the methodology used.

Moreover, it is known that NBS can reduce air pollution in different types of environments, and many ongoing projects are testing NBS impacts for different pollutant and climate conditions [19]. The implementation of NBS needs to be context-adapted, because the same solution can have a different impact depending on the specific characteristics of the urban area, like its land cover, population density and microclimate.

This work aims to assess how the impact of NBS can differ among cities with different characteristics, namely: (i) Eindhoven, The Netherlands; (ii) Tampere, Finland; and (iii) Genova, Italy. A mesoscale modelling system with a 1 km × 1 km spatial resolution was applied for two types of NBS, green roofs (GR) and green parks (GP), to assess their impact on temperature and air quality. Nature-based solutions are a common strategy to reduce urban heat in the cities, moreover, temperature changes affect directly air pollutant concentration. The designed NBS scenarios were tailored to each city following a co-creation process with the stakeholders.

2 CASE STUDIES

Three European cities were chosen due to their track record in smart and citizen-driven solutions for sustainable development, growing urbanization and different climate zones: (i) Eindhoven, The Netherlands; (ii) Genova, Italy; and (iii) Tampere, Finland.

The City of Eindhoven is one of the five largest cities in The Netherlands with 230,000 inhabitants. Eindhoven has a temperate marine climate/oceanic climate (Köppen "Cfb") with slightly warmer summers and colder winters than the coastal parts of the Netherlands. The rapid urbanisation in recent years has led to the disappearance of significant blue zones. Currently, Eindhoven environmental challenges include flooding, urban heat stress and air pollution [20], which will worsen with climate change.

Genova is the largest city in the Liguria Region, with 607,000 inhabitants. Genova is characterised by a narrow coastal zone with hills and steep mountains in the backcountry. The city climate is subtropical and Mediterranean (Köppen "Cfa" and "Csa"). Due to extreme weather and a highly urbanised landscape, the city faces environmental challenges such as water management issues, heat stress, and water and air pollution [20]. With climate change, the city will have to face hazards like wildland fires, heatwaves, landslides and surface floods.

Tampere is the third-largest city in Finland and the largest inland centre in the Nordic countries. With 238,140 inhabitants and a population density of 448 per square kilometre, Tampere is one of the three most rapidly developing regions in-country and the centre of Finnish industry. Tampere has a borderline humid continental climate/subarctic climate (Köppen "Dfb" and "Dfc"). Twenty-four per cent of Tampere's surface area is water and 76% is land. Tampere's environmental challenges are related to flooding and stormwater management, as well as water pollution and reduced biodiversity [20], [21]. With climate change, the city will also face colder winters and warmer summers.

3 METHODOLOGY

This section describes the methodology for assessing the impacts of NBS on temperature and air quality, by introducing the WRF-CHEM model and the chosen simulation setup, and by characterising each NBS scenario.

3.1 WRF-CHEM modelling system

WRF-CHEM is the weather research and forecasting (WRF) model coupled with chemistry. It is a fully coupled mesoscale online model developed by Grell et al. [22] in which the air quality component is consistent with the meteorological component. Both components use the same transport scheme (vertical and horizontal coordinates), physics schemes for sub-grid scale transport and the same time step, and therefore interpolation in space and time is not required. The WRF dynamic solver integrates the compressible, non-hydrostatic Euler equations in flux form using variables that have conservation properties [23]. Detailed information on the model structure and its underlying equations are available in Skamarock et al. [24].

Three online-nested domains with increasing resolution at a downscaling ratio of five are used–with the coarser domain of 25 km × 25 km horizontal resolution covering Europe and part of the North Atlantic Ocean, and the innermost domain of 1 km × 1 km horizontal resolution covering the area containing the case-study region. The innermost domains have the following dimensions: Eindhoven, 51 km × 41 km; Genova, 41 km × 36 km and Tampere, 46 km × 36 km. The model was forced by ERA-Interim data from the European Centre for Medium-Range Weather Forecasts (ECMWF) global analysis [25], with a horizontal resolution of 1° × 1° and with a 6 h temporal resolution. Furthermore, WRF-CHEM uses as inputs anthropogenic emissions [26], biogenic emissions (MEGAN model; [27]), initial and boundary conditions (MOZART model; [28]), and orography and land use (USGS 24 classes database). The required input data for WRF-CHEM were obtained from each city database as well as from open-source databases. WRF-CHEM output data are hourly three-dimensional fields that comprise a wide range of meteorological variables and pollutant concentrations, such as temperature (T), nitrogen oxides (NO_X) and ozone (O_3), the main variables assessed in this study. For a more detailed description of the modelling setup and its evaluation see Ascenso et al. [13].

3.2 NBS scenario description

The simulations were performed for the 7 hottest consecutive days of the representative year of the present/baseline scenario. The representative year is the one with the smallest climate anomaly regarding temperature and precipitation for the period 2012 to 2016. The hottest week was simulated to balance the biggest NBS potential impact with less computational costs. The resulting simulation periods are: Eindhoven, 1–7 August 2013; Genova, 28 July–3 August 2013 and Tampere, 25–31 July 2012.

The tested NBS scenarios were defined with the help of the municipalities during co-creation workshops, in the scope of the UNALAB project. The NBS were implemented in the model by changing land use and emissions for the grid cells (1 km × 1 km) that overlapped with the implementation areas. Note that by modifying the land use information, related variables, such as albedo, were also changed. For the GR scenarios the land use was altered to grass. For the GP scenarios, it was changed to shrubs, and the anthropogenic emissions were reduced to zero since the existing urban areas were converted into green parks. The main characteristics of the considered NBS scenarios are shown in Table 1, and their location and land use information in Fig. 1.

Table 1: Information and description of the tested NBS scenarios (the percentage of area with NBS is relative to the area inside the city boundaries). Legend: GR = green roofs; GP = green park.

City	Sigla	Description	NBS area (% | km^2)	Ref
Eindhoven	EIN_GR	GR in industrial areas	6 | 5	Fig. 1(a)
	EIN_GP	GP in zones of the city lacking in green space	11 | 10	Fig. 1(b)
Genova	GEN_GR	GR in the Lagaccio and S. Teodoro neighborhoods	0.8 | 2	Fig. 1(c)
	GEN_GP	GP on the margins of the Polcevera river	0.4 | 1	Fig. 1(d)
Tampere	TAM_GR	GR in the city centre	0.5 | 1	Fig. 1(e)

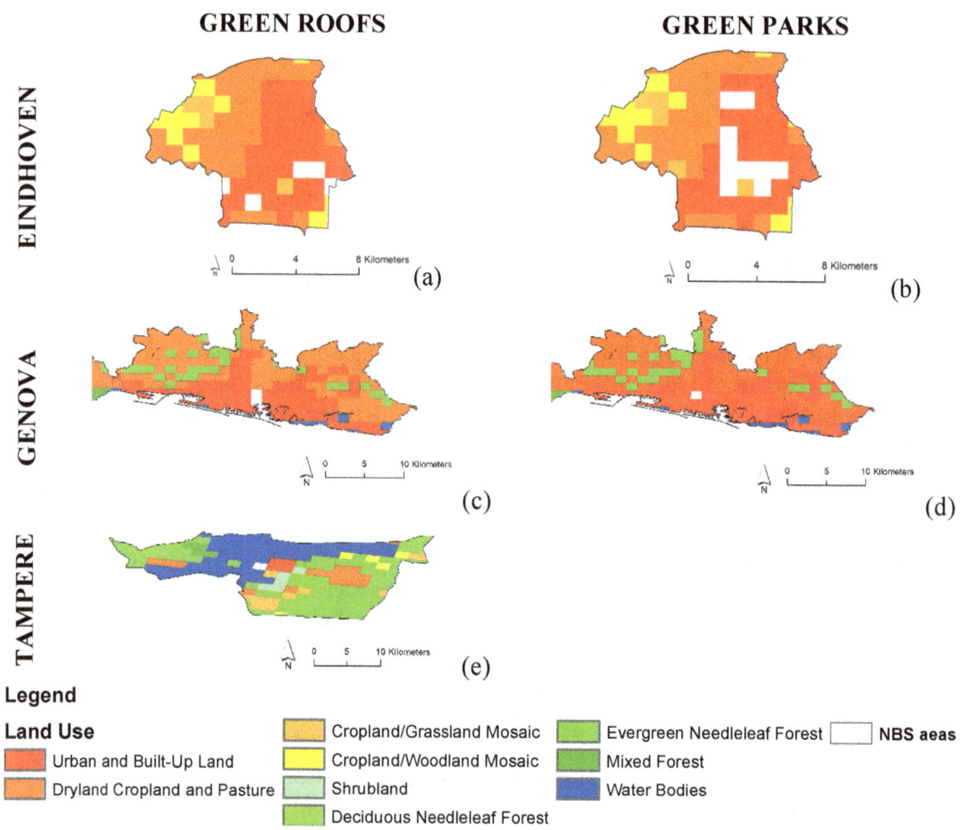

Figure 1: Location of NBS (white areas) and land use information, for each scenario and case study.

Eindhoven scenarios stand out for having the largest area covered with NBS while being the smallest municipality. Whereas for Genova and Tampere the NBS scenarios have similar dimensions that correspond to a small percentage of municipality area. Tampere has already extensive green areas and considering the city environmental challenges, the municipality did not suggest a green park scenario. The case-specific solutions are a great way to compare the different cities' potential to benefit from NBS implementation. This assessment was done based on the model grid cells average, thus the results are normalized for 1 km^2 of NBS implementation. This way it is possible to get a practical view of how effective nature-based solutions in different environments are.

4 RESULTS AND DISCUSSION

In this section the model results for T, NO_2 and O_3 concentrations are analysed and discusses based on their hourly values, spatial averages and daily profiles. This assessment is based on values from the model cells where the NBS areas were implemented (see Fig. 1).

4.1 Temperature

During the study period, the average temperature for the baseline scenario was 21.9°C in Eindhoven, 24.9°C in Genova and 20.2°C in Tampere. Table 2 presents the average temperature values for the different cities and scenarios.

Table 2: Average temperature for the case study cities and the simulated scenarios. Legend: GR = green roofs; GP = green park.

City	Average temperature (°C)		
	Baseline	GR	GP
Eindhoven	21.9	21.6	21.5
Genova	24.9	24.1	24.6
Tampere	20.2	19.3	

All the cities results indicate a temperature averaged reduction with the tested NBS.

However, the impact of NBS can vary greatly along the day. Fig. 2 shows the temperature daily profile for the baseline (°C; black line), the NBS scenarios (°C; dotted line) and the relative differences (%; grey line).

The temperature reduction occurs mainly during night. This is explained by the change from the urban land use, with low albedo characteristics, which promotes the release of energy during the night, to green surfaces that accumulate less energy. This pattern is more noticeable for the Genova and Tampere NBS scenarios; these cities had the least amount of precipitation during the study period and received the highest amount of solar radiation, which promoted more evapotranspiration and thus a larger temperature reduction. Green surfaces increase the evapotranspiration rates enhancing the latent heat flux and reducing the sensible heat flux [29], which leads to a phenomenon called the "cool-island" effect [30].

The NBS implementation resulted in an increase in temperature at dawn, especially in Genova and Tampere. At these morning hours, the reduction in the sensible heat flux that occurs during the night shifts to an increase.

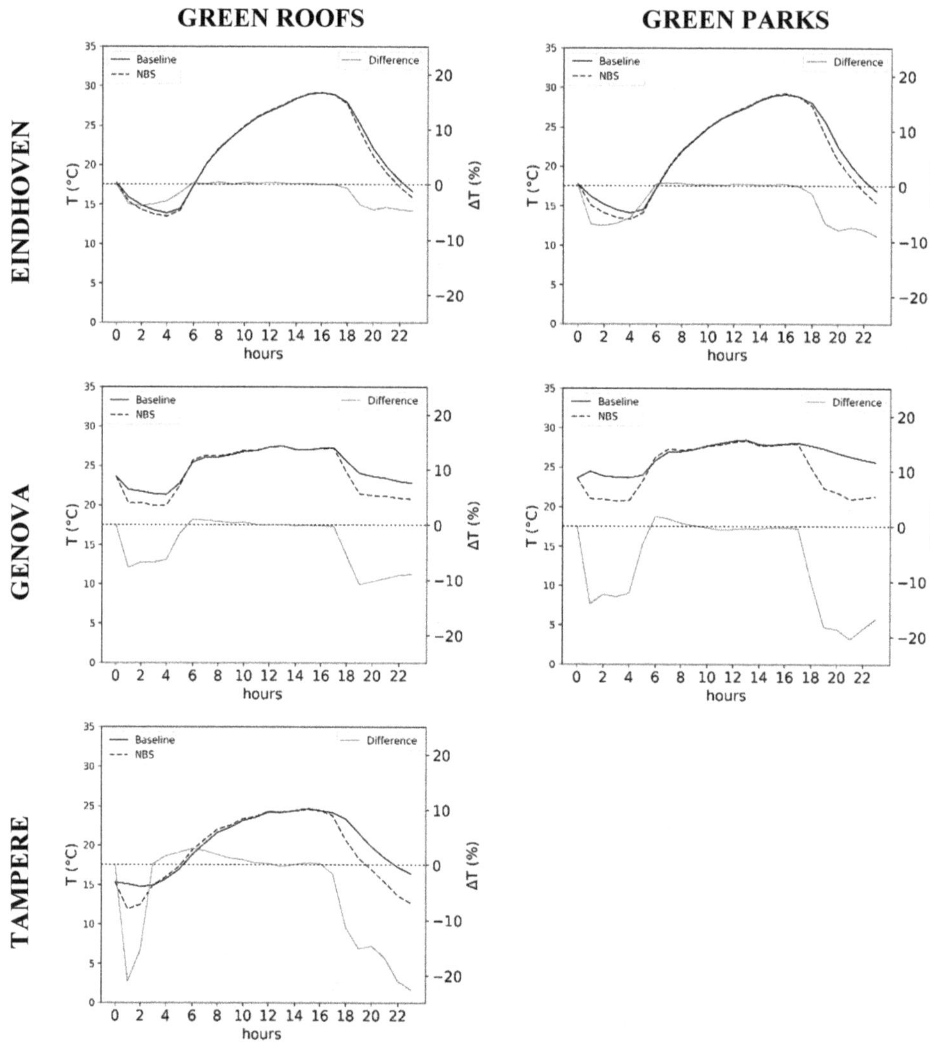

Figure 2: Temperature daily profile for baseline (°C; black line), NBS scenarios (°C; dotted line) and relative differences (%; grey line), averaged through the NBS areas (UTC time).

4.2 Air quality

Hourly NO_2 and O_3 simulated values were treated and are here presented and discussed. Among the case studies, Genova has the highest baseline NO_2 concentration values, twice exceeding the hourly limit value for the protection of human health, in one week. For Eindhoven and Tampere, NO_2 concentrations are low and do not present a risk for human health. However, Eindhoven has the highest ozone levels, with an average of 82 $\mu g \cdot m^{-3}$ maximum daily eight-hour mean and an hourly maximum of 108 $\mu g \cdot m^{-3}$. Tampere is the case study with the best air quality.

The overall NBS impacts on air quality are very small. The average difference between baseline and NBS scenarios is smaller than 1 $\mu g{\cdot}m^{-3}$, except for GEN_GP which had an averaged NO_2 reduction of 5 $\mu g{\cdot}m^{-3}$ (–12%) and for GEN_GR with a reduction in O_3 of 2 $\mu g{\cdot}m^{-3}$ (–3%). GP scenarios show the bigger impact in NO_2 concentrations due to the change in anthropogenic emissions, which were considered zero over the area where the GP was implemented.

Fig. 3 shows the NO_2 daily profile for the baseline ($\mu g{\cdot}m^{-3}$; black line), the NBS scenarios ($\mu g{\cdot}m^{-3}$; dotted line), and the relative differences (%; grey line).

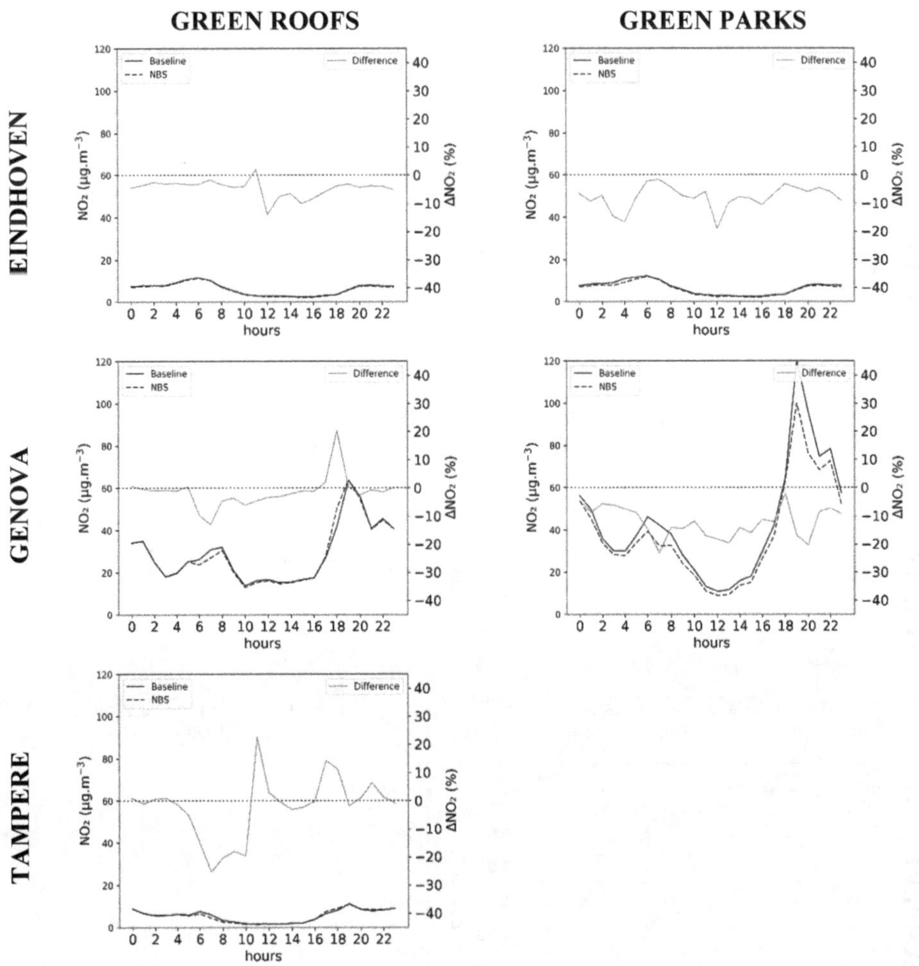

Figure 3: NO_2 daily profile for baseline ($\mu g{\cdot}m^{-3}$; black line), NBS scenarios ($\mu g{\cdot}m^{-3}$; dotted line) and relative differences (%; grey line), averaged through the NBS areas (UTC time).

Fig. 3 shows a general reduction of NO_2 concentration levels along the daily profile, in particular for the cities of Genova and Tampere. The impact for the city of Eindhoven is negligible. For the GR scenarios this reduction is larger at dawn, probably due to increase in temperature (seen in Fig. 2), which in turn promotes an increase in the boundary layer height, enabling air pollutants dispersion. Similarly, at the end of the day, the reduction in temperature led to an increase in NO_2 concentration for the GEN_GR and TAM_GR scenarios.

When considering the GP impacts for Eindhoven and Genova, the reducing trend is also present, but residual for Eindhoven. On the other hand, the reduction of NO_2 in Genova shows the effect of changing emissions along the day, with a larger reduction in the peak morning and afternoon hours; the impact of removing emissions was more relevant than the effect of NBS.

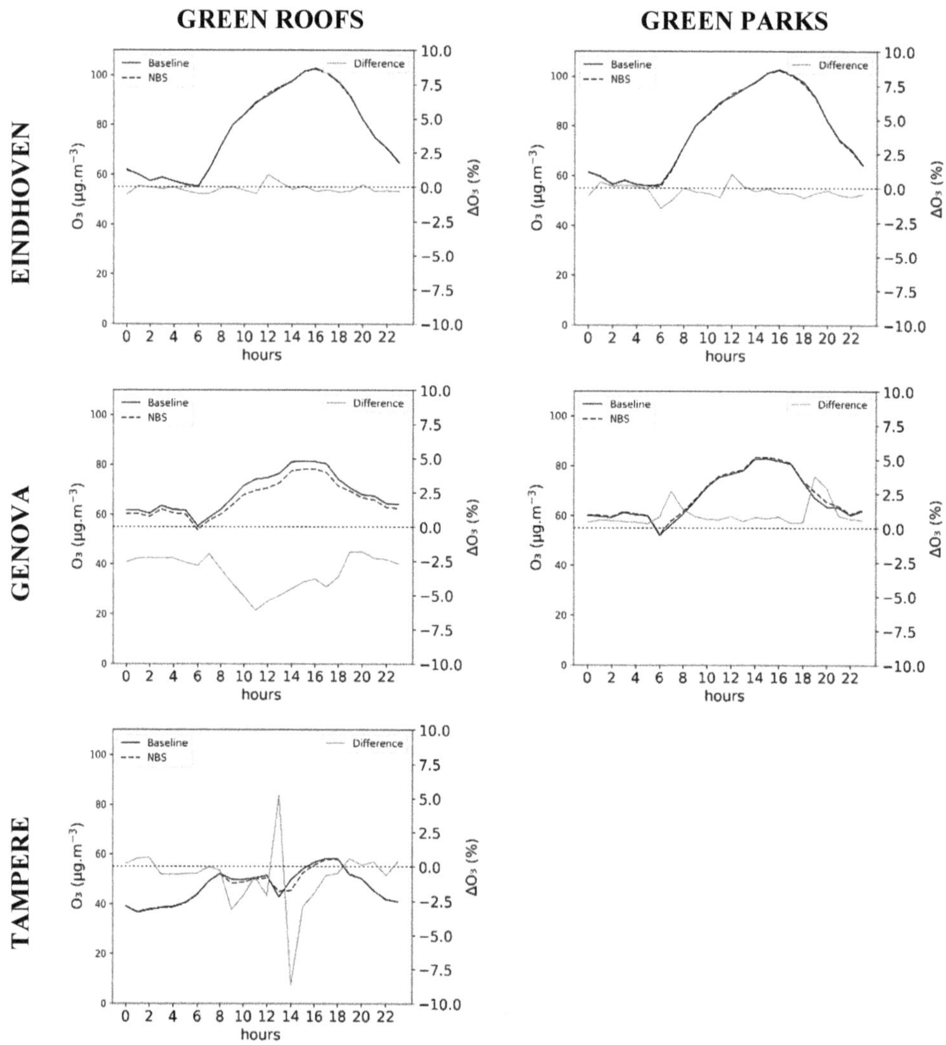

Figure 4: O_3 daily profile for baseline (µg·m^{-3}; black line), NBS scenarios (µg·m^{-3}; dotted line) and relative differences (%; grey line), averaged through the NBS areas (UTC time).

Fig. 4 shows the O_3 daily profile for the baseline (μg·m^{-3}; black line), the NBS scenarios (μg·m^{-3}; dotted line) and the relative differences (%; grey line).

O_3 concentration was the indicator who least benefited from the NBS implementation. Once again, the impact for the city of Eindhoven, which is the city where ozone levels are the main air quality problem, is minimal. This can be explained by the very low baseline NO_x values calculated by the model, resulting in small differences (<1 μg·m^{-3}), and that also originated a small impact in ozone concentrations as well.

The GEN_GP scenario resulted in an average increase of 1% in O_3 concentrations. In urban areas where NO_2 concentrations are high, such as Genova, O_3 increases with the reduction of NO_x, because of the NOx/VOC ratio in the atmosphere. Hence, the decrease in NO concentrations during the day, especially during dawn and nightfall (due to emission reduction), led to less ozone titration. The opposite effect can be seen in GEN_GR that showed the most ozone reduction with an average decrease of 3% (–2.4 μg·m^{-3}). The NO concentrations increased during the day leading to more ozone removal through NO titration.

Tampere has the lowest O_3 concentration of the three cities, and therefore the small differences can correspond to high percentages. The peaks in the averaged daily profile are related to high hourly differences reached on a particular day when the highest concentrations of the week occurred.

Table 3 contains a summary of the results based on the model cell with the highest impact along the study period.

Table 3: Summary of the NBS impact results for temperature and air quality for all cities; maximum average of temporal relative differences (%), i.e. for the model cell with the highest impact along the study period.

	GREEN ROOFS						GREEN PARKS					
	T		NO_2		O_3		T		NO_2		O_3	
	%	°C	%	μg·m^{-3}	%	μg·m^{-3}	%	°C	%	μg·m^{-3}	%	μg·m^{-3}
Eindhoven	–6	–1.27	–8	–0.47	0	–0.28	–6	–1.27	–11	–0.7	0	–0.33
Genova	–6	–1.56	–1	–0.24	–3	–2.39	–6	–1.67	–12	–5.11	1	0.64
Tampere	–5	–0.93	–1	–0.08	–1	–0.39						

Despite the different environmental characteristics, the relative differences show similar results between cities and scenarios. The impact of NBS on the temperature was approximately 5–6%. For NO_2 the GR scenario impact was the same for Genova and Tampere (–1%) and was slightly higher (–8%) for Eindhoven, where there was a constant reduction along the average daily profile. The GP scenario, for Eindhoven and Genova, resulted in a reduction of 11–12% (mainly due to emission reduction). Regarding O_3, and despite the very small changes of the concentration values, the GP scenario had the least impact 0–1%, and for GR the reduction ranged from 0–3%.

Overall, Genova is the city that will benefit the most if these solutions are implemented. Genova is challenged with poor air quality and high temperatures, and the temperature at night does not decrease much. Thus, Genova has an interesting potential for NBS implementation.

5 CONCLUSIONS

The WRF-CHEM model was applied with high-resolution to quantify the impacts of green roofs and green parks on temperature and air quality (NO_2 and O_3), for the hottest week in a reference year of the present climate on three distinct cities. The studied scenarios were proposed by the corresponding municipalities through a co-creation process. This research highlights the importance of considering the daily patterns of the NBS impact to enable more informed decisions and increase the effectiveness of these solutions.

Research results indicate that NBS can have a small impact on air quality and a bigger impact on temperature. NBS were more effective in areas with high temperatures and poor air quality. Thus, Genova is the city that will benefit the most if these solutions are implemented. However, the relative effects of NBS on temperature were very similar throughout the case studies and scenarios (–5 to –6%) suggesting that, for this particular indicator, environmental conditions are not a major influence on NBS performance. Although the impact of NBS on air quality was small for Eindhoven and Tampere, these cities face other climate and environmental-related challenges and these solutions could still be beneficial.

ACKNOWLEDGEMENTS

The authors would like to thank the municipalities of Eindhoven, Genova and Tampere for providing information on the case studies. This work was supported by the UNaLab project (Grant Agreement No. 730,052, Topic: SCC-2-2016–2017: Smart Cities and Communities Nature-based solutions) and the GENESIS project (PTDC/GESURB/29444/2017). Thanks are also due for the financial support to the PhD grants of A. Ascenso (SFRH/BD/136875/2018) and CESAM (UIDB/50017/2020+UIDP/50017/2020), to FCT/MCTES through national funds, and the co-funding by the FEDER, within the PT2020 Partnership Agreement and Compete 2020.

REFERENCES

[1] IPCC, *Climate Change 2014: Synthesis Report. Contribution of Working Groups I, II and III to the Fifth Assessment Report of the Intergovernmental Panel on Climate Change,* eds R.K. Pachauri & L.A. Meyer, IPCC: Geneva, Switzerland, 151 pp., 2014.

[2] Weilnhammer, V. et al., Extreme weather events in Europe and their health consequences: A systematic review. *International Journal of Hygiene and Environmental Health*, **233**, 113688, 2021. DOI: 10.1016/j.ijheh.2021.113688.

[3] WHO, Ambient (outdoor) air pollution. https://www.who.int/news-room/fact-sheets/detail/ambient-(outdoor)-air-quality-and-health. Accessed on: 2 Aug. 2021.

[4] EEA, Air quality in Europe – 2020 report, No. 09/2020, 162 pp., 2020.

[5] Borrego, C. et al., Air quality, urban fluxes and cities resilience under climate change: A brief overview. *International Journal of Environmental Impacts*, **1**(1), pp. 14–27, 2018. DOI: 10.2495/EI- **1**-N1-14-27.

[6] Rafael, S., Correia, L.P., Ascenso, A., Augusto, B., Lopes, D. & Miranda, A.I. Are green roofs the path to clean air and low carbon cities? *Science of the Total Environment*, **798**, 149313, 2021. DOI: 10.1016/j.scitotenv.2021.149313.

[7] Nowak, D.J., Crane, D.E. & Stevens J.C., Air pollution removal by urban trees and shrubs in the United States. *Urban Forestry and Urban Greening*, **4**, pp. 115–123, 2006. DOI: 10.1016/j.ufug.2006.01.007.

[8] Tiwari, A. et al., Considerations for evaluating green infrastructure impacts in microscale and macroscale air pollution dispersion models. *Science of the Total Environment*, **672**, pp. 410–426, 2019. DOI: 10.1016/j.scitotenv.2019.03.350.

[9] Donateo, A. et al., An evaluation of the performance of a green panel in improving air quality, the case study in a street canyon in Modena, Italy. *Atmospheric Environment*, **247**, 118189, 2021. DOI: 10.1016/j.atmosenv.2021.118189.

[10] Rodrigues, A. et al., Modelling the potential of green infrastructures to reduce the impact of climate change on air quality at microscale. *Springer Proceedings in Complexity*, pp. 147–152, 2020. DOI: 10.1007/978-3-030-22055-6_23.

[11] Augusto, B., Roebeling, P., Rafael, S., Ferreira J., Ascenso A. & Bodilis, C., Short and medium- to long-term impacts of nature-based solutions on urban heat. *Sustainable Cities and Society*, **57**, 102122, 2020. DOI: 10.1016/j.scs.2020.102122.

[12] Rafael, S., Augusto, B., Ascenso, A., Borrego, C. & Miranda, A.I., Re-naturing cities: evaluating the effects on future air quality in the city of Porto. *Atmospheric Environment*, **222**, 117123, 2019. DOI: 10.1016/j.atmosenv.2019.117123.

[13] Ascenso, A. et al., Impacts of nature-based solutions on the urban atmospheric environment: a case study for Eindhoven, The Netherlands. *Urban Forestry and Urban Greening*, **57**, 126870, 2021. DOI: 10.1016/j.ufug.2020.126870.

[14] Fallmann, J., Forkel, R. & Emeis, S., Secondary effects of urban heat island mitigation measures on air quality. *Atmospheric Environment*, **125**, pp. 199–211, 2016. DOI: 10.1016/j.atmosenv.2019.117123.

[15] Arghavani, S., Malakooti, H. & Bidokhti, A.A., Numerical evaluation of urban green space scenarios effects on gaseous air pollutants in Tehran metropolis based on WRF-Chem model. *Atmospheric Environment*, **214**, 116832, 2019. DOI: 10.1016/j.atmosenv.2019.116832.

[16] Tallis, M.J., Amorim, J.H., Calfapietra, C., Freer-Smith, P., Grimmond, S. & Kotthaus, S., The impacts of green infrastructure on air quality and temperature: Planning. *Handbook on Green Infrastructure: Planning, Design and Implementation*, pp. 30–49, 2015. DOI: 10.4337/9781783474004.00008.

[17] Rafael, S., Vicente, B., Rodrigues, V., Miranda, A.I., Borrego, C. & Lopes, M., Impacts of green infrastructures on aerodynamic flow and air quality in Porto's urban area. *Atmospheric Environment*, **190**, pp. 317–330, 2018. DOI: 10.1016/j.atmosenv.2018.07.044.

[18] Amorim, J.H., Rodrigues, V., Tavares, R., Valente, J. & Borrego, C., CFD modelling of the aerodynamic effect of trees on urban air pollution dispersion. *Science of the Total Environment*, **461–462**, pp. 541–551, 2013. DOI: 10.1016/j.scitotenv.2013.05.031.

[19] European Commission. *Evaluating the Impact of Nature-based Solutions: A Handbook for Practitioners*. Publications Office of the European Union: Luxembourg, 2021. DOI: 10.2777/244577.

[20] UNALAB, Our Cities R1 | UNaLab. https://unalab.eu/en/our-cities. Accessed on: 2 Aug. 2021.

[21] Municipality of Tampere, Information on Tampere, 2015. https://www.tampere.fi/en/city-of-tampere/information-on-tampere.html. Accessed on: 2 Aug. 2021.

[22] Grell G.A. et al., Fully coupled 'online' chemistry within the WRF model. *Atmospheric Environment*, **39**(37), pp. 6957–6975, 2005. DOI: 10.1016/j.atmosenv.2005.04.027.

[23] Ooyama, K.V., A thermodynamic foundation for modeling the moist atmosphere. *Journal of the Atmospheric Sciences*, **47**(21), pp. 2580–2593, 1990. DOI: 10.1175/1520-0469(1990)047<2580:ATFFMT>2.0.CO;2.

[24] Skamarock, C. et al., A description of the advanced research WRF version 3 (No. NCAR/TN-475+STR). University Corporation for Atmospheric Research, 2008. DOI: 10.5065/D68S4MVH.
[25] Dee, D.P. et al., The ERA-Interim reanalysis: configuration and performance of the data assimilation system. *Quarterly Journal of the Royal Meteorological Society*, **137**(656), pp. 553–597, 2011. DOI: 10.1002/qj.828.
[26] EMEP/CEIP, Emissions as used in EMEP models, 2015. www.ceip.at/webdab-emission-database/emissions-as-used-in-emep-models. Accessed on: 2 Aug. 2021.
[27] Guenther, A., Karl, T., Harley, P., Wiedinmyer, C., Palmer, P.I. & Geron, C., Estimates of global terrestrial isoprene emissions using MEGAN (Model of emissions of gases and aerosols from nature). *Atmospheric Chemistry and Physics*, **6**(11), pp. 3181–3210, 2006. DOI: 10.5194/acpp–6-3181-2006.
[28] Emmons, L.K. et al., Description and evaluation of the model for ozone and related chemical tracers, version 4 (MOZART-4). *Geoscientific Model Development*, **3**(1), pp. 43–67, 2010. DOI: 10.5194/gmd-3-43-2010.
[29] Rafael, S., Martins, H., Sá, E., Carvalho, D., Borrego, C. & Lopes, M. Influence of urban resilience measures in the magnitude and behaviour of energy fluxes in the city of Porto (Portugal) under a climate change scenario. *Science of the Total Environment*, **566–567**, pp. 1500–1510, 2016. DOI: 10.1016/j.scitotenv.2016.06.037.
[30] Santamouris, M., Cooling the cities: A review of reflective and green roof mitigation technologies to fight heat island and improve comfort in urban environments. *Solar Energy*, **103**, pp. 682–703, 2014. DOI: 10.1016/j.solener.2012.07.003.

MODEL-BASED ESTIMATION AND PREDICTION OF ATMOSPHERIC POLLUTANTS IN THE ENVIRONMENT

UROOSA TAGAR & RANJAN VEPA
School of Engineering and Material Science, Queen Mary University of London, UK

ABSTRACT

Air pollution in urban areas is mainly due to the intense use of motorized transport for travelling. To monitor the change in the concentration of pollutants in the atmosphere on a real time basis, we may use any control-theory engendered filter such as a Kalman filter, an extended Kalman filter, an unscented Kalman filter or a particle filter, which are statistically based methods that rely on some form of dynamic model of the changes in the concentration, and are therefore suitable for monitoring concentrations of carbon monoxide (CO), carbon dioxide (CO_2), nitric oxide (NO), unburnt hydrocarbons, nitrogen dioxide (NO_2) and other pollutants from cars and jet engines. In this study air quality monitoring data for the years 2013 to 2017 (5 years) was obtained from the website of Department for Environment, Food and Rural Affairs (DEFRA), UK. A formal statistical model that can be used to predict the dynamics of pollutants in the atmosphere or any other similar pollutant is not available and it is really necessary that such a model is developed based on the observed data. A simple second order model of this data was established using MATLAB software. After that the concentrations of NO, NO_2, ozone and particulate matter (PM_{10}) in the atmosphere at a particular location within London city, were estimated using certain techniques namely, Least-squares estimation technique, Theil–Sen regression technique and an auto-regressive predictive filter technique also known as the linear predictive coding (LPC) estimator. Based on the second order model, a Kalman filter technique was implemented to predict the concentration of the pollutants. Results obtained from these techniques were compared with each other and generally indicate a steady decline of the pollutant concentrations while the Kalman filter and LPC estimator predict a periodic reduction in the concentration of the same pollutants over the 5 year period.

Keywords: air pollution, concentration estimation, prediction, nitric oxide, nitrogen dioxide, particulate matter, ozone.

1 INTRODUCTION

Various species of pollutants and toxic chemicals are characterised by several emission pathways into the atmosphere. They could be emitted by point sources in relation to the size of large cities due to relatively rare events in time such as, accidental release at nuclear power plants or volcanic eruptions or from regionally distributed sources such as emission of photochemical smog generators on roads and highways as well as forest fires. Moreover, time interval of such releases could quite varied. These air pollutants generally are known to be carried hundreds or thousands of kilometres from wherever they are released to distant points on the surface of the globe depending on the nature and composition of pollutants. Consequently they could affect the health of several groups of humans (see for example, Murray and Lipfert [1]) and result in a long-term damage to both the population and the environment. Such incidents can also have a huge economic impact due to a variety of reasons. For example the volcanic eruption of Eyjafjallajökull in Iceland over a period of six days in April, 2010 caused enormous disruption to air travel to and from London Heathrow, following the cancellation of several hundreds of flights because of the damage volcanic ash that was ejected to the atmosphere could cause to aircraft. Consequently the high episodic concentrations of air pollutants (e.g., nitrogen oxides, particulates, and carbon monoxide) in major population centres across the world, is a matter of deep concern due to their harmful effects on human health and the environment.

WIT Transactions on Ecology and the Environment, Vol 252, © 2021 WIT Press
www.witpress.com, ISSN 1743-3541 (on-line)
doi:10.2495/AIR210021

Statistical analysis (see for example Wilcox [2]) of air pollution data has been extensively advanced in the recent decade (see for, e.g., Gyarmati-Szabo et al. [3] and references therein). For short term predictions of pollution concentration, statistical methods are traditionally more suitable than physically based deterministic models [4]. The long-term goals of the research in this area, were succinctly stated by Gyarmati-Szabo et al. [3], as:

i. prediction of critical levels of pollutants to give out health warnings to local populations;
ii. identifying and predicting changing trends in high concentration levels;
iii. assessing temporal changes (secular or periodic) in air pollution levels due to the impact of human activities on the environment, either directly due to changing emission patterns or indirectly due to climate change.

In a recent study, Gyarmati-Szabo et al. [3], a novel approach to extreme value modelling is proposed and applied to the problem predicting pollutant concentrations in the atmosphere over the city of Leeds (West Yorkshire, UK), on both long (e.g., daily or yearly) and short (e.g., quarter of an hour) time scales. Although estimation methods based on linear Kalman filtering were introduced several decades ago (see for example Desalu et al. [5]), recent applications (Samaranayake et al. [6]) have focussed on nonlinear Kalman filtering techniques such as extended Kalman filtering and large scale ensemble Kalman filtering based on the use of a variety of nonlinear models. Advanced regression based methods have also been pursued as evidenced by the work of Ortiz and Friedrich [7]. A detailed review of the theory and application of those forecasting models have been presented by Bai et al. [8]. A host of different modelling approaches have also been developed, ranging from single species continuous time models characterised by ordinary differentials for the evolution of the state (see for example Liu et al. [9]) to dispersive and distributed models characterised by partial differential equations for the evolution and spread of a species in time and space, as in Leelőssy et al. [10].

In this paper, an approach similar to that proposed by Gyarmati-Szabo et al. [3] is adopted, to study pollutant concentrations in the atmosphere over the city of London at one location (London, Bloomsbury), on long time scales such as yearly and 5-yearly periods, based of available hourly measurements of pollutant concentrations from 2013 to 2017. The concentrations of NO, Nitrogen dioxide (NO_2), Ozone and particulate matter (PM_{10}) in the atmosphere at a particular location within London city, were estimated using different techniques namely, Least-squares estimation technique, Theil–Sen regression technique and an auto-regressive predictive filter known as the linear predictive coding (LPC) estimator technique. Based on the second order model, a Kalman filter was implemented to predict the concentration of the pollutants. Results obtained from these filters were compared with results of least-squares estimation method and Theil–Sen regression are similar and generally indicate a steady decline of the pollutant concentrations while the Kalman filter and LPC estimator predict a periodic reduction in the concentration of the same pollutants over the 5 year period. It is expected that a physical model would be developed and used for prediction over a longer time scale and for estimation and classification of the pollution components in the atmosphere.

2 METHODOLOGY

Air quality measurements give only quantitative information about ambient concentrations without identifying causes of air quality problems. There are over 1,500 sites across the UK that monitor air quality. They are organised into networks that gather a particular kind of information, using a particular method. There are two major types – automatic and non-automatic networks that gather air pollution information.

For monitoring and reporting air pollution the UK has been divided into two zones: non-agglomeration zones and agglomeration zones. There are fifteen non-agglomeration zones which match:

i. The boundaries of England's Government Offices for the Regions; and
ii. The boundaries agreed by the Scottish Government, Welsh Government and Department of the Environment in Northern Ireland.

Based on the above zones, there are 16 regions and 16 urban areas which are used for reporting latest real-time monitoring data of air pollutants. These are a subset of the UK agglomeration and non-agglomeration zones.

This section provides details of the study area including geometric characteristics and air quality measurements data of a five year time frame for above mentioned pollutants.

2.1 Case study

This case study concerns the estimation and prediction of the concentrations of nitric oxide (NO), nitrogen dioxide (NO_2), ozone and particulate matter, (PM_{10}) at a particular site in city of London.

2.2 Air quality monitoring site

The study area is Bloomsbury in the city of London. Bloomsbury has no official boundaries, but can be roughly defined as the square of territory bounded by Tottenham Court Road to the west, Euston Road to the north, Gray's Inn Road to the east, and either High Holborn or the thoroughfare formed by New Oxford Street, Bloomsbury Way and Theobalds Road to the south. Bloomsbury merges gradually with Holborn in the south, with St Pancras and King's Cross in the north-east and with Clerkenwell in the south-east.

2.3 Data collection and analysis

In this research study, the air quality measurement data was collected from the website of UK Air Information Resource, Department for Environment, Food and Rural Affairs (DEFRA) UK with their permission. The data that has been used is the hourly concentration of NO nitrogen dioxide (NO_2), ozone and particulate matter (PM_{10}) in a central corridor of London, called Bloomsbury, from 2013 to 2017 years.

2.4 Concentration of the pollutants

Concentration data of the pollutants from year 2013 to year 2017 is collected from the UK Air Information Resource [11], Department for Environment, Food and Rural Affairs (DEFRA).

A preliminary inspection of the data generally not only indicated several season variations on a yearly basis but also variations over 5 year period. These variations dictated the choice of the time scales and the methods of analysis that could be used. The data analysis methods used in this study are the recursive least squares method, the Theil–Sen linear regression estimator, the linear predictive coding estimator and the Kalman filter.

2.5 Data analysis techniques

2.5.1 Least squares method

The least squares method (Simon [12]) is a form of mathematical regression analysis that finds the line of best fit for a set of data, providing a visual demonstration of the relationship between the data points. Each point of data is representative of the relationship between a known independent variable and an unknown dependent variable. In our work the variables used for least square estimation are pollutant concentration measured in units (μg/m^3) and time in hours but represented as a whole year.

2.5.2 Theil–Sen estimator technique

Theil–Sen estimator is a method for robustly fitting a line to sample points in the plane (simple linear regression) by choosing the median of the slopes of all lines through pairs of points. As defined by Theil [13], Theil–Sen estimator of a set of two-dimensional points (x_i, y_i) is the median (m) of the slopes $\left(y_j - y_i\right)/\left(x_j - x_i\right)$ determined by all pairs of sample points. Sen [14] extended this definition to handle the case in which two data points have the same x coordinate. In Sen's definition, one takes the median of the slopes defined only from pairs of points having distinct x coordinates. The Theil–Sen estimator has several important features such as unbiasedness (Wilcox [15], Wang and Yu [16]) and most suitable for prediction and forecasting applications.

2.5.3 Linear predictive coding estimator technique

Linear predictive coding (O'Shaugnessy [19]) is a tool used mostly in audio signal processing and speech processing for representing the spectral envelope of a digital signal of speech in compressed form, using the information of a linear predictive model. It is one of the most powerful speech analysis techniques for encoding good quality speech at a low bit rate and provides extremely accurate estimates of speech parameters.

In linear predictive coding, the signal of interest is modelled as,

$$y_{n+1} = \sum_{i=1}^{n} a_i y_i + e_i \,, \tag{1}$$

So as make the error e_i a minimum. An estimate of the prediction of the measurement y_i , $i = 1, 2 \ldots n$, is then given by,

$$\hat{y}_{n+1} = \sum_{i=1}^{n} a_i y_i \,. \tag{2}$$

2.5.4 Kalman filter estimation technique

The Kalman filter model (Brown and Hwang [20]) assumes that the state of a system, represented as a state vector, at a time t evolved from the prior state at time $t-1$ according to the below equation:

$$\mathbf{x}_{i+1} = \mathbf{A}_i \mathbf{x}_i + \mathbf{B}_i \mathbf{u}_i \,, \tag{3}$$

where $\mathbf{x}_i$ is the state vector containing the terms of interest for the system (concentration and rate of change of concentration) at time t, $\mathbf{u}_i$ is the vector containing any control or disturbance inputs, $\mathbf{A}_i$ is the state transition matrix which applies the effect of each system state parameter at time $t-1$ on the system state at time t and $\mathbf{B}_i$ is a distribution matrix that distributes the inputs to the various individual state equations. The measurement model takes the form,

$$\mathbf{y}_i = \mathbf{H}_i \mathbf{x}_i + \mathbf{v}_i , \tag{4}$$

where $\mathbf{y}_i$ is the measurements vector that is linearly related to the state vector $\mathbf{x}_i$ and $\mathbf{v}_i$ is the vector containing disturbance inputs to the measurements.

The Kalman filter which is used to obtain an estimate of the state vector given by $\hat{\mathbf{x}}_i$, then takes the form,

$$\hat{\mathbf{x}}_{i+1} = \mathbf{A}_i \hat{\mathbf{x}}_i + \mathbf{B}_i \hat{\mathbf{u}}_i + \mathbf{K}_i \left(\mathbf{y}_i - \mathbf{H}_i \hat{\mathbf{x}}_i \right), \tag{5}$$

where $\mathbf{K}_i$ is the optimum Kalman gain multiplying an estimate of the measurement error.

3 RESULTS

3.1 Pollutants analysis using the least square technique and the Theil–Sen estimator technique

In the first instance, the Nitric Oxide (NO) concentration in the atmosphere was estimated using least square estimation method (LSEM).

Following the application of the LSEM, the pollutant concentrations were further analysed using Theil–Sen estimator technique. These results were also compared with LSEM results. Figs 1–5 show the comparative results of NO pollution data using LSEM and Theil–Sen estimation techniques of the years 2013 to 2017 (5 years) of London Bloomsbury.

In Fig. 1, data analysed was of the period from January 2013 to January 2014, allowing an extra month time to show the trend of pollution. The pollutant concentration was expressed in μg/m³ unit and time in terms of year. The scale bar is kept generalized for each analysed graph keeping the same number of days for each year but the data point used are

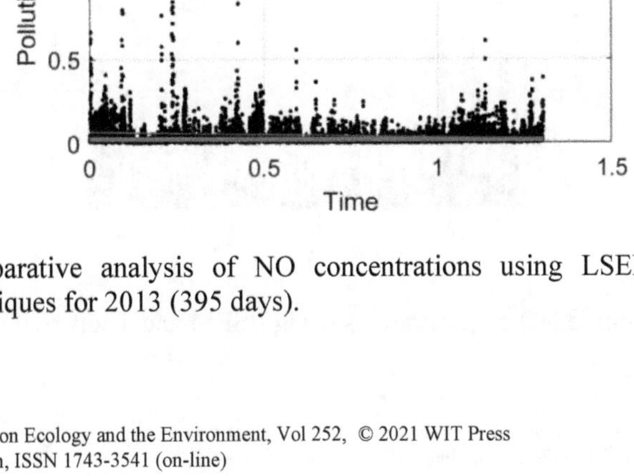

Figure 1: Comparative analysis of NO concentrations using LSEM and Theil–Sen techniques for 2013 (395 days).

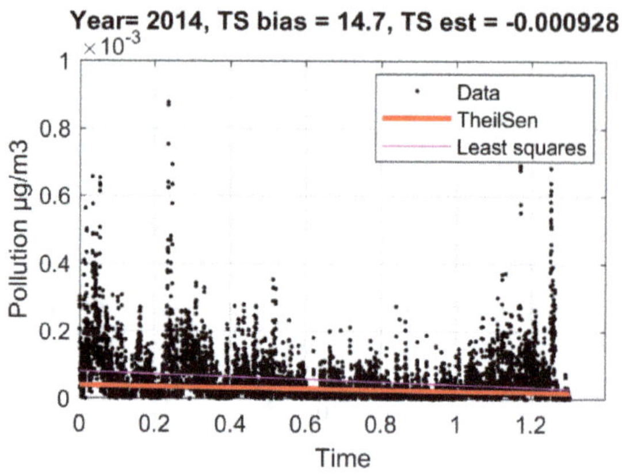

Figure 2: Comparative NO concentrations using LSEM and Theil–Sen, 2014 (395 days).

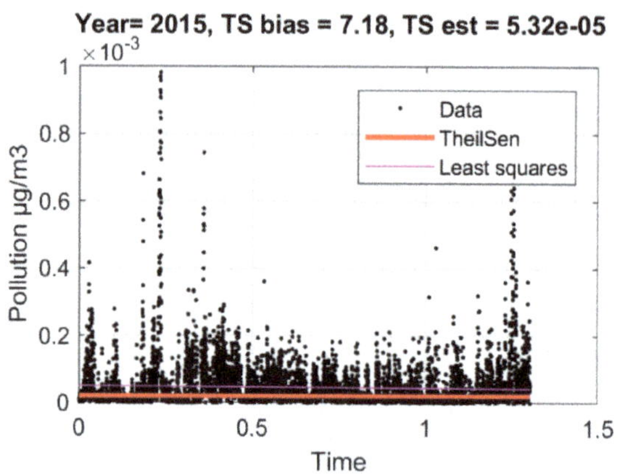

Figure 3: Comparative NO concentrations using LSEM and Theil–Sen, 2015 (395 days).

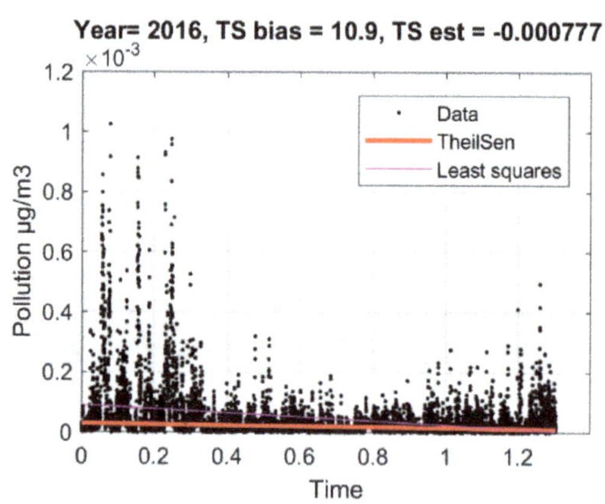

Figure 4: Comparative NO concentrations using LSEM and Theil–Sen, 2016 (395 days).

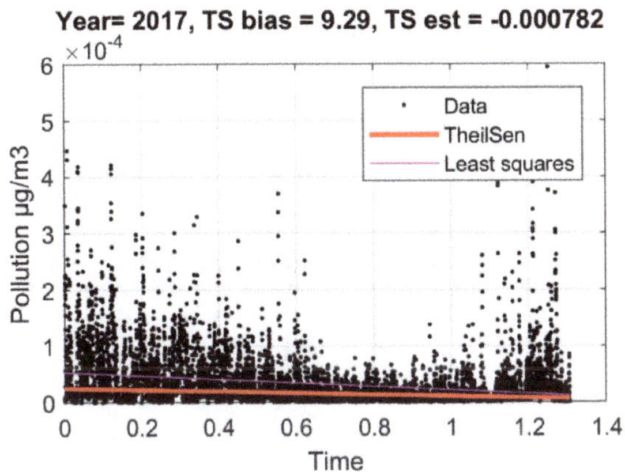

Figure 5: Comparative NO concentrations using LSEM and Theil–Sen, 2017 (395 days).

hourly collection of pollutant concentration. The graphical representation of the concentration of nitric oxide (NO) shows that there are major three peaks over a year, and the concentration was much lower during the summer period of June and July (0.6–0.8 time scale bar). The data also shows that in year 2014, NO concentration was much less than the year 2013. A major rise in the NO peaks during the winter months of January and February, was also observed. Comparing with the other years data obtained, overall concentration was much higher in 2014. The highest concentration peak is above 500 µg/m^3. The concentration in the year 2015 was found to be much more compared to previous year but the major increase in concentration was in the winter period.

The slope of the least-squares regression line is the average change in the predicted values of the response variable when the explanatory variable increases by 1 unit. Fig. 6 refers to the slope and bias of the Theil–Sen estimator for NO pollution concentration over a 5 year time frame. Bias means that the expected value of the estimator is not equal to the pollution parameter. Intuitively in a regression analysis, this would mean that the estimate of one of the parameters is too high or too low. In 2017, the highest concentration peak was less than 300 µg/m^3. Figs 7–9 refer respectively to the Theil–Sen estimators for nitrogen dioxide (NO_2) concentration, ozone concentration and particulate matter (PM_{10}) in the atmosphere. The magnitudes of the slopes and biases of the Theil–Sen estimators in each are of similar order. This means these three techniques predicted the different pollutant's behaviour in a same manner over a fixed period of time. The continuous negative slope over 5 years for NO, NO_2 and ozone shows the negative correlation of pollutant concentration over time. It means the concentration was decreased with increasing time period and Theil–Sen estimator technique correlated same as the least square technique.

The slope has been changed for PM_{10} concentration and continuous positive trend in the slope of Theil–Sen over 5 years shown in Fig. 9 proves the positive correlation between concentration and time.

3.2 Pollutants estimation using the Kalman filter

The available data from the air quality measurements was also analysed using a lightly damped second order model Kalman filter, to capture the seasonal and annual variations of

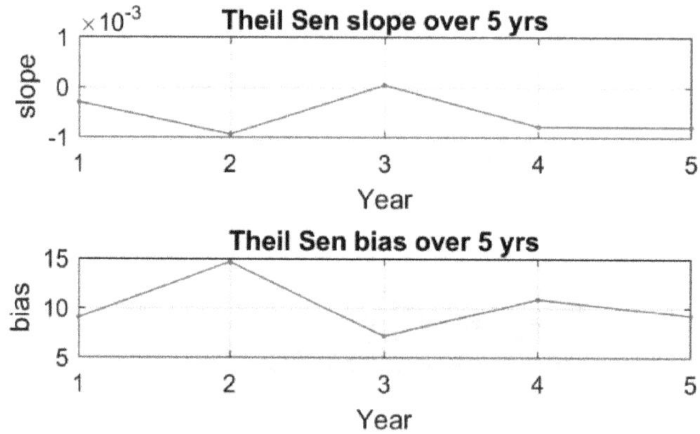

Figure 6: The Theil–Sen slope and bias over 5 years of NO concentration.

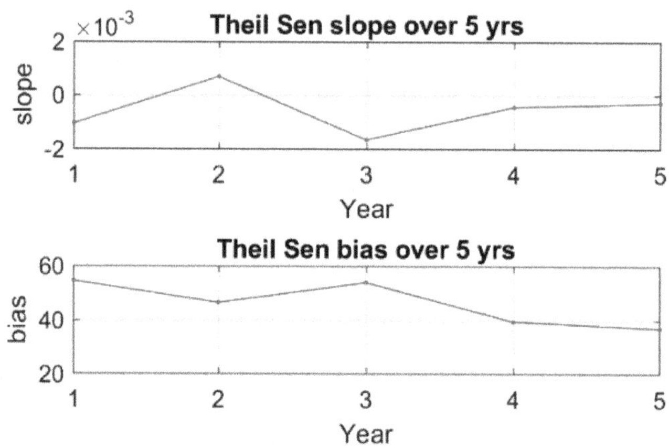

Figure 7: The Theil–Sen slope and bias over 5 years of NO_2 concentration.

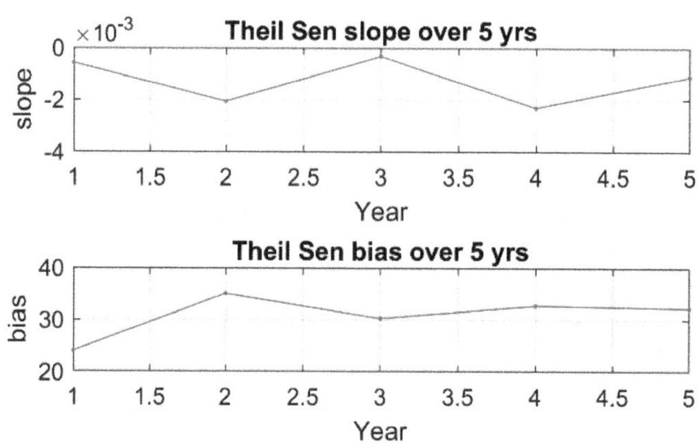

Figure 8: The Theil–Sen slope and bias over 5 years of ozone concentration.

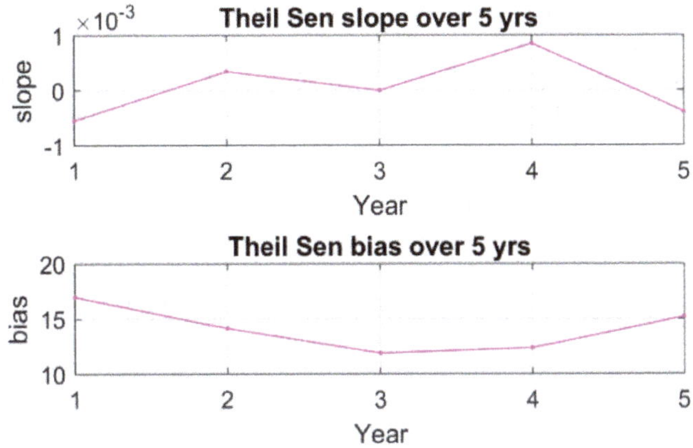

Figure 9: The Theil–Sen slope and bias over 5 years of PM_{10} concentration.

the pollutant concentration over the five year period. Typical estimates of NO concentrations are compared with the measurement and model predicted data ass shown in Fig. 10. Similar results were obtained for nitrogen dioxide (NO_2) concentration, ozone concentration and particulate matter (PM_{10}) in the atmosphere. We can see that filter estimate (Kalman filter) is predicting the pollutants in the same manner but with reduced error.

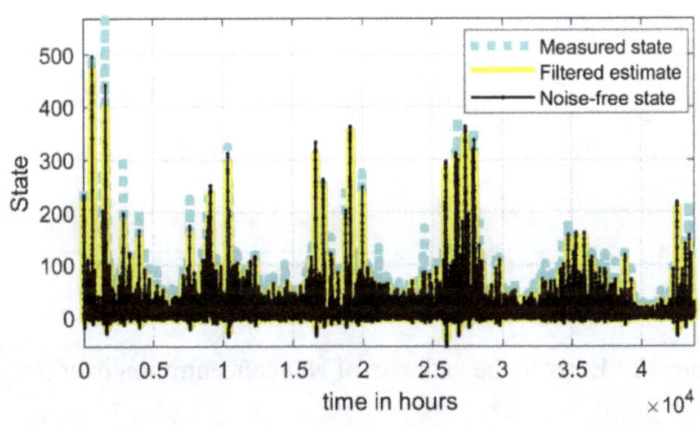

Figure 10: Comparison of measured, model predicted and Kalman filter estimated NO concentrations.

The noise free data represents the smooth data. The data points of pollutants are named as state, a general term to show al data. It can be seen that our model can easily be fit into all points and gives a smooth signal to easily predict the behaviour.

3.3 Pollutants analysis using the LPC estimator technique

The LPC estimation method was also applied to the pollutant concentrations. Typical estimates of NO concentrations are compared with the original measurement (labelled as the “original signal” in the figure) in Fig. 11. The error in the LPC estimate is shown in Fig. 12. It can be seen that LPC estimator predicted and followed the original data points in a very same manner. It means LPC estimator can predict the pollution data points efficiently. The amplitude shows the frequency of signals produced by pollutant data points named sample number.

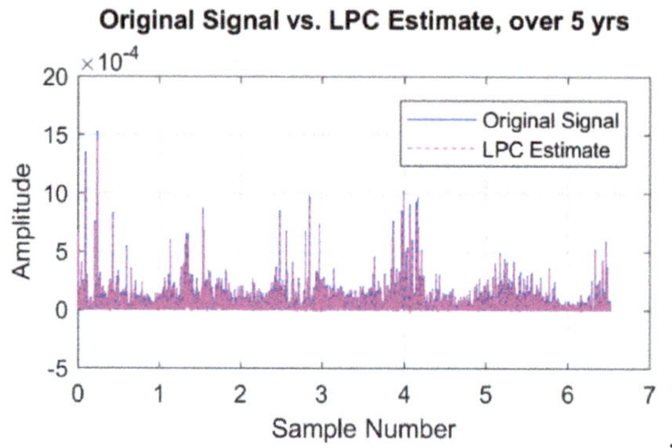

Figure 11: Comparison of the LPC estimate with the original measurement (labelled as the “original signal” in the figure).

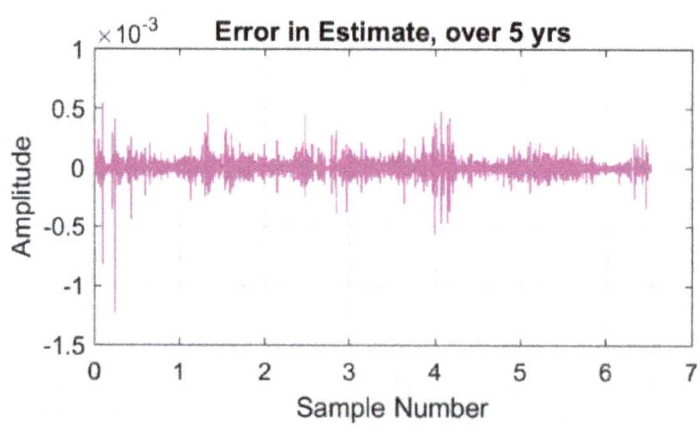

Figure 12: Error in the estimate of NO concentration over 5 years.

4 DISCUSSION AND CONCLUSIONS

In this study, application of the least square method, the Theil–Sen estimator, the Kalman filter and the LPC estimator have been successfully used to predict the above mentioned pollutant’s concentrations and thereby forecasted the changes at this particular urban site called London Bloomsbury. The results generally indicate that over the 5 year time period, the overall concentration of the pollutants was predicted which was steadily declining with increase in time. However it is also possible to understand the limitations of the predictions. The linear fit of least square estimate and Theil–Sen estimate showed the same behaviour over 5 year data points. Negative slope of Theil–Sen estimate showed a declining trend of pollutants concentration over 5 year time slot. Furthermore, results of Kalman filter predicted the same behaviour as of measured state and noise free data (pollutant concentration). The frequency (amplitude) spectra of pollutant concentration by LPC estimator follows the same signal of original data. A very low error estimate of LPC ranging between -0.5×10^{-3} to 0.5×10^{3} proves the efficiency of LPC and makes this technique suitable for predicting the behaviour of atmospheric pollutants for future. It can be concluded that these techniques can be efficiently used to predict the pollutant concentrations. Primarily, although the estimates are quite accurate, the prediction is only valid in the short term, which could be typically less than a year. It is desirable that predictions can be made in the longer term as well.

5 FUTURE WORK

It is planned to use more physically meaningful models to predict and forecast the changes in the pollution concentration over a longer prediction period. Air pollution dynamics can be modelled using mathematical representations to describe the novel relationship between emissions, meteorology, atmospheric concentrations, surface depositions and other factors [21]. A dynamic model of pollution generation, particularly gaseous pollutants may be developed based on system dynamic modelling considerations. The system dynamics models enable the quantity of each generated component such as nitric oxide to be estimated and predicted so effective and timely methods of pollution control may be developed. Furthermore, pollution generation in general is directly related to the growth of the population and hence can be related to the birth and death dynamics of the population. These basic models of the behaviour of simple populations utilise the standard tools of dynamical systems as outlined by Metz and Diekmann [22], McCauley and Murdoch [23] and Benenson [24]. Typical examples of population growth and death are the Logistic growth model, Predator–Prey models, and Lokta–Volterra competition. The Logistic growth model can and has been effectively used to model the generation of pollution. Finally, it is expected that the dynamic model developed will all be used for prediction, estimation and classification of the pollution generated and compared with the actual data.

ACKNOWLEDGEMENT

The first author wishes to acknowledge the financial support from the Pakistan Higher Education Council.

REFERENCES

[1] Murray, C.J. & Lipfert, F.W., Revisiting a population-dynamic model of air pollution and daily mortality of the elderly in Philadelphia. *Journal of the Air and Waste Management Association*, **60**(5), pp. 611–628. DOI: 10.3155/1047-3289.611, 2010.

[2] Wilcox, R.R., *Fundamentals of Modern Statistical Methods: Substantially Improving Power and Accuracy*, 2nd ed., Springer: New York, 2010.

[3] Gyarmati-Szabo, J., Bogachev, L.V. & Chen, H., Nonstationary POT modelling of air pollution concentrations: Statistical analysis of the traffic and meteorological impact. *Environmetrics*, **28**(5), 2017. DOI: 10.1002/env.2449.

[4] Catalano, M., Galatioto, F., Bell, M. & Namdeo, A., Improving the prediction of air pollution peak episodes generated by urban transport networks. *Journal of Environmental Science and Policy*, **60**, pp. 69–83.

[5] Desalu, A.A., Gould, L.A. & Schweppe, F.C., Dynamic estimation of air pollution. *IEEE Transactions on Automatic Control*, **AC19**(6), pp. 904–910, 1974.

[6] Samaranayake, S., Glaser, S., Holstius, D., Monteil, J., Tracton, K. & Bayen, A., Real-time estimation of pollution emissions and dispersion from highway traffic. *Computer-Aided Civil and Infrastructure Engineering*, **29**, pp. 546–558. DOI: 10.1111/mice.12078, 2014.

[7] Ortiz, S.T. & Friedrich, R., A modelling approach for estimating background pollutant concentrations in urban areas. *Atmospheric Pollution Research*, **4**, pp. 147–156. DOI: 10.5094/APR.2013.015, 2013.
[8] Bai, L., Wang, J., Ma, X. & Lu, H., Air pollution forecasts: An overview. *International Journal of Environmental Research and Public Health*, **15**(780), pp. 1–44, 2018. DOI: 10.3390/ijerph15040780.
[9] Liu, B., Luan S. & Gao, Y., Modeling the dynamics of a single-species model with pollution treatment in a polluted environment. *Discrete Dynamics in Nature and Society*, pp. 1–8, 2013. DOI: 10.1155/2013/412409.
[10] Leelőssy, A., Molnár, Jr. F., Izsák, F., Havasi, A., Lagzi, I. & Mészáros, R., Dispersion modeling of air pollutants in the atmosphere: A review. *Central European Journal of Geosciences* **6**(3), pp. 257–278, 2014. DOI: 10.2478/s13533-012-0188-6.
[11] UK Air Information Resource, Department for Environment, Food and Rural Affairs. https://uk-air.defra.gov.uk/air-pollution/. Accessed on: 2 Apr. 2019.
[12] Simon, D., *Optimal State Estimations*, John Wiley: Hoboken, NJ, 2006.
[13] Theil, H., A rank-invariant method of linear and polynomial regression analysis. *Proceedings, Koninklijke Nederlandse Akademie van Wetenschappen*, **A53**, pp. 386–392, 1950.
[14] Sen, P.K., Estimates of the regression coefficient based on Kendall's tau. *Journal of the American Statistical Association*, **63**, pp. 1379–1389, 1968.
[15] Wilcox, R.R., Some results on extensions and modifications of the Theil–Sen regression estimator. *British Journal of Mathematical and Statistical Psychology*, **57**(2), pp. 265–280, 2004.
[16] Wang, X. & Yu, Q., Unbiasedness of the Theil–Sen estimator. *Journal of Nonparametric Statistics*, **17**(6), pp. 685–695, 2005.
[17] Fernandes, R. & Leblanc, S., Parametric (modified least squares) and non-parametric (Theil–Sen) linear regressions for predicting biophysical parameters in the presence of measurement errors. *Remote Sensing of Environment*, **95**, pp. 303–316, 2005.
[18] Peng, H., Wang, S. & Wang, X., Consistency and asymptotic distribution of the Theil–Sen estimator. *Journal of Statistical Planning and Inference*, **138**(6), pp. 1836–1850, 2008.
[19] O'Shaugnessy, D., *Speech Communication: Human and Machine*, Addison Wesley, 1978.
[20] Brown, R.G. & Hwang, P.Y.C., *Introduction to Random Signals and Applied Kalman Filtering*, John Wiley, 1992.
[21] Daly, A. & Zannetti, P., Air pollution modelling: An overview, *Ambient Air Pollution*, Arab School for Science and Technology and The Enviro-Comp Institute: USA, 2007.
[22] Metz, J.A.J. & Diekmann, O., The dynamics of physiologically structured populations, 68. *Lecture Notes in Biomathematics*, Springer–Verlag: Heidelberg, 1986.
[23] McCauley, E. & Murdoch, W.W., Predator–prey dynamics in rich and poor environments. *Nature*, **343**, pp. 455–457.
[24] Benenson, I., Modeling population dynamics in the city: From a regional to a multi-Agent approach. *Discrete Dynamics in Nature and Society*, **3**, pp. 149–170, 1999.

ACEMIS: SOFTWARE FOR THE COMPUTATION OF AIRCRAFT FLIGHT EMISSIONS BASED ON FUEL CONSUMPTION

SUDEEP ARYAL SHARMA, HAROLD BARBEY, REMI JULIO, KEVIN COCQ, JEREMY CURE, ARNAUD GLARIA & VALERIE LAVASTE
TEC (Technology and Engineering Center), Capgemini Engineering, France

ABSTRACT

The world has witnessed a significant growth in air traffic over the past 50 years. While this may be a good thing for a country's economy, increase in pollution and rise in temperature, due to the growth of greenhouse gas (GHG) emissions are worrying factors. Thus the methods to track pollutants emitted by flights, in order to mitigate the pollution, have become of utmost importance. ACEMIS (AirCraft EMISsion), presented in this paper, is a tool developed for the calculation of emissions based on fuel consumption of flights. The input data is one of the most important factors to create a realistic simulation and this paper explains how several influencing factors and actual flight data are considered for computation. It also explains the computational methodology to calculate the fuel consumption of a flight and simulate realistic emissions of not only carbon dioxide CO_2 but other gaseous emissions such as nitric oxides/nitrogen dioxide NO_x, water H_2O, carbon monoxide CO, hydrocarbons HC, and sulfur dioxide SO_2. This novel tool enables the user to simulate chemical emissions of a flight or a fleet between two or more airports, then displays and analyzes the simulation results. Finally, this calculation method is encoded using Python programming language to create a graphical friendly user interface.

Keywords: environment, aircraft, aeronautics, Python, greenhouse gas emission, GHG, pollution, carbon dioxide.

1 INTRODUCTION

The economic growth in the aeronautical sector, illustrated by novel airports creation (among others), has resulted in 8% increase in air traffic between 2014 and 2017, with a forecast of 42% increase from 2017 to 2040 [1]. The amplification of air traffic proportionally increases emissions of pollutants as fuel burn process in flight generates byproducts such as greenhouse gases (GHG). The significant emissions are carbon dioxide CO_2 (70% of the exhaust), water vapor H_2O (30%), sulfur dioxide SO_2, nitric oxide/nitrogen dioxide NO_x, carbon monoxide CO and hydrocarbons HC (around 1% of the exhaust combined). There are many studies focusing on the impact of CO_2 onto the environment, whereas other flight emissions have been historically ignored. This has to change in order to act significantly against the negative impact of the aeronautic sector on the environment. The United Nations' Intergovernmental Panel on Climate Change (IPCC) published one of the first reports on the impact of non-CO_2 emissions in 1999 highlighting the contribution of NO_x at high altitudes in the depletion of ozone layer [2]. Water vapor forms cirrus-clouds at high altitude, which has a serious warming effect despite its short life span because of the collective effect of thousands of continuous flights [3]. The climate impacts of NO_x, contrails, and clouds due to H_2O emitted at high altitude is so strong that they can eat the planet more in one day than all the aviation's CO_2 accumulated in the atmosphere since the 1940s [4]. Hence, it has become increasingly important today and in the future to keep a record on GHG and/or pollutant emissions like H_2O, SO_2, NO_x, CO and HC, in addition to CO_2.

WIT Transactions on Ecology and the Environment, Vol 252, © 2021 WIT Press
www.witpress.com, ISSN 1743-3541 (on-line)
doi:10.2495/AIR210031

CO_2 emissions from aviation have been included in the EU emissions trading system (EU ETS) since 2012. Under the EU ETS, all European and non-European airlines operating in Europe are required to monitor, report, and verify their emissions, then they may have to pay allowances against those emissions. Every year, they receive tradeable allowances covering a certain level of emissions from their flights [5]. In 2016, the International Civil Aviation Organization (ICAO) agreed on a resolution for a global market-based measure to address CO_2 emissions from international aviation as of 2021 [6]. The agreed resolution sets out the objective and key design elements of the global scheme, as well as a roadmap for the completion of the work on implementing modalities. The Carbon Offsetting and Reduction Scheme for International Aviation (CORSIA) program aims to stabilize CO_2 emissions at 2019 levels by requiring airlines to offset the growth of their emissions after this date [6]. It however takes into account only CO_2 emission.

To respect the regulation as set out by ETS or CORSIA, airlines must keep tracking all types of emissions contributing to global warming. Even though methods are proposed to gather the in-flight emissions data [7], there is currently no complete tool for the calculation of flight emissions other than CO_2 [8]. To fill this gap and anticipate the future of aviation, the Capgemini Engineering group envisioned an emission and fuel consumption-tracking graphics friendly user interface (GUI) tool called ACEMIS (AirCraft EMISsions). It uses scientific calculation methodologies to simulate the fuel consumption and flight emissions (CO_2, NO_x, H_2O, CO, HC, SO_2) in two ways, one by using MOZAIC or IAGOS data [9] from flights that have already taken place ("MOZAIC/IAGOS Fleet" part in the tool), and the other by building personalized flight routes ("Flight Profile" part in the tool, Fig. 1).

MOZAIC (Measurements of OZone, water vapor, carbon monoxide and nitrogen oxides by in-service AIrbus airCraft) is an observation program that took place between Aug. 1994 and Nov. 2014 [9]. Its purpose was to measure the composition of the atmosphere through commercial aircraft of various airlines. Several chemical components have been recorded: ozone (O_3), carbon monoxide (CO), water vapor (H_2O) and total nitrogen oxides (NO_y). IAGOS (In-service Aircraft for a Global Observing System), which started in Jul. 2011 until today, is the continuation of the MOZAIC program and includes among others measurements of ozone (O_3), carbon monoxide (CO), water vapor (H_2O), nitrogen oxides (NO_x, NO_y) and greenhouse gases (CO_2, CH_4) [9]. All measurements are geo-localized (latitude, longitude, altitude…) and supplemented by meteorological data (temperature, pressure, humidity…).

2 METHODOLOGY

ACEMIS is a GUI software that calculates the fuel consumption and pollutant emissions. The user can enter flight information such as flight name, aircraft type, fuels used (and their concentration), weather conditions, departure and arrival airports, and passenger fill rate of the aircraft (Fig. 1). Realistic simulation can be done in the MOZAIC/IAGOS module using real flight data received from MOZAIC or IAGOS database [9], whereas personalized flight routes are simulated in the FLIGHT PROFILE module. The right part (red dashed line, Fig. 1) displays useful information such as fleet, progress of the simulation (inside the red dotted box) and the information of the selected flight.

In this section, the methodologies behind the fuel consumption and emissions calculation are explained.

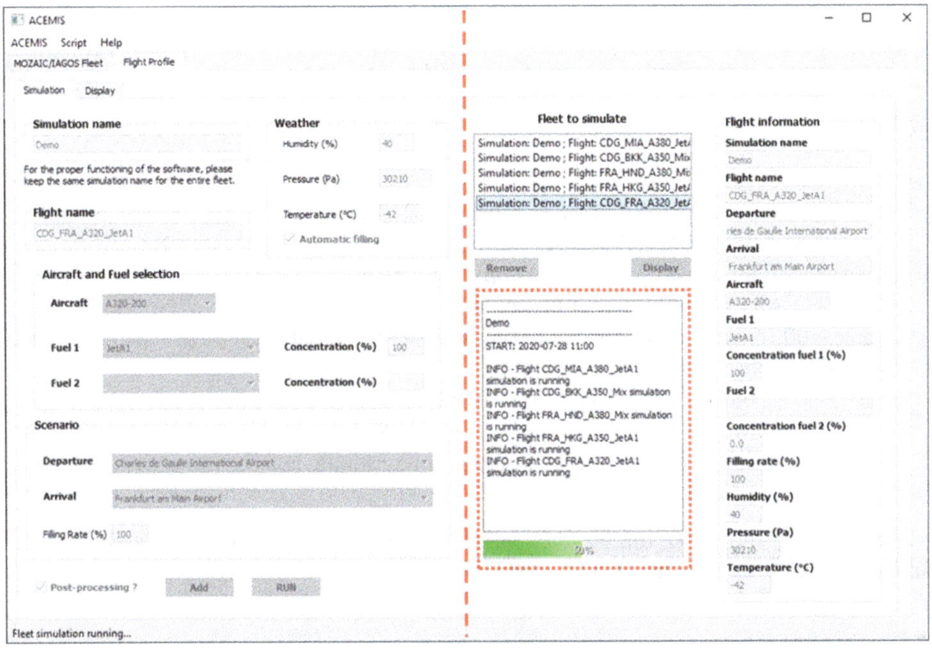

Figure 1: Graphical user interface of ACEMIS – version 2.1 – Flight Profile part.

2.1 Calculation of fuel consumption

The fuel consumption calculation requires data from the Auxiliary Power Unit (APU) and the engine: the standard thrust on the ground Tr_{00}, the specific consumption SFC_{00}, the fuel flow at idle speed FF_{IDLE}, and the fuel flow of the APU FF_{APU}. These data, together combined with the "aircraft" parameters and ambient conditions, make possible to calculate the rate of fuel burned by the aircraft at any point.

The fuel flow is the product of the specific fuel consumption and the thrust. Thrust is calculated differently depending on whether the aircraft is stabilized in the cruise phase, the climb phase, or the braking phase (this is referred as back thrust, eqns (1)–(3)). In the first case, the thrust is determined from the balance of forces acting on the aircraft; in the second case, an empirical method is used.

$$\begin{cases} SFC(z,M) = SFC_{00} \times \sqrt{T(z)/T_{00}} \times (1+M). \\ Tr(z,M) = Tr_{00} \times (0.568 + [0.25 \times (1.2 - M)^3]) \times [P(z)/P_{00}]^{0,6}. \\ FF(z,M) = SFC(z,M) \times Tr(z,M). \end{cases} \tag{1}$$

$$\begin{cases} SFC(z,M) = SFC_{00} \times \sqrt{T(z)/T_{00}} \times (1+M). \\ Tr(m_{A/C}) = (m_{A/C} \times g_0) \times [\cos\beta/f - \sin\beta]/nb_{eng}. \\ FF(z,M) = SFC(z,M) \times Tr(z,M). \end{cases} \tag{2}$$

$$\begin{cases} SFC(z,M) = SFC_{00} \times \sqrt{T(z)/T_{00}} \times (1+M). \\ Tr(m_{A/C}) = m_{A/C} \times (V_{INV}/t_{T-D})/nb_{eng}. \\ FF(z,M) = SFC(z,M) \times Tr(z,M). \end{cases} \quad (3)$$

where:

- Atmospheric parameters: T (temperature; K), P (pressure; Pa), RH (relative humidity);
- Aircraft parameters: FF (fuel flow; kg.s^{-1}), Tr (thrust; N), SFC (specific consumption), M (Mach number), m (mass; kg), nb_{eng} (number of engines), V_{INV} (reverse speed; m.s^{-1}), t_{T-D} (duration of braking phase; s);
- Indices: 00 (ground condition), *eng* (engine), *APU* (Auxiliary Power Unit), *IDLE* (idle phase), *A/C* (aircraft);
- Other: g_0 (acceleration due to gravity = 9.80665 m.s^{-2}), γ (isentropic coefficient of air = 1.4), r (specific ideal gas constant = 287.058 J.K^{-1}.kg^{-1}), f (finesse = ration of coefficient of lift and drag; four possible configurations, noted from f_0 to f_3), β (descent slope of the aircraft).

When using the *APU* during the parking phase, the fuel flow denoted FF_{eng} and expressed in kg/s, can be summed up as follows in eqn (4):

$$FF_{eng} = FF_{APU}. \quad (4)$$

Similarly, at idle speed, the fuel flow is provided by the ICAO data (eqn (5)):

$$FF_{eng} = FF_{IDLE}. \quad (5)$$

2.2 Calculation of emissions

Calculation of emissions requires the fuel flow value. The product of the fuel flow and the emission indices (ratio of mass of pollutant emitted and mass of fuel consumed) allows the calculation of the emission of the pollutant considered per unit of time during a flight. These calculations are done based on two methods: first, the complete combustion assumption for CO_2, H_2O and SO_2 emissions, and second the Boeing Fuel Flow Method (BFFM2 method) for NO_x, CO and HC [10].

$$\begin{cases} EI(CO_2) = m_C \times (M_C + 2.M_O)/(M_C). \\ EI(H_2O) = m_H \times (M_O + 2.M_H)/(2.M_H). \\ EI(SO_2) = m_S \times (M_S + 2.M_O)/(M_S). \end{cases} \quad (6)$$

where:

- fuel parameters: m (mass percentage of a compound; kg/kg), M (molar mass of a compound; g.mol^{-1});
- Indices: C (carbon), H (hydrogen), O (oxygen), S (sulfur).

The BFFM2 method consists in extrapolating the emission indices at an altitude from the emission value on the ground affected by the ambient conditions – temperature, pressure and humidity [10]. This is done in three steps:

1. The calculation of fuel flow rate on the ground determined from the previously calculated consumption (FF_{eng}), the ambient conditions (T, P) and the speed of the aircraft (M), eqn (7):

$$FF_{00} = FF_{eng} \times \frac{(T/T_{00})^{3,8}}{(P/P_{00})} \times e^{0,2\times M^2}. \quad (7)$$

2. Linear interpolation of ICAO data using above value to associate it with emission indices $EI_{NOx,00}$, $EI_{CO,00}$, and $EI_{HC,00}$
3. Calculation of the aircraft's emission indices (eqn (8)) using the values obtained from step 2 above (in the absence of relative humidity data, the value of 50% is applied).

$$\begin{cases} EI_{NOx}(z) = EI_{NOx,00} \times ((T/T_{00})^{3.3}/(P/P_{00})^{1.02})^{-0.5} \times e^{-19(q-0.00634)}. \\ EI_{CO}(z) = EI_{CO,00} \times \sqrt{((T/T_{00})^{3.3}/(P/P_{00})^{1.02})}. \\ EI_{HC}(z) = EI_{HC,00} \times \sqrt{((T/T_{00})^{3.3}/(P/P_{00})^{1.02})}. \end{cases} \quad (8)$$

where:

- Aircraft parameters: *EI* (emission indice of a pollutant; kg/kg), *FF* (fuel flow rate; $kg.s^{-1}$), *M* (Mach number at an altitude z);
- Atmospheric parameters: *T* (temperature; K), *P* (pressure; Pa), *q* (specific humidity);
- Indices: 00 (ground parameter), *eng* (in-flight parameter).

3 SIMULATION RESULTS

In ACEMIS, the results could be analyzed under the "Display" tab. Within the sub-tab of "Flight" the result of one single flight can be displayed (Fig. 2). For the purpose of demonstration, the results of the flight trajectory, fuel flow rate, and emissions graph of CO_2, CO, and NO_x in Fig. 2 are displayed for a flight from Paris Charles de Gaulle International Airport to Miami International Airport with an A380-800 aircraft, 100% JetA1 fuel and 99% passenger fill rate. The weather conditions are the same as the cruise conditions.

From Fig. 2(b), the maximum fuel burn (ca. 7 $kg.s^{-1}$) is reached during take-off and landing phases, which in turn produces more emissions (concentrated on a brief period of time, compared to the total time flight). At cruise, the fuel consumption is ca. 2 $kg.s^{-1}$. From Fig. 2(c), the CO_2 emissions are ca. 22 $kg.s^{-1}$ during the take-off and landing phases, while it stabilizes at ca. 7 $kg.s^{-1}$ at cruise, correlated with the fuel consumption profile (Fig. 2(b)). Both fuel consumption and CO_2 emissions are about three times more important in the take-off and landing phases. Based on Fig. 2(d), the CO emission is ca. 0.0085 $kg.s^{-1}$ for the take-off and landing phases, while it stabilizes at ca. 0.0022 $kg.s^{-1}$ at cruise. Here again, CO emission during take-off and landing is nearly four times more than the cruise level. The same observation is made from Fig. 2(e), where NO_x emissions are ca. 0.5 $kg.s^{-1}$ during take-off and landing phases, while it stabilizes at ca. 0.04 $kg.s^{-1}$ at cruise. Thus, NO_x emissions during take-off and landing are about 12 times increased compared to the cruise level.

Within the sub-tab "Fleet", the results of all the simulations at the same time are depicted if the post-processing option was ticked. Then, the consumption and emissions statistics of the entire fleet simulated are also depicted. For the purpose of demonstration, a fleet of five flights was created, depicted in Table 1 (refer to Fig. 1 to see the ACEMIS graphic interface while fleet simulation is running).

The fuel consumption and CO_2 emissions distribution of the above fleet are shown in Fig. 3: there is a high fuel consumption and CO_2 emission region near departure and arrival airports (green region, Fig. 3(a) and (b)), corresponding to the take-off and landing phases.

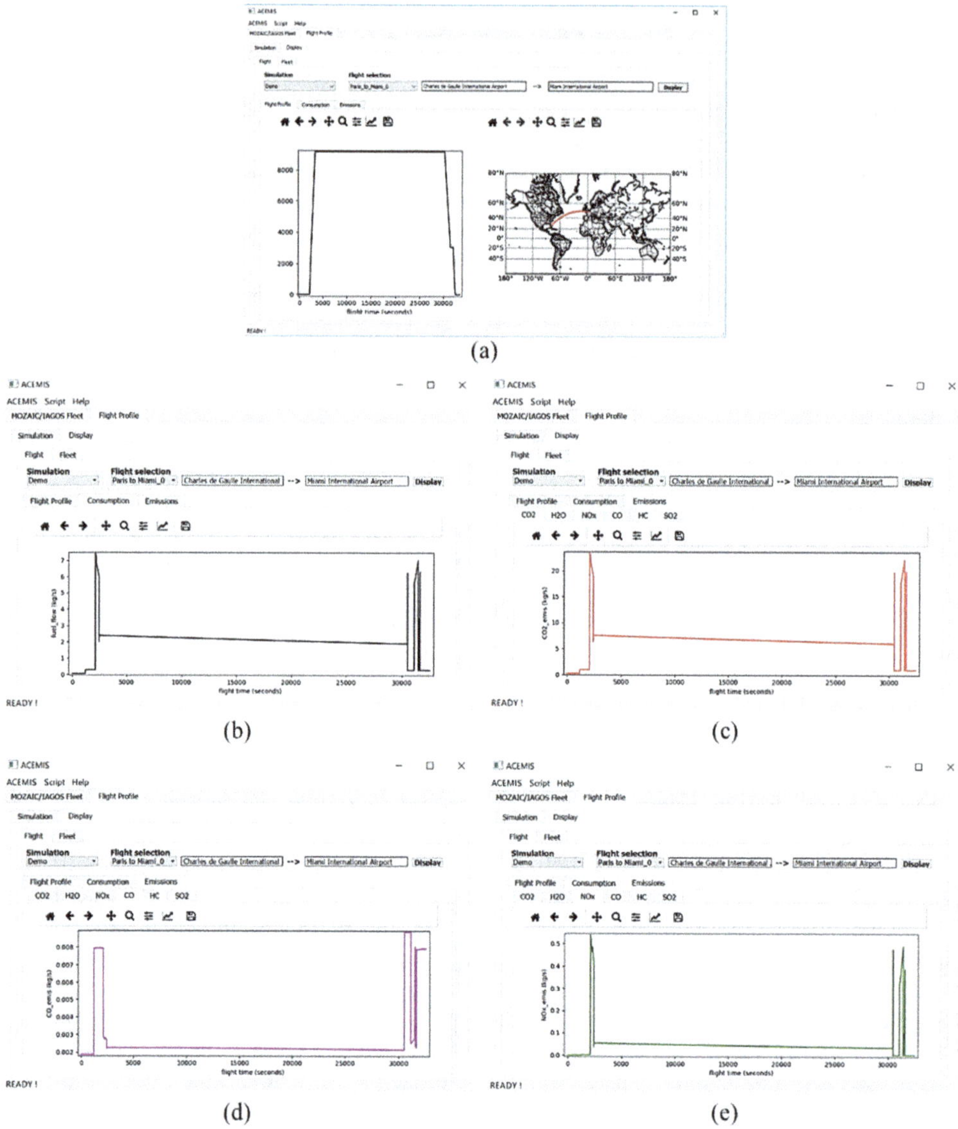

Figure 2: Display menu of “Flight Profile” part. (a) Flight altitude (left part) and orthodromic projection of the flight route on the map (right part); (b) Fuel consumption; (c) CO_2 emissions; (d) CO emissions; and (e) NO_x emissions, as a function of flight time.

The potential of the ACEMIS tool is depicted Fig. 4 hereafter: it shows the geographical distribution and intensity of CO_2 emissions over a sample of approximately 10,000 flights from Frankfurt International Airport in Germany (based on the MOZAIC/IAGOS database).

Table 1: Fleet simulation of five flights in ACEMIS.

	Flight 1	Flight 2	Flight 3	Flight 4	Flight 5
Simulation name	Demo_10	Demo_10	Demo_10	Demo_10	Demo_10
Flight name	CDG_MIA_A380_JetA1_10	CDG_BKK_A350_Cat1_10	FRA_HND_A380_Cat1_10	FRA_HKG_A350_JetA1_10	CDG_FRA_A320_JetA1_10
Aircraft	A380-800	A350-900	A380-800	A350-900	A320-200
Fuel 1	JetA1	–	–	JetA1	JetA1
Concentration fuel 1 (%)	100	–	–	100	100
Fuel 2	–	Catalytic cracking bio oil n°1	Catalytic cracking bio oil n°1	–	–
Concentration fuel 2 (%)	–	100	100	–	–
Departure Airport	Charles de Gaulle	Charles de Gaulle	Frankfurt-am-Main Airport	Frankfurt-am-Main Airport	Charles de Gaulle
Arrival airport	Miami International	Suvarnabhumi Airport	Tokyo Haneda International	Hong Kong International	Frankfurt-am-Main Airport
Filling rate	99%	90%	99%	90%	100%
Humidity (%)	32.3	40	35	38.5	40
Pressure (Pa)	26,838	28,439	31,892	27,485	30,210
Temperature (°C)	–48	–51	–46	–49	–42
Post-process?	Yes	Yes	Yes	Yes	Yes

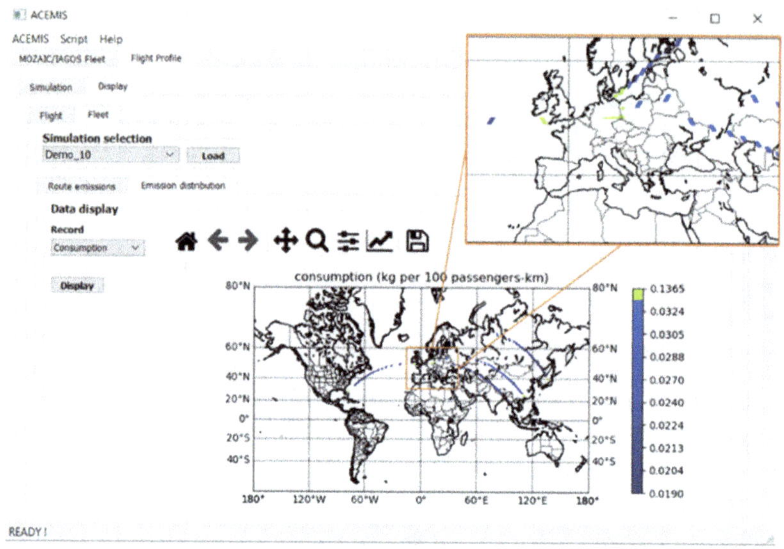

(a)

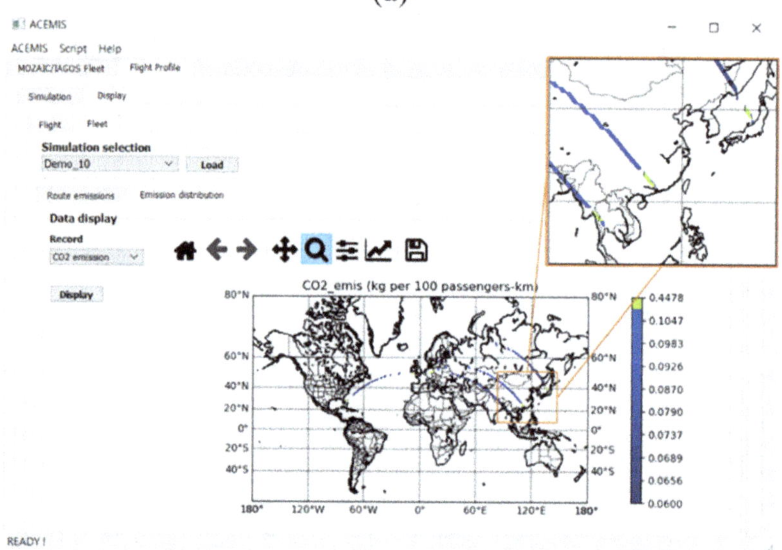

(b)

Figure 3: Display menu of "Flight Profile" part – Geographical distribution of (a) fuel consumption; (b) CO_2 emissions along the fleet routes.

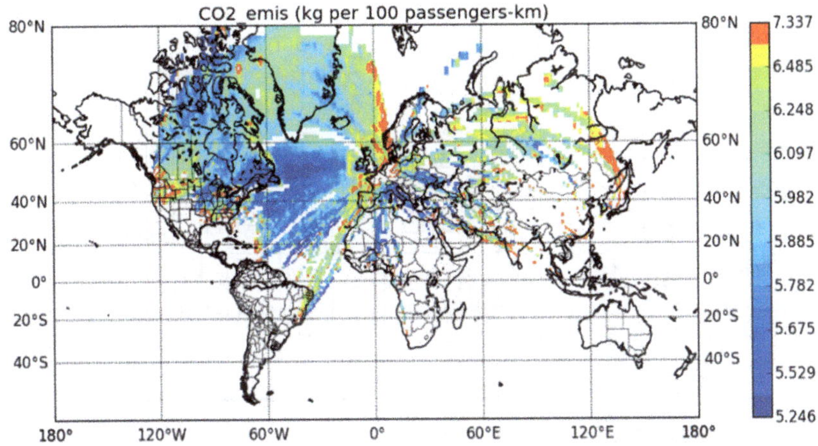

Figure 4: Display menu – Geographical distribution of CO_2 emission over a sample of approximately 10,000 flights.

4 DISCUSSION

In order to validate the ACEMIS emissions model, a comparison with free and available calculators were done. ICAO has provided an online carbon emission calculator tool for CO_2 emissions of a flight [11]. Furthermore, the CORSIA program has developed a tool called CO_2 Estimation and Reporting Tool (CERT) [12], with more options about aircrafts, fuels and other parameters compared to the ICAO online simulator. Similarly, in France, the *Direction Générale de l'Aviation Civile* (DGAC) put in place an online carbon emission calculator known as TARMAAC (which stands for *Traitements et Analyses des Rejets éMis dans l'Atmosphère par l'Aviation Civile*) [13].

Hereafter, Table 2 represents the comparison between ICAO online calculator, CERT tool from CORSIA program, and ACEMIS for the fuel consumption and CO_2 emissions related to the same flight profiles. The results show the same degree of magnitude for the total fuel consumption and the CO_2 emissions. The ICAO calculator does not take all the parameters for the flight simulation, such as the aircraft used, the type of fuel, the weather data or the passenger-filling rate of the aircraft [11], whereas the CERT from CORSIA and ACEMIS do. In addition, ICAO calculator assumes that all aircrafts are entirely configured with economic seats. The values calculated are an average of the type of aircrafts, fuel used, and passenger fill rate information, available in the ICAO database [11]. Since JetA1 fuel is the most common fuel used today, it is assumed that the ICAO simulations are based on flights operating with 100% JetA1, so the same conditions were chosen with the CERT and ACEMIS. Therefore, ICAO results are approximated via several factors. On the other hand, the CERT seems to approximate the fuel used during the flights, as well as the CO_2 emissions calculation (Table 2). ACEMIS results may be closer to reality, thanks to the calculations done in each flight segment, and experienced with the MOZAIC/IAGOS database [9]. Each ACEMIS simulation is specific to a single aircraft and engine, depending on various parameters. The results therefore are varying realistically, depending on the type of aircraft and its characteristics.

Table 2: Results of CO_2 emissions calculated with ICAO online simulator, CERT_CORSIA (simulator tool from CORSIA program), and ACEMIS v2.1 about multiple flights comparison. Fuel: 100% JetA1; Paris stands for Charles de Gaulle International Airport, New York for John F. Kennedy International Airport, Shanghai for Shanghai Pudong International Airport, Vienna for Vienna International Airport, and Dubai for Dubai International Airport.

Flight	Aircraft	ICAO online simulator			CERT_CORSIA			ACEMIS v2.1		
		Total fuel consumption (kg)	Total emissions CO_2 (kg)	Emissions CO_2/Fuel consumption	Total fuel consumption (kg)	Total emissions CO_2 (kg)	Emissions CO_2/Fuel consumption	Total fuel consumption (kg)	Total emissions CO_2 (kg)	Emissions CO_2/Fuel consumption
Paris–New York	A330-300	55,014.7	98,193.6	1.78	41,000.0	131,000.0	3.20	49,239.7	155,449.8	3.16
Paris–New York	A380-800	55,014.7	178,057.7	3.24	88,000.0	279,000.0	3.17	74,356.0	234,741.8	3.16
Paris–Shanghai	A330-200	88,673.9	97,953.6	1.10	63,000.0	200,000.0	3.17	63,099.3	199,204.6	3.16
Paris–Shanghai	A380-800	88,673.9	215,735.9	2.43	142,000.0	449,000.0	3.16	116,958.2	369,237.1	3.16
Paris–Vienna	A320-200	5,414.5	16,922.6	3.13	8,000.0	26,000.0	3.25	5,274.7	16,652.2	3.16
Paris–Vienna	E190	5,414.5	10,830.5	2.00	3,000.0	9,000.0	3.00	2,968.1	9,370.3	3.16
Paris–Dubai	A330-200	78,847.6	82,321.8	1.04	36,000.0	113,000.0	3.14	35,457.9	111,940.5	3.16
Paris–Dubai	A380-800	78,847.6	181,308	2.30	80,000.0	252,000.0	3.15	67,496.6	213,086.9	3.16

5 CONCLUSION

There is today a global consensus that climate change can be largely imputed to human activities, and that our society should move towards sustainable and renewable energy sources. ACEMIS tool is contributing towards aircraft traffic related pollution reduction. ACEMIS can be useful to airlines in many ways:

- for diagnosis purpose and continuous improvement of their environmental footprint, not only based on CO_2 emissions, but also considering other pollutants to estimate the global impact of their fleet activity;
- to enable airlines to not only comply with the environmental requirements (such as the CORSIA program), but also to prevent further environment restriction based on the other chemical pollutants.

This is equally helpful to the regulatory bodies in the aviation sector (e.g. DGAC/DSNA in France), engineers, and environmental auditors (among others) to analyze the emissions levels. It also leads to the awareness of interested party about the most/the least polluted regions, along the flight path. This in turn could help to schedule future routes as avoiding climate sensitive regions might be the most promising approach to reduce the climate impact of non-CO_2 emissions [14].

ACEMIS is a scalable tool developed using Python programming language. Due to the independent nature of the modules, the future aircraft technologies affecting the simulation can be easily integrated within existing or novel modules. With a simple update of the database files, it is possible to integrate new aeronautics developments, such as disruptive aircraft design, novel engines, biofuels, electricity and hydrogen technologies… The Capgemini Engineering group is currently working on new features such as simulation of noise around airports, inclusion of emission simulation of cargo flights, 3D results display and inclusion of electrical and hydrogen powered engines, in order to provide one of the first pollutants simulation of these novel technologies. The final objective is to get a solution that is beneficial to future users, thanks to quick updates and easy integration of new technologies.

6 SOFTWARE AVAILABILITY

Name: ACEMIS.
Developer: Capgemini Engineering.
Contact: Capgemini Engineering (www.altran.com).
Programming framework: Python.
Language: English.
Year first available: 2021.

ACKNOWLEDGEMENT

ACEMIS is the result of the collective discussions of numerous people of the Technology and Engineering Center of the Capgemini Engineering Group who worked in this project, at various time and interval. We would like to thank everyone who contributed, and particularly Sébastien Garnier and Thomas Noiret who initiated this work.

MOZAIC/CARIBIC/IAGOS data were created with support from the European Commission, national agencies in Germany (BMBF), France (MESR), and the UK (NERC), and the IAGOS member institutions (http://www.iagos.org/partners). The participating airlines (Lufthansa, Air France, Austrian, China Airlines, Iberia, Cathay Pacific, Air Namibia, Sabena) supported IAGOS by carrying the measurement equipment free of charge since 1994. The data are available at http://www.iagos.fr thanks to additional support from AERIS.

REFERENCES

[1] European Aviation Environmental Report 2019. https://www.easa.europa.eu/eaer/system/files/usr_uploaded/219473_EASA_EAER_2019_WEB_HI-RES_190311.pdf.

[2] Aviation and the Global Atmosphere, IPCC special report, 1999. https://www.ipcc.ch/site/assets/uploads/2018/03/av-en-1.pdf.

[3] The Growth in Greenhouse Gas Emissions from Commercial Aviation, Part 1 of a series on airlines and climate changes, 2019. https://www.eesi.org/papers/view/fact-sheet-the-growth-in-greenhouse-gas-emissions-from-commercial-aviation. Accessed on: 19 Apr. 2021.

[4] Aviation: 2 to 3 times more damaging to the climate than industry claims, 2018. https://www.transportenvironment.org/news/aviation-2–3-times-more-damaging-climate-industry-claims. Accessed on: 19 Apr. 2021.

[5] Reducing emissions from aviation. https://ec.europa.eu/clima/policies/transport/aviation_en#tab-0-0. Accessed on: 19 Apr. 2021.

[6] International Civil Aviation Organization (ICAO), Annex 16 to the Convention on International Civil Aviation, Environmental Protection, vol. IV, Carbon Offsetting and Reduction Scheme for International Aviation (CORSIA), 1st ed., Oct. 2018 (full report).

[7] Schaefer, M., Methodologies for aviation emission calculation: A comparison of alternative approaches towards 4D global inventories. Thesis, Berlin University of Technology, 2006. DOI: 10.14279/depositonce-1476.

[8] Air Transport Action Group, Facts and Figures. https://www.atag.org/facts-figures.html. Accessed on: 19 Apr. 2021.

[9] In-service Aircraft for a Global Observing System. https://www.iagos.org/, and data http://www.iagos-data.fr/. Accessed on: 19 Apr. 2021.

[10] DuBois, D. & Paynter, G.C., Fuel flow Method2 for estimating aircraft emissions. *SAE Transactions*, **115**(1), pp. 1–14. https://www.jstor.org/stable/44657657.

[11] International Civil Aviation Organization (ICAO), ICAO Carbon Emissions Calculator. https://www.icao.int/environmental-protection/CarbonOffset/Pages/default.aspx. Accessed on: 19 Apr. 2021.

[12] International Civil Aviation Organization (ICAO), Carbon Offsetting and Reduction Scheme for International Aviation (CORSIA), CO_2 Estimation and Reporting Tool (CERT). https://www.icao.int/environmental-protection/CORSIA/Pages/CERT.aspx. Accessed on: 19 Apr. 2021.

[13] Calculateur d'émissions de gaz à effet de serre de l'aviation, Direction Générale de l'Aviation Civile. https://eco-calculateur.dta.aviation-civile.gouv.fr/. Accessed on: 19 Apr. 2021. (In French.)

[14] Climate Change 2013, The Physical Science Basis, Intergovernmental Panel on Climate Change, Report, 2013. https://www.ipcc.ch/site/assets/uploads/2018/02/WG1AR5_all_final.pdf.

EFFECTS OF AEROSOLS FEEDBACKS IN MODELING METEOROLOGY AND AIR QUALITY IN THE ANDEAN REGION OF SOUTHERN ECUADOR

RENE PARRA
Instituto de Simulación Computacional (ISC-USFQ), Colegio de Ciencias e Ingeniería,
Universidad San Francisco de Quito, Ecuador

ABSTRACT

Feedbacks are essential in atmospheric modeling. In this arena, aerosols promote direct and indirect effects. Direct effects are related to the scattering and absorption of solar radiation. Indirect effects correspond to cloud formation and cloud properties, which change the vertical temperature profile, relative humidity, and atmospheric stability. Modeling with direct and indirect effects should improve numerical performance based on a more detailed description of interactions. Benefits for air quality modeling were identified in the city Cuenca (Andean region of Southern Ecuador, 2,500 masl) when working with direct effects, compared to modeling without aerosols feedbacks. These benefits created expectations about the potential improvement when including direct and indirect effects. Therefore, we used the Weather Research and Forecasting with Chemistry Version 3.2 (WRF-Chem 3.2) model to assess aerosols feedbacks in modeling the meteorology and transport of air pollutants in Cuenca during September 2014. We considered the following scenarios: (1) No aerosols interaction (NI), (2) direct effects (DE), and (3) direct and indirect effects (DIE). Records were considered positive modeled if the maximum deviation between observed and modeled values agreed with prescribed accuracies (100% means all the records were captured by modeling). The NI scenario captured 82.7, 74.8, and 75.6% of short-term air quality, long-term air quality, and meteorological records. The DE scenario kept the performance for short-term (82.7%) and long-term (74.8%) air quality and increased to 76.2% for meteorological records. However, the DIE scenario decreased to 75.3% for short-term air quality, although it increased to 76.8% for meteorology. The inclusion of indirect feedbacks decreased the $PM_{2.5}$ modeling performance. WRF-Chem allows selecting numerous physics and scheme options that need to be assessed to identify a suitable combination for each region. Results can be notably affected by the selected parameters. Other aerosols schemes, alternatives for generating initial and boundary conditions, data assimilation, and updated emission inventories need to be assessed as components directly involved in modeling performance.

Keywords: WRF-Chem, Cuenca, modeling performance, direct effects, indirect effects.

1 INTRODUCTION

Air pollution is the result of complex interactions between atmospheric emissions and meteorology. Feedbacks are essential in atmospheric modeling. Recent research has shown that meteorology and chemistry feedbacks are essential for numerical weather and air quality forecasting [1]. In this arena, aerosols promote direct and indirect effects. Direct effects are related to the scattering and absorption of solar radiation [2]. Indirect effects correspond to cloud formation and cloud properties, which change the vertical temperature profile, relative humidity, and atmospheric stability. Therefore and based on detailed interactions, it is expected that including direct and indirect effects will improve atmospheric modeling performance.

Modeling both meteorology and air quality is challenging, especially in regions with complex topography, as the Andean region of South America. Based on the Eulerian Weather Research and Forecasting with Chemistry model (WRF-Chem), promising results were obtained for Cuenca [3], [4], a high (2,500 masl) Andean city of Southern Ecuador (Fig. 1). WRF-Chem is an online chemical transport model that includes treating the

WIT Transactions on Ecology and the Environment, Vol 252, © 2021 WIT Press
www.witpress.com, ISSN 1743-3541 (on-line)
doi:10.2495/AIR210041

aerosols direct and indirect effects [5] with numerous options and parameters. WRF-Chem was used to assess the influence of planetary boundary layer schemes when modeling meteorology and air quality [3] and assessing the influence on air quality due to the shift from diesel to electric buses in Cuenca [4].

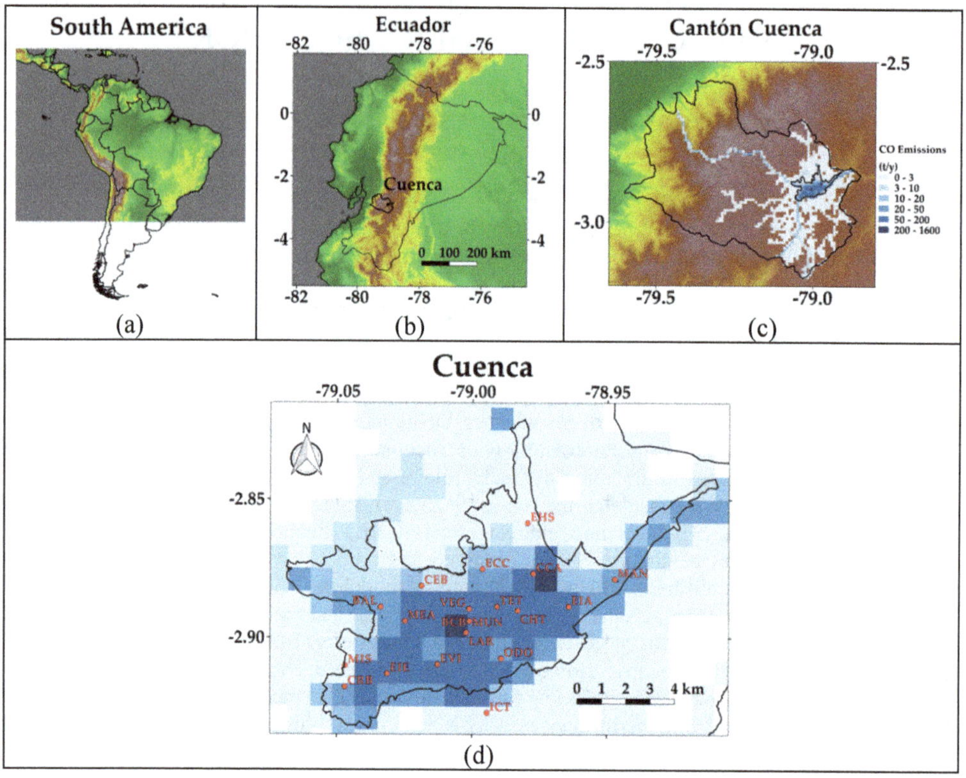

Figure 1: Location of: (a) Ecuador; (b) Cuenca; (c) CO emissions (t^{y-1}); (d) Urban area of Cuenca and the air quality stations (red dots). MUN station (2,500 masl) provides short-term air quality levels and meteorology. The others are passive stations for monitoring monthly-mean air quality levels.

Benefits for air quality modeling were identified in Cuenca when working with direct effects, compared to modeling without aerosols feedbacks [3]. These benefits created expectations about the potential improvement of a complete description of aerosols feedbacks through direct and indirect effects. Therefore, the main objective of this contribution focused on a preliminary assessment of the effects of a complete inclusion of aerosols feedbacks in modeling meteorology and air quality in this city.

2 METHOD

2.1 Emission inventory of Cuenca

The last emission inventory of Cuenca was made for the year 2014 [6], including the contribution from on-road traffic, vegetation, industries, a power facility, use of solvents, service stations, domestic GLP consumption, air traffic, landfills, handcrafted production of bricks, dust erosion, and mining. On-road traffic was the main source of most of the pollutants (94.9% of carbon monoxide (CO); 71.2% of nitrogen oxides (NO_x), 42.4% of fine particulate matter ($PM_{2.5}$), 39.6% of non-methane volatile organic compounds (NMVOC). Industries were the most important source (60.1%) of sulfur dioxide (SO_2). The power facility located at the NE of the urban area was also a significant source of SO_2 (35.1%) and NO_x (18.5%). About 600 artisanal producers of bricks located at the NW of the city were significant contributors of $PM_{2.5}$ (38.5%). Fig. 1 depicts the spatial distribution (1 km × 1 km of resolution) of CO emissions during 2014.

From this emission inventory, we generated hourly speciated emissions data for this contribution.

2.2 Modeling approach

We used the WRF-Chem Version 3.2 for modeling meteorology and air quality in Cuenca during September of 2014 for the following three scenarios: (1) No aerosols interaction (NI), (2) direct effects (DE), and (3) direct and indirect effects (DIE). September was selected because emissions from on-road traffic and other sources are representative for other months. Additionally, O_3 levels can be higher during this month than the World Health Organization guideline (100 µg m^{-3}, maximum 8 h mean). Meteorological simulations were carried through a master domain of 70 × 70 cells (27 × 27 km each) and three nested sub-domains. The third (100 × 82 cells of 1 km each and 35 vertical levels) covered the region of Cuenca (Fig. 1). Initial and boundary conditions were generated from the final NCEP FNL Operational Global Analysis data [7]. For the inner sub-domain, the chemistry option of WRF-Chem was activated, using the CBMZ [8] for gaseous pollutants and the MOSAIC [9] for aerosols. Table 1 indicates the schemes and options used for modeling.

Table 1: Schemes and options for modeling meteorology and the air quality in Cuenca (WRF-Chem V3.2) [10].

Component	Modeling approach			Scheme/model
	NI	DE	DIE	
Microphysics	2	2	2	Lin et al. [11]
Longwave radiation	1	1	1	RRTM [12]
Shortwave radiation	2	2	2	2 Goddard [13]
Surface layer	1	1	1	MM5 similarity [14]
Land surface	2	2	2	Unified Noah land surface [15]
Planetary boundary layer	1	1	1	Yonsei University [16]
Cumulus parameterization	5	5	5	Grell 3D Ensemble [17]
Chem_opt	7*	7*	9**	
Wet scavenging	Off	Off	On	
Cloud chemistry	Off	Off	On	
Aerosols to radiation feedback	Off	On	On	

*CBMZ chemical mechanism and MOSAIC using 4 sectional aerosol bins. **CBMZ chemical mechanism using 4 sectional aerosols bins including aqueous reactions.

2.3 The air quality network of Cuenca

Since 2012, the city has one automatic station in the historic center (MUN station, Fig. 1), monitoring meteorology and short-term air quality. There are also about 20 stations with passive sensors for measuring monthly-mean air quality levels of NO_2 and O_3. Monitoring is based on methods established in the Ecuadorian air quality regulation, under the responsibility of the local government.

Meteorological and air quality records were considered positive modeled if the maximum deviation between observed and modeled values agreed with the accuracies (100% means all the records were captured by modeling) of Table 2. Also, modeled meteorological and short-term air quality values were assessed through the indicators of Table 2, corresponding to the following expressions:

$$GE = \frac{1}{N}\sum_{i=1}^{N}|Pi - Oi|,\ BIAS = Pm - Om,\ RMSE = \sqrt{\frac{1}{N}\sum_{i=1}^{N}(Pi - Oi)^2},$$

where GE = gross error; RMSE = root mean square error; N = number of values; Pm = mean modeled value; Om = mean observed value; Pi = modeled value; and Oi = observed value.

Table 2: Accuracy and indicators for meteorological and air quality modelling [18].

Parameter	Accuracy	Indicator	Benchmark	Ideal value
Meteorology:				
Hourly surface temperature	±2°C	GE	< 2 K	
		BIAS	< ± 0.5 K	
Hourly wind speed at 10 m above the surface	± 1 m s^{-1}	RMSE	< 2 m s^{-1}	
		BIAS	< ± 0.5 m s^{-1}	
Short-term air quality:				
Maximum 8 h CO mean	± 50% mg m^{-3}	RMSE		0
Maximum 8 h O_3 and 24 h $PM_{2.5}$ means	± 50% µg m^{-3}	RMSE		0
Long-term air quality:				
Monthly NO_2 and O_3 means	± 30% µg m^{-3}			

3 RESULTS AND DISCUSSION

Fig. 2 compares the observed and modeled temperatures. The NI and DE scenarios reached performances of 75.9 and 76.2%, respectively (Table 3). The DIE scenario increased the performance to 77.0%. Both GE and BIAS were in the benchmark range for temperature (Table 4). The observed wind rose indicated wind coming mainly from the east–northeast (ENE) direction (Fig. 3). However, modeled results indicated the east (E) as the main direction of the coming wind. The NI scenario captured 75.3% of observed wind speed values, The DE and DIE scenarios increased to 76.1 and 76.6%, respectively. GE and BIAS were in the benchmark range for wind speed (Table 4).

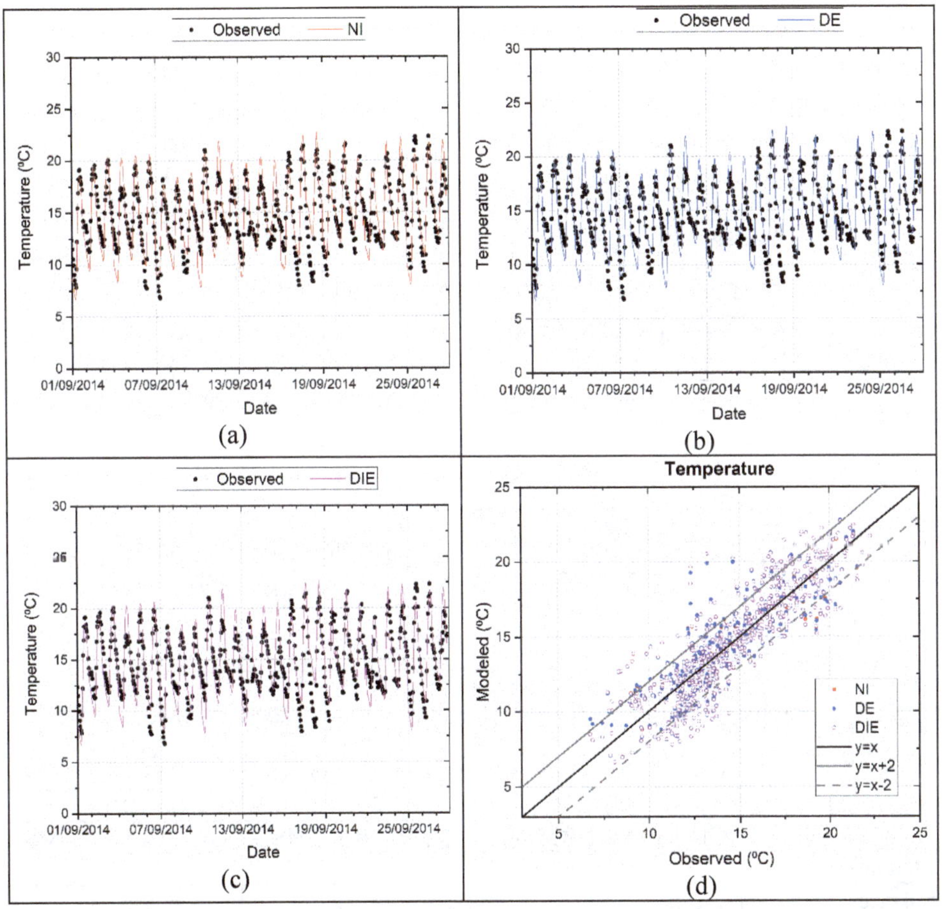

Figure 2: Observed and modeled temperature. (a) No aerosol interactions (NI); (b) Direct effects (DE); (c) Direct and indirect effects (DIE); and (d) Observed vs. modeled temperatures.

The DE mean daily profile for temperature was similar to NI (not showed). The DIE hourly mean values showed variations between –0.1 to 0.2°C compared to NI. The mean daily modeled profiles of global solar radiation and wind speed overestimated the observed values during daylight hours (Fig. 4). The DE mean profile for solar radiation showed reductions up to –4 W m^{-2} than NI. On contrary, the DIE increased hourly mean values up to 56 W m^{-2}. The DE mean profile for wind speed was similar to NI (Fig. 4), although the DIE showed variations between –0.1 to 0.1 m s^{-1} compared to NI. The average performances for modeling temperature and wind speed for the NI, DE, and DIE scenarios were 75.6, 76.2, and 76.8%, respectively.

Due to the numerous interactions involved, it is challenging the interpretations of results when modeling with aerosols feedbacks [19]. Direct effects can reduce surface temperature due to the reflection of solar radiation. However, the corresponding layer of the atmosphere can be warmed simultaneously due to the absorption of solar radiation by black carbon. The

Table 3: Percentage of records positively captured in modeling meteorology and air quality.

	Modeling approach		
Meteorology:	NI	DE	DIE
Surface temperature	75.9	76.2	77.0
Wind speed at 10 m	75.3	76.1	76.6
Average performance	75.6	76.2	76.8
Short-term air quality:			
CO, daily maximum 8 h mean	100.0	100.0	100.0
$PM_{2.5}$, daily 24 h mean	63.0	63.0	40.7
O_3, daily maximum 8 h mean	85.2	85.2	85.2
Average performance	82.7	82.7	75.3
Long-term air quality:			
NO_2, monthly mean	93.3	93.3	93.3
O_3, monthly mean	56.3	56.3	56.3
Average performance	74.8	74.8	74.8
Global performance	77.7	77.9	75.6

Table 4: Indicators for modeling meteorology and air quality.

	Modeling approach								
	NI			DE			DIE		
	GE	BIAS	RMSE	GE	BIAS	RMSE	GE	BIAS	RMSE
Meteorology:									
Surface temperature	1.3	0.1	–	1.3	0.1	–	1.3	0.0	–
Wind speed at 10 m	–	0.3	1.0	–	0.3	1.0	–	0.3	1.0
Short-term air quality:									
CO, daily maximum 8 h mean	–	–	0.19	–	–	0.19	–	–	0.18
$PM_{2.5}$, daily 24 h mean	–	–	3.05	–	–	3.04	–	–	5.19
O_3, daily maximum 8 h mean	–	–	19.1	–	–	19.1	–	–	19.4

increased solar radiation values provided by the DIE approach can be attributed to the increased light scattering in partly cloudy conditions.

Modeling performances and RMSE values for short-term air quality were equals or similar for CO (Table 3, Table 4, Fig. 5(a)) and O_3 (Fig. 5(c)).

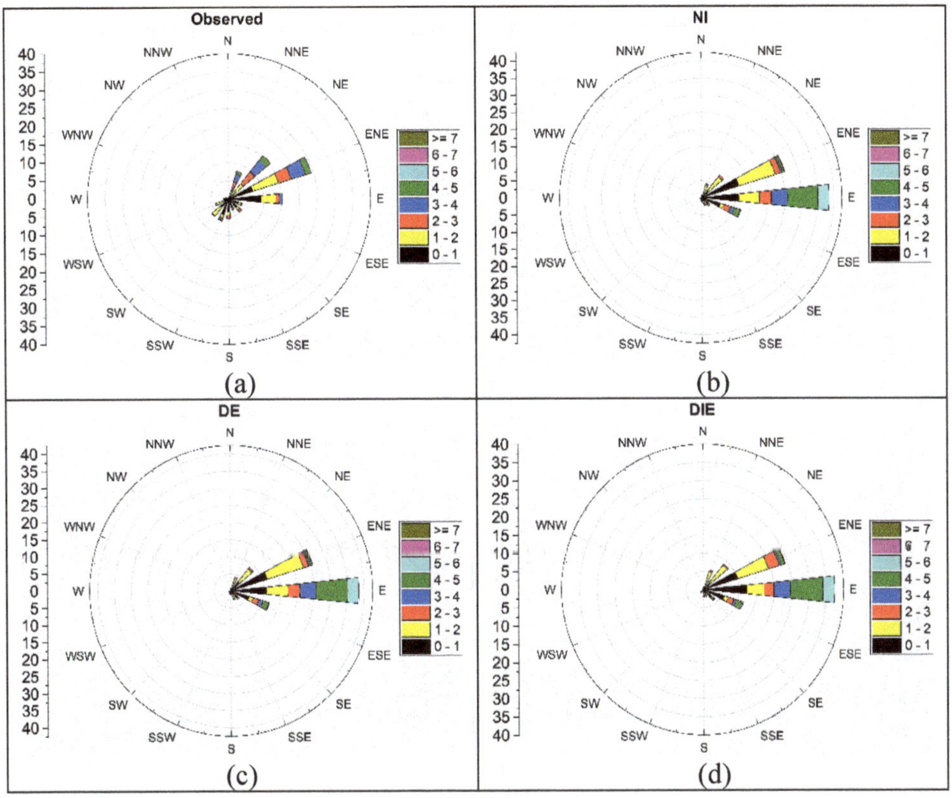

Figure 3: Wind roses. (a) Observed; (b) No aerosol interactions (NI); (c) Direct effects (DE); and (d) Direct and indirect effects (DIE).

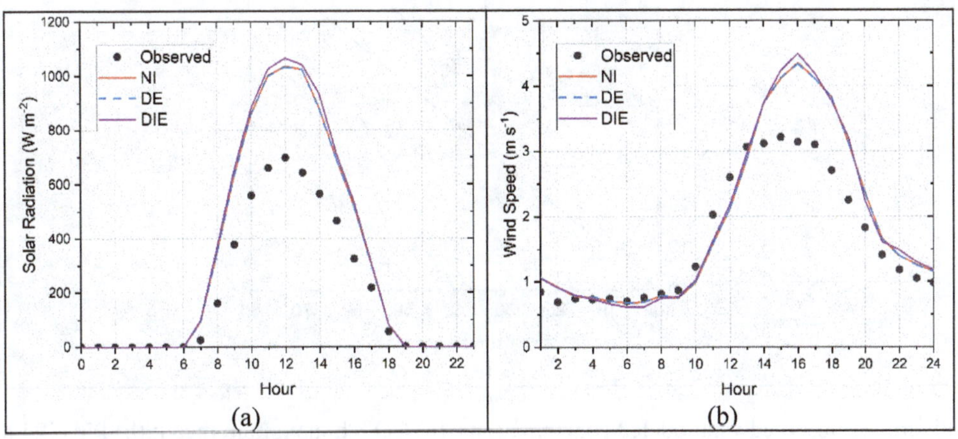

Figure 4: Mean daily profiles. (a) Global solar radiation; and (b) Wind speed.

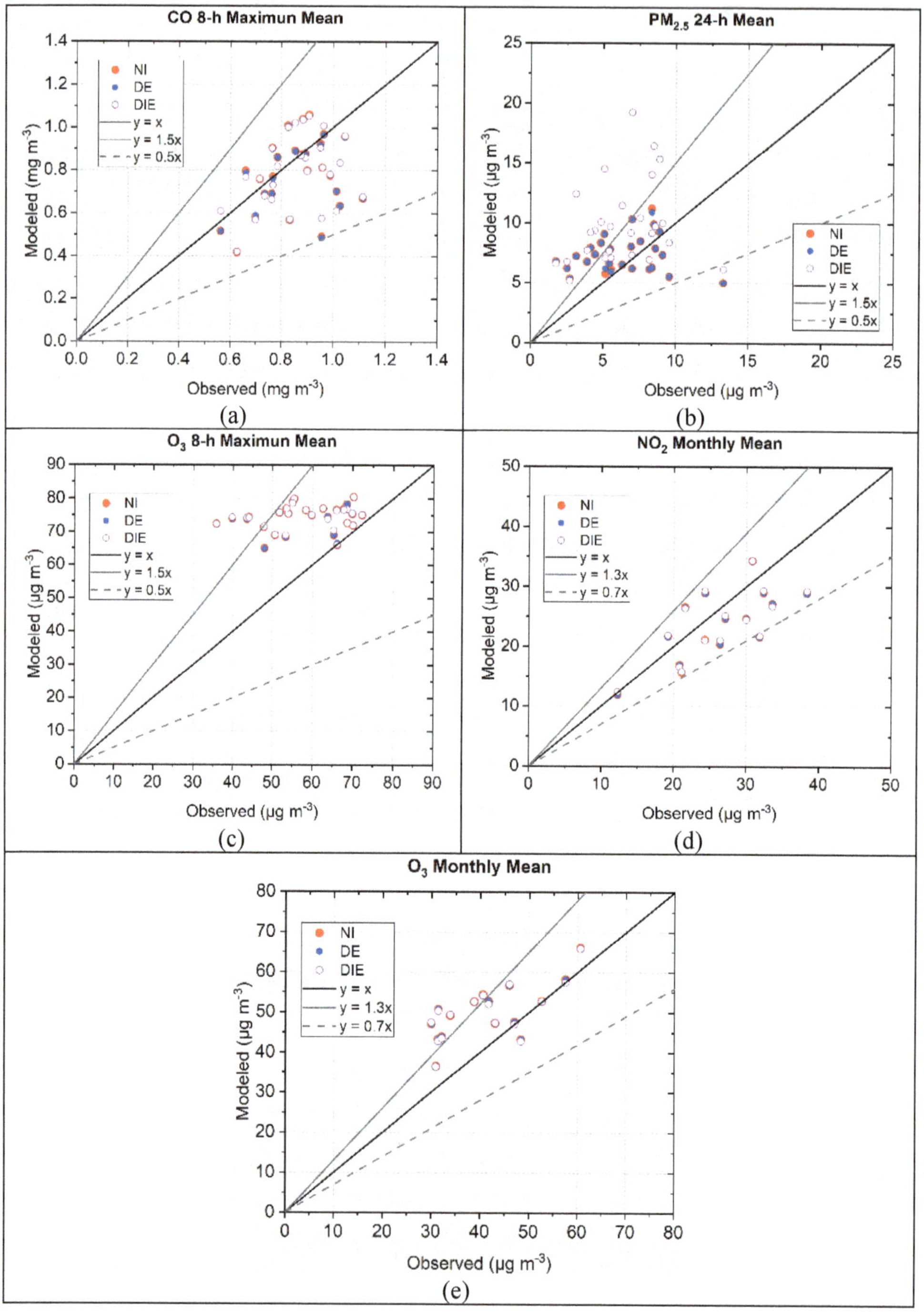

Figure 5: Observed vs. modeled concentrations. (a) CO 8 h maximum mean; (b) $PM_{2.5}$ 24 h mean; (c) O_3 8 h maximum mean; (d) NO_2 monthly mean; and (e) O_3 monthly mean.

For $PM_{2.5}$, the NI and DE reached modeling performances of 63.3%. However, it decreased to 40.7% when using the DIE approach (Fig. 5(b)). The RMSE parameter was similar for NI and DE (3 $\mu g\ m^{-3}$, Table 4), but increased to 5.2 for DIE.

NI and DE captured 82.7% of records for short-term air quality, which was decreased to 75.3% by DIE. For all the approaches, the performances for long-term air quality were the same (74.8%)(Table 3, Fig. 5(d) and 5(e)).

Results indicated a minute improvement performance when modeling with direct effects than without aerosols interactions. An additional improvement was obtained for meteorology using direct and indirect effects, although this approach increased the $PM_{2.5}$ concentrations. This increase resulted in higher modeled values than $PM_{2.5}$ records, decreasing the modeling performance as a consequence.

This result is consistent with the decrease of performance reported by Makar et al. [20] when modeling particulate matter using direct and indirect effects over domains in North America and Europe (Table 5). In Cuenca, on average, $PM_{2.5}$ mean daily concentrations were increased up to 2.5 $\mu g\ m^{-3}$ when considering indirect effects. For example, Fig. 6 shows the modeled $PM_{2.5}$ concentrations by the NI and DIE approaches on 21 September 2014, which depicts higher values for DIE in the historic center. This increase is consistent with the increase between 1 to 10 $\mu g\ m^{-3}$ for monthly PM_{10}, reported by Forkel et al. [21] for July 2006 over large parts of Continental Europe. In the same way, Tucella et al. [22] reported an overestimation of particle number concentration when modeling an event over Europe with direct and indirect feedbacks. However, Liu et al. [23] reported good performance for most meteorological fields and chemical concentrations when modeling over East Asia (Table 5).

The increase in $PM_{2.5}$ concentrations, when including indirect effects poises whether $PM_{2.5}$ emissions were overestimated or if this increase was excessive due to inadequate modeling of the aerosols indirect effects.

The treatment of indirect effects is one of the critical uncertainties. A better representation of aerosol processes and feedbacks, their interactions with clouds and radiation; currently, components need to be improved [24], especially for regions of complex topography as the Andean zone.

WRF-Chem allows selecting numerous physics and scheme options that need to be assessed to identify a suitable combination for the region. Results can be notably affected by the selected parameters. Land surface models and schemes of cumulus parameterizations deserve a special effort as components affecting atmospheric modeling.

Updated aerosols schemes, initial and boundary conditions, data assimilation, fusion with machine learning and artificial intelligence approach, and even the emission inventory need to be assessed as components directly involved in modeling performance.

4 CONCLUSIONS

To our knowledge, we made the first exercise of modeling meteorology and air quality in Cuenca, considering direct and indirect aerosols interactions.

Although being a preliminary assessment, our results suggested that modeling meteorology and air quality for the Andean region of Ecuador can be improved using direct aerosols effects. The parameterizations used for indirect feedbacks did not improve or even decreased the $PM_{2.5}$ modeling performance, which poises whether $PM_{2.5}$ emissions were overestimated or if this increase was excessive due to inadequate modeling of the aerosols indirect effects.

Table 5: Comparison with other studies on the influence of aerosols feedbacks modeling.

	Case or reference				
	This assessment	Makar et al. [20]	Forkel et al. [21]	Tucella et al. [22]	Liu et al. [23]
Region	Cuenca, Ecuador	North America and Europe	Europe	Europe	East Asia
Period	September 2014	2006 and 2010 for North America 2010 for Europe	June and July 2006	15 May 2008	January, April, July, and October 2008
Parameters assessed	Meteorology and air quality	Meteorology and air quality	Meteorology and air quality	Aerosols and cloud droplets number concentrations (CNDC)	Meteorology and air quality
Models	WRF-Chem 3.2	GEM-MACH, WRF-Chem 3.4.1 and WRF-CMAQ	WRF-Chem 3.3	WRF-Chem 3.4	WRF-Chem and MADRID
Spatial resolution	1 km	0.25° (≈27.8 km)	22.5 km	30 km	36 km
Summary of results reported about modeling performance	Higher values of $PM_{2.5}$. Higher values of solar radiation at surface. Modeling performance can be improved using direct aerosols effects. Using direct and indirect effects did not improve or even degraded modeling performance.	Summertime model performance for O_3 and other gases improved, although the particulate matter performance was degraded. Aerosol indirect effect dominated feedbacks compared to the direct effect. Direct and indirect effects competed.	Higher mean PM_{10} near the ground was simulated for significant parts of continental Europe. Significantly higher values of solar radiation at the surface for a large part of the model domain.	Overestimation of particle number concentration by a factor of 2 to 2.5 CDNC was overestimated by a factor of 5. Probably there was an excessive nucleation rate.	The model reproduced reasonably well the spatial and seasonal variations of most meteorological fields and chemical concentrations. Limitations by uncertainties in emissions and/or imperfectness in modeling aerosol processes.

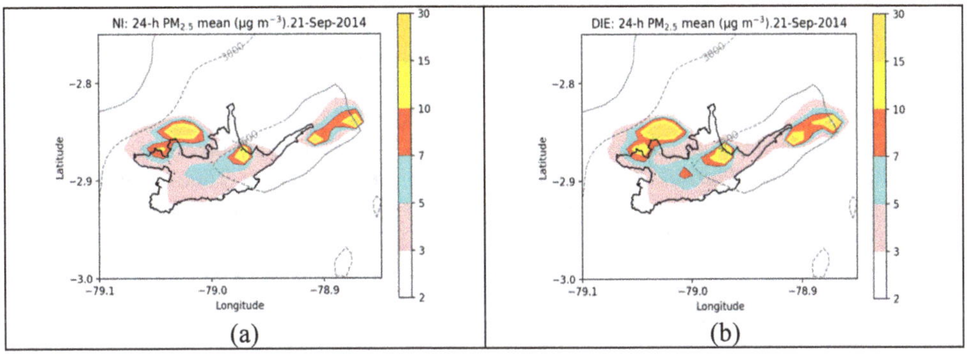

Figure 6: Modeled 24 h $PM_{2.5}$ mean at 21 September 2014: (a) No interaction (NI), (b) Direct and Indirect Effects (DIE).

ACKNOWLEDGEMENT

This research is part of the project "Emisiones y Contaminación Atmosférica en el Ecuador 2021–2022". Simulations were done at the High Performance Computing system at the USFQ.

REFERENCES

[1] Baklanov, A. et al., Online coupled regional meteorology chemistry models in Europe: Current status and prospects. *Atmospheric Chemistry and Physics*, **14**(1), pp. 317–398, 2014.
[2] Boucher, O., Atmospheric aerosols: Properties and climate impacts. Springer Netherlands, 2015. https://www.springer.com/gp/book/9789401796484. Accessed on: 14 Apr. 2021.
[3] Parra, R., Performance studies of planetary boundary layer schemes in WRF-Chem for the Andean region of southern Ecuador. *Atmospheric Pollution Research*, **9**(3), pp. 411–428, 2018.
[4] Parra, R. & Espinoza, C., Insights for air quality management from modeling and record studies in Cuenca, Ecuador. *Atmosphere*, **11**(9), p. 998, 2020.
[5] WRF-Chem, https://ruc.noaa.gov/wrf/wrf-chem/. Accessed on: 2 May 2021.
[6] EMOV EP, Inventario de Emisiones Atmosféricas del Cantón Cuenca 2014, p. 88. https://www.researchgate.net/publication/311789933_Inventario_de_Emisiones_Atmosfericas_del_Canton_Cuenca_2014. Accessed on: 2 May 2021.
[7] Commerce NC for EPWSSD of NCEP FNL Operational Model Global Tropospheric Analyses, continuing from Jul. 1999. UCAR/NCAR: Research Data Archive, 2000. https://rda.ucar.edu/datasets/ds083.2/. Accessed on: 14 May 2021.
[8] Zaveri, R.A. & Peters, L.K., A new lumped structure photochemical mechanism for large-scale applications. *Journal of Geophysical Research: Atmospheres*, **104**(D23), pp. 30387–30415, 1999.
[9] Zaveri, R.A., Easter, R.C., Fast, J.D. & Peters, L.K., Model for simulating aerosol interactions and chemistry (MOSAIC). *Journal of Geophysical Research*, **113**(D13), D13204, 2008.
[10] Skamarock, W.C. et al., A description of the advanced research WRF version 3:125.

[11] Lin, Y.-L., Farley R.D. & Orville H.D., Bulk parameterization of the snow field in a cloud model. *Journal of Climate and Applied Meteorology*, **22**(6), pp. 1065–1092, 1983.

[12] Mlawer, E.J., Taubman, S.J., Brown, P.D., Iacono, M.J. & Clough, S.A., Radiative transfer for inhomogeneous atmospheres: RRTM, a validated correlated-k model for the longwave. *Journal of Geophysical Research*, **102**(D14), pp. 16663–16682, 1997.

[13] Chou, M.-D. & Suarez, M.J., A solar radiation parameterization (CLIRAD-SW) for atmospheric studies, 48.

[14] Paulson, C.A., The mathematical representation of wind speed and temperature profiles in the unstable atmospheric surface layer. *Journal of Applied Meteorology and Climatology*, **9**(6), pp. 857–861, 1970.

[15] Tewari, M.F. et al., Implementation and verification of the unified NOAH land surface model in the WRF model. *20th Conference on Weather Analysis and Forecasting. 16th Conference on Numerical Weather Prediction*, Seattle, pp. 11–15, 2004.

[16] Hong, S.-Y., Noh, Y. & Dudhia, J., A new vertical diffusion package with an explicit treatment of entrainment processes. *Monthly Weather Review*, **134**(9), pp. 2318–2341, 2006.

[17] Grell, G.A., Prognostic evaluation of assumptions used by cumulus parameterizations. *Monthly Weather Review*, **121**, pp. 764–787, 1993.
DOI: 10.1175/1520-0493(1993)121<0764:PEOAUB>2.0.CO;2.

[18] European Environment Agency, *The Application of Models Under the European Union's Air Quality Directive : A Technical Reference Guide [Internet]*, Publications Office: LU, 2011. https://data.europa.eu/doi/10.2800/80600. Accessed on: 9 Apr. 2021.

[19] Rosenfeld, D. et al., Global observations of aerosol-cloud-precipitation-climate interactions. *Reviews of Geophysics*, **52**, pp. 750–808, 2014.

[20] Makar, P.A. et al., Feedbacks between air pollution and weather, Part 2: Effects on chemistry. *Atmospheric Environment*, **115**, pp. 499–526, 2015.

[21] Forkel, R. et al., Effect of aerosol-radiation feedback on regional air quality: A case study with WRF-Chem. *Atmospheric Environment*, **53**, pp. 202–211, 2012.

[22] Tuccella, P., Curci, G., Crumeyrolle, S. & Visconti, G., Modeling of aerosol indirect effects with WRF-Chem over Europe. *Air Pollution Modeling and its Application XXIII*, eds D. Steyn & R. Mathur, Springer: Cham, pp. 91–95, 2014.
DOI: 10.1007/978-3-319-04379-1_15.

[23] Liu, X.-Y., Zhang, Y., Zhang, Q. & He, K.-B., Application of online-coupled WRF/Chem-MADRID in East Asia: Model evaluation and climatic effects of anthropogenic aerosols. *Atmospheric Environment*, **124**, pp. 321–336, 2016.

[24] Baklanov, A. & Zhang, Y., Advances in air quality modeling and forecasting. *Global Transitions*, **2**, pp. 261–270, 2020.

SECTION 2
INDOOR AIR POLLUTION

SEASONAL CONCENTRATIONS OF ELEMENTS IN BOTH INDOOR AND OUTDOOR HOUSING IN $PM_{2.5}$

AMGALAN NATSAGDORJ[1], BULGANKHANGAI TUGSBAYAN[1], TSELMUUNGARAV BATTUMUR[1], BOLORMAA OYUNTSETSEG[1], JIYI LEE[2], YONG PYO KIM[3] & BATTSETSEG USUHBAYAR[4]
[1]Department of Chemistry, School of Arts and Sciences, National University of Mongolia, Mongolia
[2]Department of Environmental Science and Engineering, Ewha Womans University, Republic of Korea
[3]Department of Chemical Engineering and Materials Science,
Graduate Program in System Health Science and Engineering, Ewha Womans University, Republic of Korea
[4]Department of Chemistry, School of Biomedicine,
Mongolian National University of Medical Sciences, Mongolia

ABSTRACT

Young children and elders, who spend most of their time at home, are at high risk of indoor air pollution. Therefore, it is important to take appropriate measures to reduce the levels of air pollutants, in particular, PM. This study determined the elemental characteristics of indoor $PM_{2.5}$ concentrations in ger, house, and apartments where fuel types were different. The indoor air samples of $PM_{2.5}$ and elemental chemical species in each ger were collected and measured from December 2019 to August 2020 in 18 ger houses and apartments in the residential area (Songino-Khairkhan district) and city center area (Bayanzurkh district) of Ulaanbaatar. In this study, we used inductively coupled plasma mass spectrometry (ICP-OES) to evaluate the $PM_{2.5}$ bound elemental composition. Average indoor $PM_{2.5}$ concentrations were 94.24 and 42.07 $\mu g/m^3$ in residential and city centers for the winter, and 12.06 and 23.51 $\mu g/m^3$ for the summer period, respectively. In the $PM_{2.5}$ samples, high concentrations of ambient elements were observed for Ca, Al, K, Zn, and Pb. The concentration of toxic heavy metals such as Pb was 1.4 times higher than the Mongolian National Standard (MNS 4585:2016, which is 1 $\mu g/m^3$). In conclusion, the seasonal variation of the elements' concentrations mostly depended on consumption, such as coal and wood consumption (W/S > 1). Crustal matter elements depended on dusting and vacuuming activities (W/S < 1). In addition to elements, there are other air pollutants in indoor that should be identified and quantified to estimate the health effects due to indoor air pollution, such as elemental carbon (EC) and organic compounds. Thus, it is warranted to further study the levels of air pollutants in indoor, especially in the residential area in Ulaanbaatar.

Keywords: air pollution, fine particulate matter, indoor and outdoor housing, elemental composition, indoor stove, clean fuel, lead.

1 INTRODUCTION

Mongolia's population reached 3.1 million in 2016 [1], one of the least densely populated countries in the world, yet its two-thirds live in the capital, Ulaanbaatar [2]. According to Volodya et al., the main reasons for the inflow of people to the capital are environmental and economic problems. Increased anthropogenic disturbances such as mining, overgrazing, pasture land, inappropriate usage of crop lands, and deforestation have caused extreme soil destruction, intensified land degradation, and frequent sand storms [2]. Several of Ulaanbaatar`s current socio-economic issues, especially environmental issues, are closely related to the impact from that inflow of people. Due to the housing deficit, most people that move to Ulaanbaatar have become residents of what is known as the residential area (Ger District). The ger district is home to approximately 736,000 residents. Ger districts, consisting of gers (Mongolian yurt) and other dwellings such as houses made of brick or wood, are distributed mainly in the northern valleys and western suburbs around the urban area (marked in yellow in Fig. 1). For heating and cooking, each household living in gers and houses uses coal and wood [3]. Most of the coal used in the ger district are not treated and emit a large amount of air pollutants compared to treated coal. In the ger district, there are

WIT Transactions on Ecology and the Environment, Vol 252, © 2021 WIT Press
www.witpress.com, ISSN 1743-3541 (on-line)
doi:10.2495/AIR210051

two prominent dwelling types: ger and house. Houses are generally constructed from wood or brick. It is common for plots to host more than one household, which often leads to disputes regarding the redevelopment of ger areas to apartment blocks. It is generally agreed that one of the major sources of PM in Ulaanbaatar is the burning of coal and wood in individual residences in ger area [3]. As at 2015, there were 184,000 stoves and 3,341 hot water boilers used to heat these households in the ger district and raw untreated coal was the main fuel source for the stoves and hot water boilers. Ger district households consume approximately 800,000 metric tons of raw coal annually [4]. The vast majority of gers and houses in peri-urban areas are heated with raw coal lit by small amounts of wood in small chimney stoves, while apartment households in outside of the ger district almost exclusively employ clean indoor use (CIU) heat that creates no indoor emissions. Therefore, Ulaanbaatar faces a challenge to ban coal consumption and promote the use of processed coal and clean fuels to mitigate air pollution. The geographical characteristics and regional climate are also not favorable for air quality management in Ulaanbaatar. Ulaanbaatar is located in a basin surrounded by mountains at high altitude in 47° 55' 16.4" N and 106° 55' 6.8" E [5] which leads to a long and cold winter. Those unique geographical conditions provide suitable conditions for the formation of thick inversion layers and low mixing height in winter. According to Wang and Kai, maximum mixing height was below 500 m and thickness of inversion layer was 500 to 700 m in winter and emphasized that temperature inversion layer became thicker and stronger in winter (from September to January). That low mixing height, thick inversion layer forming a stable atmosphere, and continuous emitting air pollutants which resulted in severe particulate pollution.

Due to air pollution, pneumonia is leading among the respiratory diseases making up 20.2% of the patient population and it is especially high among children under 5 years old, making 70.9% of patients [8]. The Mongolian government for the first time has decided to ban the use of raw coal in Ulaanbaatar in 2018. According to the Government resolution No. 62 of 2018 [9], the use of raw coal has been banned since 15 May 2019, except entities and organizations that provide electricity and heat in the territory of Songino Khairkhan and Bayankhoshuu districts of Ulaanbaatar. Household indoor air quality can directly affect the health conditions of people dwelling in the home. However, there are only a few studies on the indoor air quality in ger [10], [11]. Furthermore, these studies were limited to one district in winter time. It is important to understand the relationship between the outdoor and indoor air quality to develop effective air quality management directions. Moreover, it is important to understand the different air quality characteristics depending on the household types, ger, house, and apartment. These differences combined with variations in outdoor particle infiltration between building types likely result in substantially different indoor concentrations between them. Indoor concentrations of $PM_{2.5}$ should be estimated by housing type, household heating source, presence of second-hand tobacco smoke (SHS), and season.

The overall objective of this study is to understand the PM pollution in a household in Ulaanbaatar. To do that: (1) $PM_{2.5}$ samples in both indoor and outdoor sites were collected at both residential and downtown district with different household types; (2) concentrations of their elemental species were analyzed; and (3) major sources of the observed elements were identified. To our knowledge, this is the first study in Mongolia to simultaneously analyze both indoor and outdoor concentrations of elements and their relationship.

2 STUDY AREA

This study was conducted in two areas in Ulaanbaatar, Mongolia, and in each area both indoor and outdoor sampling were carried out. Ulaanbaatar is divided into nine districts, each

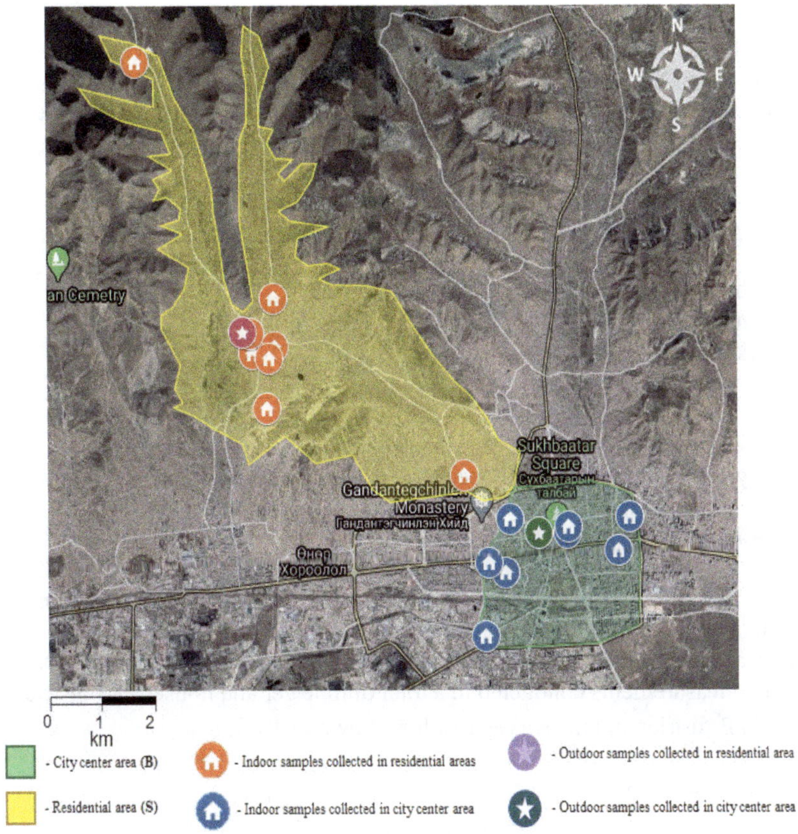

Figure 1: Indoor and outdoor PM sampling points in Songinokhairkhan (S: ger area, colored in yellow), and Bayanzurkh (B: city center, colored in green) districts in Ulaanbaatar. *(Source: Modified from Google Maps.)*

of which is subdivided into subdistricts (khoroo). There are a total of 121 khoroo. As shown in Fig. 1, the study areas are Songinokhairkhan (residential area, denoted S), and Bayanzurkh (city center, denoted B) districts. Songinokhairkhan is located in northeast part of Ulaanbaatar and one of the biggest districts by area in Ulaanbaatar and contains ger district. Bayanzurkh district is located in the center of the city and has a high population density (272 persons per km^2) [12]. Demographical information on the sampling areas are given in Table 1. The $PM_{2.5}$ sampling was performed between 2 December and 9 March (winter) 2019 and between 28 June and 28 August (summer) 2020 to evaluate the differences between summer and winter concentrations of air pollutants. We measured concentrations in three household types, ger, house and apartment. The locations of the sampling points are shown in Fig. 1. The selection of the sampling points was made in consideration of the possible domestic heating sources and a comparison was made by selecting sampling points in both residential (S) and city center (B) areas. Residential sampling area is affected mainly by domestic heating emissions and by traffic. It is expected that the city center sampling area is less affected by heating related emissions. In winter, samples were collected at four houses and two gers in S and four apartments in B, respectively. In summer, samples were collected at three houses in S and five apartments in B, respectively. Since in summer there is no heating in gers, sampling was not carried out in the ger. The participating households were recruited by doctors at a primary healthcare center. Outdoor samplings were carried out at the APRD five monitoring sites at S and NAMEM B districts, respectively, both of which were operated by Government.

Table 1: Demographical information of the sampling areas [12].

Sampling sites	Household	Population	Apartment	House	Ger
Songinokhairkhan district (S, residential area)	5,200	327,580	886	1,951	1,900
Bayanzurkh district (B, city area)	4,274	361,689	4,274	–	–

In addition to the sampling, a survey was carried out to identify the indoor characteristics such as the type and amount of fuel used, frequency of fuel addition, the number of meals cooked, the number of family members and indoor smoking status during the study period. Specially designed housing questionnaires were collected from members of each household (aged over 11 years old). The survey was collected from a total of 119 people. In addition, major vitals such as blood pressure, heart rate, and blood oxygen levels were measured in these 119 individuals during indoor measurements. The survey was conducted through a cross-sectional design covering two districts (Table 1) of the Ulaanbaatar city. Data analysis was based on measurements conducted in a total of nine ger and houses in S district and nine apartments in B district and the survey result is shown in Table 1.

2.1 Particulate matter sampling

The indoor particle concentration in each household was measured using two different $PM_{2.5}$ sampling instruments for 8 hours (10:00–18:00) during daytime and 12 hours during night time. Each household was measured once in each season. Two samplers were used for indoor sampling. For daytime sampling, the MiniVol TAS Tactical Air Sampler was placed in the center of each microenvironment on a table roughly 1.5 m from the floor level in order for the sampling to occur within the breathing zone of a seated individual. Due to the noise problem, a real-time measuring Portable AsLung V2 sampling instrument based on light-scattering technology was used during night-time. Thus, indoor elemental data are for the daytime data only. In addition to $PM_{2.5}$, temperature, pressure, and moisture were also recorded hourly. The Partisol™ 2000i was used for collection of outdoor $PM_{2.5}$. The outdoor samples were sampled for 23 h at the same day as the indoor samples. The filter cartridge was equipped with a 2 μm pore sized 47 mm PTFE membrane disk filter (Teflon® with support ring, MTL) with integrated filter collection. Samples were weighed by a six-digit electronic microbalance (Ba-C, BIOBASE, China). Quality control in filter weighting was ensured by using appropriate laboratory and field blanks. Each filter was weighed at least three times to ensure that the difference of each measurement was within 0.1 mg. After weighing, the sampled filters were stored in a refrigerator at 4°C for subsequent analysis.

2.2 Elemental analysis

External calibration was performed before and after analysis. Before the analysis, the samples are divided into four equal parts and one piece has been used for elemental analysis. Then 34 elements were analyzed by inductively coupled plasma mass spectrometry (ICP-OES), but 18 elements were detected (see Table 3). Sixteen elements (As, Be, Bi, Ga, Mo, Ni, Re, Sb, Se, Ta, Th, U, V, Y, Yb, Zr) were below limit of detection (LoD) and not reported in this study. A quarter of PTFE filters were used for each sample digestion and placed 40 ml vials. The filter was extracted with 20 mL 2% nitric acid for 30 minutes six times under heated (69°C) ultrasonication. The heated digestion was cooled down to room temperature. Then the digestions were filtrated on 0.45 μm pore size filter (MCE, 25 mm syringe filter, JetBiofil) and the solution became ready for analysis. The chemical compositions of the $PM_{2.5}$ filter samples were analyzed using ICP/OES (iCAP7000, Thermo Fisher Scientific, USA). The certified reference materials were used for calibration purposes in the analyses of elements. Eleven certified reference materials were used for calibration purposes in the analyses of major elements such as: GNM-SET-001-0-5, GNM-SET-002-0-3, GNM-SET-003-0-3, GNM-SET-004-0-4, GNM-SET-014-0-4, GNM-SET-015-0-4, GNM-SET-030-0-3 and GNM-M33198. Analysis of blank samples, determination of detection limits, and recovery tests were also performed for the quality control of analytical procedures (MNS 17:025) used in analysis studies. The LoD for an element is determined as three times the standard deviation of 10 analyses of the blank sample.

2.3 Statistical analysis

The varimax rotated factor analysis (FA) was performed separately on indoor and outdoor datasets for households to show possible element sources for $PM_{2.5}$. The factor analysis identifies groups of elements whose concentrations fluctuate together from one sample to another and separates those elements into factors. Elements were included in FA only if the observations had data for at least 30% of the samples. Values reported below the LoD were included in the analysis to create the largest matrix possible. In this study, one-half of the LoD of elements was used for the treatment of missing values.

3 RESULTS AND DISCUSSION

3.1 Indoor and outdoor $PM_{2.5}$ characteristics

Health effects of ambient $PM_{2.5}$ have been widely reported in terms of mortality and morbidity of cardiopulmonary diseases, the prevalence of respiratory symptoms, and hospital admission. Smoking rates in Mongolia are among the highest in the world.

Recent nationwide bans on public indoor smoking suggest indoor SHS may only make considerable contributions to exposure in personal, private indoor environments. Indoor concentration estimates thereby conservatively assume that SHS occurred only indoors at home and thus contributions from SHS were applied only to night-time indoor concentration estimates. For simplicity and due to limited information on Mongolian workplace environments, the concentration profiles of the indoor environments in which the population spends their time away from home were assumed the same as those of their home indoor environments [16]. The 24 hour profile of $PM_{2.5}$ both indoor, outdoor in winter and summer summarized in Fig. 2. In winter $PM_{2.5}$ concentration increase at 6:00 a.m. and high concentration were observed from 10:00 a.m. to 11:00 a.m. in both indoor and outdoor. The concentration decreased continuously from 12 p.m. to 4 p.m., increased again beginning at 4 p.m. and remained high until after 3 a.m., when it decreased slowly until 6 a.m. Highest and lowest average 1 hour average concentrations both indoor and outdoor were 177 μg/m^3 at 11:00 a.m. and 270 μg/m^3 10:00 a.m. Ger areas tended to have higher hour average concentration than city center. Outdoor $PM_{2.5}$ concentrations much higher than indoor during winter. In summer $PM_{2.5}$ concentration increase at 12:00 p.m. and highest daytime concentration were observed at 2 p.m. indoor and high until 12 a.m. At 2:00 a.m., unexpected $PM_{2.5}$ high concentration observed in both indoor and outdoor. Highest and lowest average 1 hour average concentrations both indoor and outdoor were 59 μg/m³ at 2:00 a.m. and 11 μg/m³ 11:00 a.m. City center tended to have higher hour average concentration than ger area. Indoor $PM_{2.5}$ concentrations much higher than outdoor during summer.

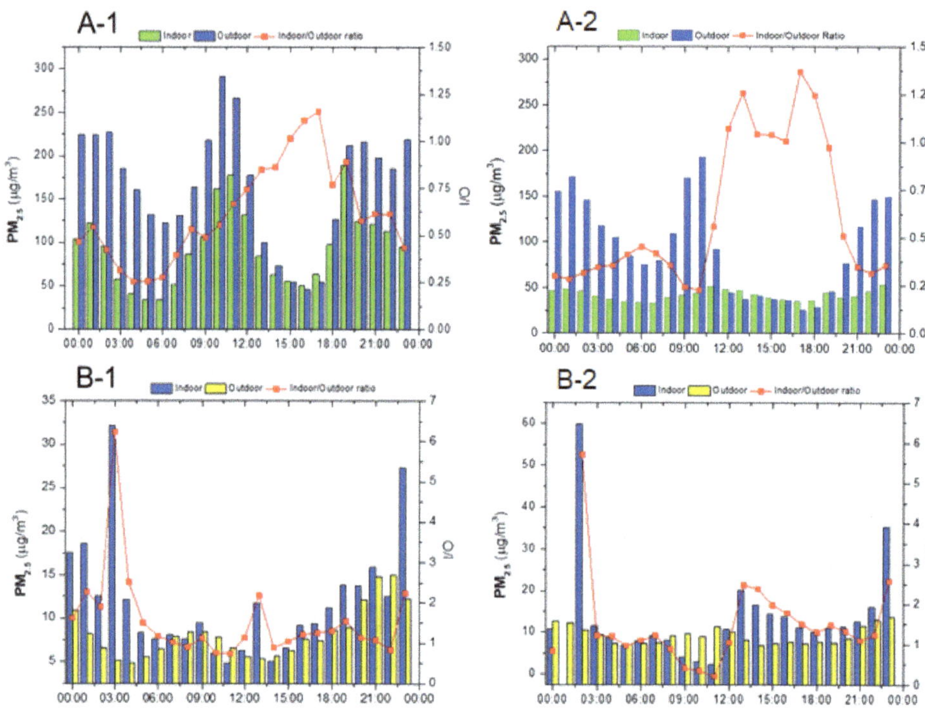

Figure 2: Indoor and outdoor 24 h profiles of $PM_{2.5}$ (A-1 Winter ger area, A-2 Winter city center, B-1 Summer ger area, B-2 Summer city center).

The observed mass concentration of $PM_{2.5}$ in each type of houses is summarized in Fig. 3. As shown in Fig. 3, the 24 hour average $PM_{2.5}$ concentrations were 95.8–44.1 μg/m³ in houses, 93.4–42.4 μg/m³ in gers and 42.1–6.29 μg/m³ in apartments. The $PM_{2.5}$ concentrations of households living in gers and houses in ger areas (S) were 3.7–3.8 times higher than the WHO recommended concentration, and 1.8–1.9 times higher than the Mongolian national standard (tolerance level). For those living in apartments downtown, it was 1.7 times higher than the recommended concentration and less than the allowable concentration of the Mongolian national standard. For each element, generally the concentrations were higher in houses and gers than apartments as shown in Fig. 2.

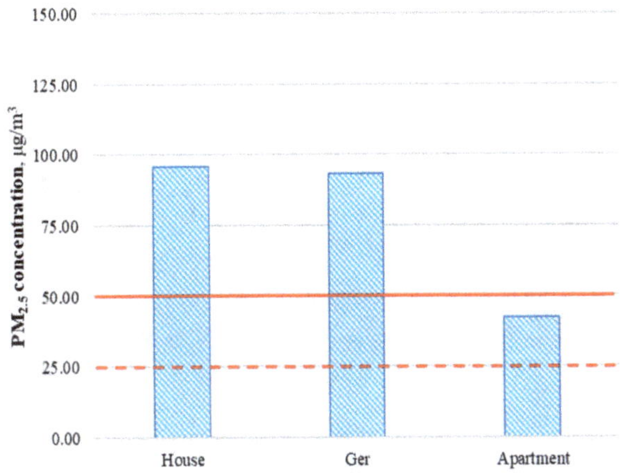

Figure 3: 24 hour average concentration of indoor $PM_{2.5}$ by type of housing (Red line: Mongolian national standard, MNS 4585:2016. Dashed red line: WHO Air Quality Guideline).

3.2 Elemental composition in $PM_{2.5}$

The mass concentrations of $PM_{2.5}$ were 95.8–44.1 μg/m^3 in houses, 93.4–42.4 μg/m^3 in gers, and 42.1–6.29 μg/m^3 in apartments. In the $PM_{2.5}$ samples, high concentrations of ambient elements as shown in Fig. 4, were observed for Ca (28.6 ± 67.1 μg/m^3), Al (4.238 ± 1.116μg/m^3), K (3.808 ± 1.663μg/m^3), Zn (1.831 ± 3.148μg/m^3) and Pb (1.411 ± 1.441 μg/m^3). The concentrations of toxic heavy metals such as Pb was 1.4 higher than the Mongolian National Standard (MNS 4585:2016, which is 1 μg/m^3). Seasonal variations of the mean and standard deviation of the element concentrations in the indoor and outdoor samples in winter and those in the indoor samples in summer are shown in Table 3 along with the seasonal ratio values, i.e., the ratio of the concentration in winter to that of summer (W/S). The concentrations of crustal elements such as Ca and Mg were lower in winter compared to that in summer. Generally, the seasonal variation of the elements' concentrations mostly depended on consumption, such as coal and wood consumption (W/S > 1). Crustal matter elements depended on dusting and vacuuming activities (W/S < 1). As shown in Table 3, the concentration patterns are similar for both summer/winter and indoor/outdoor conditions. When the trace element concentrations were analyzed, the highest concentrations were found to be Ca, Al, K, and Ba and the lowest concentrations were found to be Cr, Co, and Cu in winter. The highest concentrations were found to be Ca, Al, K, and Zn and the lowest concentrations were found to be Cr, Ti and Mn in summer. Comparisons of the elemental concentrations measured in residential (S) and city center (B) in winter and summer periods are seen in Fig. 4. With intense combustion of coal and wood in the gers and housed in residential area (S) compared to the apartments in city center (B) where coal or wood combustion activities were negligible, the concentrations of most elements in S were higher than B. In summer, as shown in Fig. 4, the pattern was reversed.

The composition of the elements in the indoor air has been extensively studied in other countries. However, in 2007, the first major study of indoor air quality in Mongolia determined the element concentration in the aerosols. Table 2 compares the previous study results from the United States, Qatar, China, Austria, and India.

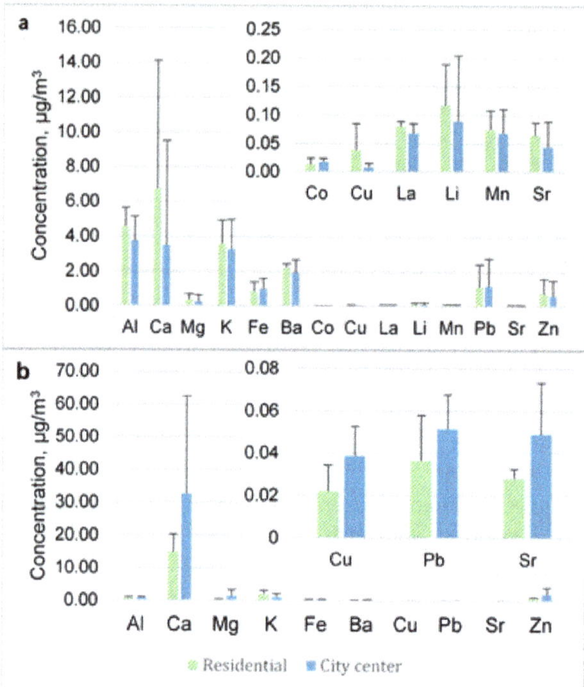

Figure 4: Mean concentrations and standard deviations of the elements in indoor samples in residential (S) and city center (B) and during (a) winter; and (b) summer (µg/m³).

Table 2: Comparison of heavy metal concentration with the results of other researchers.

Research object	Concentration, ng/m³							Reference
	Co	Ni	Cu	Zn	Cr	Cd	Pb	
House (winter)	**20**	**< LoD**	**42**	**1831**	**12**	**76**	**1411**	This study
House (summer)	**< LoD**	**< LoD**	**89**	**4420**	**45**	**< LoD**	**113**	
House	–	4	21	77	–	–	53	Shanghai, China
Apartment	0.23	4.55	6.92	19.5	4.3	0.16	6.06	Doha, Qatar [13]
House	–	–	–	–	–	0.57	23.9	Wien, Austria [14]
House	16.3	27	33	2810	7.1	9.3	821	Deli, India [15]
Standards								
MNG							1000	MNS 4585:2016
WHO		1000				5	1000	WHO, 2000
US EPA		20				5	150	US EPA, 1997

As shown in Table 2, the concentrations of heavy metals in the indoor air vary depending on the lifestyle of the country and the pollution of the environment. In the United States, Austria, and Qatar, where coal combustion was rarely used in the home, the heavy metal concentration is below the WHO standards. On the contrary, in China and India, which use raw coal, the heavy metal concentration exceeds the WHO standards. Our study found a Pb concentration of 1,411 ng/m^3 and a Zn concentration of 1,831 ng/m^3 in winter, which is close to the Pb concentration of 821 ng/m^3 and the Zn concentration of 2,810 ng/m^3, the results of a study conducted in India. India has a high concentration of outdoor air pollution and a raw indoor environment, which is similar to Ulaanbaatar, so the results are likely to be similar. The concentration of 20 elements in $PM_{2.5}$ was determined to be 1.5 times higher than the MNS (1,000 ng/m^3) in winter, 5.1 times higher than the US EPA standard (5 ng/m^3) in summer and winter. The nickel (Ni) concentration increased 1.8 times higher than the US standard (20 ng/m^3). Carcinogenic heavy metals such as cadmium, lead, and nickel may have been contaminated by human activities.

The concentrations in houses were slightly higher than those in gers as shown in Fig. 5, but the difference was small. These suggest that the combustion of coal and wood in houses and gers in residential area (S) was the dominant factor, which determined the concentration of $PM_{2.5}$ and elements' concentration in the homes at both residential area (S) and city center (B).

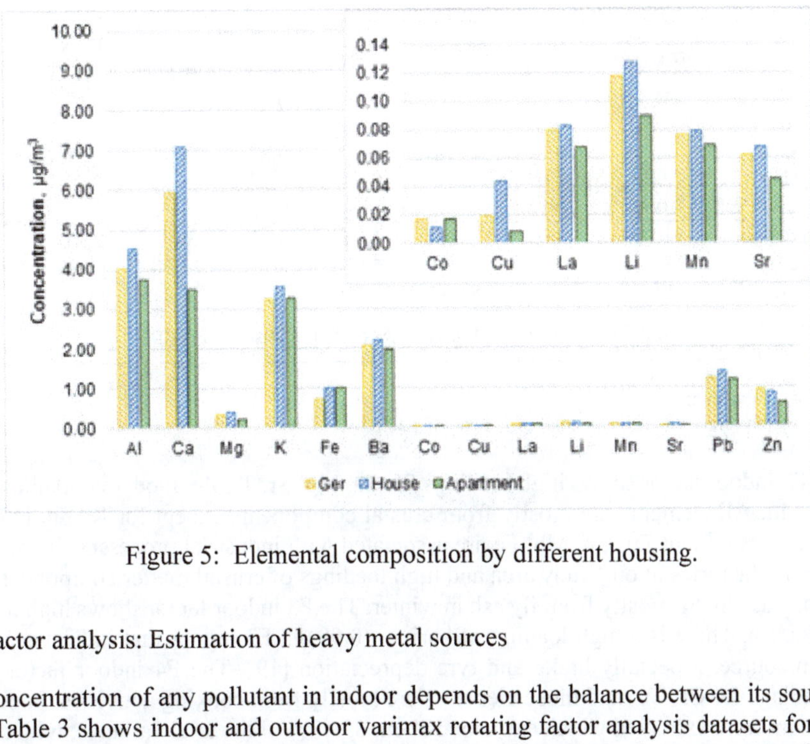

Figure 5: Elemental composition by different housing.

3.3 Factor analysis: Estimation of heavy metal sources

The concentration of any pollutant in indoor depends on the balance between its source and sink. Table 3 shows indoor and outdoor varimax rotating factor analysis datasets for homes to identify possible heavy metal sources in winter. The eigenvalue was set to 1.0 as a threshold limit on the number of extracted factors. Factor loadings equal to or greater than 0.3 are presented because those < 0.3 were considered insignificant. The chemicals introduced with ambient air depend on the location and leakiness of the houses, gers and apartments. Four factors were extracted from the indoor data which explained about 95.09% of the total variance, with the first factor (F1 indoor) accounting for 56.56%, the second (F2 indoor) for 16.97%, the third (F3 indoor) for 11.41% and the fourth (F4 indoor) for 5.97%. The F1 indoor factor shows high loadings for K, Fe, Cd, Co, Li, Mn, Pb and moderate loadings for Zn, Sr. All the elements in the first factor except Fe, Sr, Mn are mostly from combustion of coal or wood. The presence of Fe, Mn suggests the mixing of some fly ash components into this factor. The first factor probably represents coal and wood combustion.

Table 3: Factor analysis of indoor and outdoor in winter.

Components	Indoor				Outdoor			
	F1	F2	F3	F4	F1	F2	F3	F4
Al		0.05	0.82	0.24	0.77	0.40	0.41	
Ca	0.36	0.93		0.02	0.78		0.17	0.26
Mg	0.49	0.86			0.91		0.08	
K	0.84	0.46			0.11	0.63		0.72
Fe	0.83	0.47		0.05	0.62		0.71	0.07
Ba		0.02	0.99	0.04	0.30	0.38	0.75	
Cd	0.89	0.32						
Co	0.78	0.21						
Cr				0.89			0.94	
Cu		0.45	0.20	0.82		0.86		0.05
La	0.08		0.78			0.93	0.23	
Li	0.90	0.31	0.06		0.13		0.08	0.31
Mn	0.82	0.54			0.92	0.13	0.17	0.20
P	0.33				0.82	0.02		
Pb	0.91	0.33						0.85
Sr	0.56	0.82	0.08		0.82	0.34	0.09	0.39
Ti	0.27	0.95		0.04	0.94		0.22	0.11
Zn	0.49	0.86			0.42			0.79
Eigenvalue	10.54	3.05	2.05	1.08	5.81	3.69	2.41	2.39
% of Variance	**58.56**	**16.97**	**11.41**	**5.97**	**36.34**	**23.08**	**15.05**	**14.96**
Cumulative %	58.56	75.53	86.94	92.91	36.34	59.41	74.46	89.43
Possible source type	Com-2[a]	Ash	Traff[c]	Act-C[d]	Dust	Pet-C[b]	Traff[c]	Com-2[a]

[a] combustion of coal and wood;
[b] petroleum combustion;
[c] traffic emission (brake and tire depreciation), d cooking activity.

The F2 indoor factor shows high loadings for Ca, Mg, Sr, Ti, Zn, moderate loading for K, Fe, Cu, Mn. All elements are mostly from crustal components except for K, Zn. [17], [18] studies suggested that Zn, Cu, Mn, Fe are associated with industrial processes. In our study, there are no factories in our study area and high loadings of crustal matter components seem to be Zn, Cu, Mn, Fe mostly from fly ash in winter. The F3 indoor factor shows high loadings of Al, Ba, La. There is a high loading of barium (0.99) in F3 indoor that indicates a traffic emission source, especially brake and tyre depreciation [19]. The F4 indoor factor shows high loadings for Cr, Cu, and those elements are mostly from cooking activities. In outdoor, four factors were extracted which explained about 89.43% of the total variance, with the first factor (F1 outdoor) accounting for 36.34%, the second factor (F2 outdoor) for 23.08%, the third factor (F3 outdoor) for 15.05% and the fourth factor (F4 outdoor) accounting for 14.96% in outdoor $PM_{2.5}$. The F1 outdoor factor had high loads for Mg, Al, Ca, Ti, Mn, P and Sr. All elements in this factor area were most often found in crustal matter. The F2 outdoor factor had high loadings of La, Cu, K and moderate loading of Ba, Al, Sr. Ba, Cu, La are known to be the markers for traffic emission and, especially, Cu and La were known to be mainly from vehicle exhaust. The F3 outdoor factor was similar to the F3 indoor factor and had additionally Fe, Al and Ti which could be explained by road dust resuspension mixed with vehicular emission [20]. The F4 outdoor factor had high loading for Pb, Zn, K and moderate loadings for Sr, Li. The main source of Pb is known to be combustion, especially coal combustion, and K comes mostly from biomass burning (wood, incense etc). This source was also mixed with fly ash elements Zn, Sr, Li and suggested the combustion of coal and wood. Four factors were extracted from the summer indoor data, which explained about 89.96% of the total variance, with the first factor occupying 36.33%, the second 29.25%, the third 15.41% and the fourth 8.97% of the total variance. The first factor had high loading for Ca, Mg, Sr, Pb, Mn and moderate loadings for Cu. First factor's elemental composition was most often found in the crustal matter (except Pb).

The presence of Pb suggests the mixing of some anthropogenic components into this factor. The second factor elemental fractions are most often found in traffic exhaust and crustal matter.

Table 4: The factor analysis of indoor in summer.

Components	Indoor			
	F1	F2	F3	F4
Al	0.08	0.78	0.55	
Ca	0.94			
Mg	0.92			
K		0.34	0.39	0.79
Fe	0.21	0.09	0.93	0.13
Ba		0.94	0.24	
Co			0.05	0.93
Cr	0.34			
Cu	0.51	0.05	0.30	
Li		0.94		0.07
Mn	0.72		0.37	
Pb	0.80		0.47	0.26
Sr	0.96	0.01	0.12	
Ti	0.29			
Zn	0.40		0.32	
Eigenvalue	5.45	4.39	2.31	1.34
% of Variance	**36.33**	**29.25**	**15.41**	**8.97**
Cumulative %	36.33	65.57	80.99	89.95
Possible source type	Act-D[b]	Road	Ash	Act-C[a]

[a] cooking activity;
[b] dusting activity.

The third factor had high loadings for Al, Fe and moderately loadings for K, Mn, Pb, Zn that is most often found in fly ash elemental composition in the indoor factor 2. The fourth factor, the high loadings for Co and K are mostly from biomass burning [21]. Most residents (in a residential area) use wood for cooking in the warm season and the suggested factor 4 probably represents cooking activities.

4 CONCLUSIONS

Air quality in Ulaanbaatar is poor mainly due to the combustion of coal and wood in the residential area. To quantify and determine major sources of both indoor and outdoor concentrations of elements in ambient $PM_{2.5}$, $PM_{2.5}$ samples were collected in various housing types; ger and houses (dwelling made of wood or brick) where coal and wood are used for cooking and heating, and apartments (housing complexes with central heating) in summer and winter. Thirty-four elements were analyzed by using an ICP/OES, but the levels of 16 elements were below the LoD.

Generally, the seasonal variation of the elements' concentrations mostly depended on consumption, such as coal and wood consumption (W/S > 1). Crustal matter elements depended on dusting and vacuuming activities (W/S < 1). Concentration of toxic heavy metals such as lead (Pb) was 1.4 times higher than the Mongolian national standard, 1 $\mu g/m^3$ (MNS 4585:2016). It was found out that the concentrations of most elements in the residential area where coal and wood combustion activities were carried out were higher than the city center where these combustion activities were not carried out in winter. Compared with the results observed in other countries, the concentrations in Ulaanbaatar were higher than the countries with little coal or wood combustion in the home while comparable to the countries with coal and wood combustion activities in home.

Major factors affecting the elements' concentrations in both indoor and outdoor were studied by using the factor analysis method. Four factors were identified for both indoor and outdoor sources in winter. The dominant factor was coal and wood combustion for indoor, followed by the mixture of soil and fly ash, traffic and cooking. For outdoor, soil was the dominant factor, followed by traffic, a mixture of road dust and traffic resuspension, and coal and wood combustion. In summer, four factors were also identified for indoor soil, mixture of traffic and soil, mixture of soil and fly ash and cooking.

This study is the first study on the $PM_{2.5}$ bound elemental compositions of indoor during winter and summer in Ulaanbaatar after the import of raw coal was banned in Ulaanbaatar. Thus, this study may provide valuable information of indoor air quality in Ulaanbaatar. In addition to elements, there are other air pollutants in indoor that should be identified and quantified to estimate the health effects due to indoor air pollution, such as elemental carbon (EC) and organic compounds. Thus, it is warranted to further study the levels of air pollutants in indoor, especially in residential area in Ulaanbaatar.

ACKNOWLEDGEMENT

This study is supported by the National Strategic Project of Fine Dust, Republic of Korea 2020–2021.

REFERENCES

[1] Mitchell, M.J., BTI 2018 country report: Mongolia. *Asian Secur. Handb. Assess. Polit. Issues Asia-Pacific Reg.* pp. 184–190, 2019.

[2] Volodya, E., Yeo, M.J. & Kim, Y.P., Trends of ecological footprints and policy direction for sustainable development in Mongolia: A case study. *Sustain.,* **10**, 2018. DOI: 10.3390/su10114026.

[3] Guttikunda, S.K. et al., Particulate pollution in Ulaanbaatar, Mongolia. *Air Qual. Atmos. Heal.,* **6**, pp. 589–601, 2013.

[4] Engel, D., *Ulaanbaatar's Ger District Issues: Changes and Attitudes.*

[5] Ganbat, G. & Baik, J.-J., Wintertime winds in and around the Ulaanbaatar metropolitan area in the presence of a temperature inversion. *Asia-Pacific J. Atmos. Sci.,* **52**, pp. 309–325, 2016.

[6] Wang, M. et al., Meteorological factors affecting winter particulate air pollution in Ulaanbaatar from 2008 to 2016. *Asian J. Atmos. Environ.,* **12**, pp. 244–254, 2018.

[7] Wang, M. et al.. Air particulate pollution in Ulaanbaatar, Mongolia: Variation in atmospheric conditions from autumn to winter. *Sola,* **13**, pp. 90–95, 2017.

[8] Urangoo, G. et al., *Outdoor Air Pollution and Health in Ulaanbaatar City*, Ulaanbaatar, 2019.

[9] Resolution of the Government of Mongolia, *On the Prohibition of the Consumption of Raw Coal*, 2018.

[10] Lim, M. et al., Characteristics of indoor $PM_{2.5}$ concentration in gers using coal stoves in Ulaanbaatar, Mongolia. *Int. J. Environ. Res. Public Health*, **15**, 2018. DOI: 10.3390/ijerph15112524.

[11] Ahn, J.Y. et al., Determinants of indoor $PM_{2.5}$ concentrations in ger, a traditional residence, in Mongolia. *Korean J. Public Heal.*, 55, pp. 22–30, 2019.

[12] Mongolian Statistical Information Service, Population.

[13] Saraga, D. et al., Chemical characterization of indoor and outdoor particulate matter ($PM_{2.5}$, PM_{10}) in Doha, Qatar. *Aerosol Air Qual. Res.*, 17:, pp. 1156–1168, 2017.

[14] Komarnicki, K., Lead and cadmium in indoor air and the urban environment. **136**, 2005. DOI: 10.1016/j.envpol.2004.12.006.

[15] Lai, A.M. et al., Impacts of stove/fuel use and outdoor air pollution on chemical composition of household particulate matter. *Indoor Air,* **30**, pp. 294–305, 2020.

[16] Hill, L.D. et al., Health assessment of future $PM_{2.5}$ exposures from indoor, outdoor, and secondhand tobacco smoke concentrations under alternative policy pathways in Ulaanbaatar, Mongolia. *PLoS One,* **12**, pp. 1–26, 2017.

[17] Gunchin, G. et al., Three-year long source apportionment study of airborne particles in ulaanbaatar using x-ray fluorescence and positive matrix factorization. *Aerosol Air Qual. Res.,* **19**, pp. 1056–1067, 2019.

[18] Davy, P.K. et al., Air particulate matter pollution in Ulaanbaatar, Mongolia: Determination of composition, source contributions and source locations. *Atmos. Pollut. Res.,* **2**, pp. 126–137, 2011.

[19] Goddard, S.L. et al., Determination of antimony and barium in UK air quality samples as indicators of non-exhaust traffic emissions. *Environ. Monit. Assess.,* **191**, p. 641, 2019.

[20] Amgalan, N., Narantsetseg, T. & Shagjjamba, D., Valuations of elemental concentrations of particle matter in Ulaanbaatar, Mongolia. *Open J. Air Pollut.,* **5**, pp. 160–169, 2016.

[21] Yu, J. et al., Potassium: A tracer for biomass burning in Beijing? *Aerosol Air Qual. Res.,* **18**, pp. 2447–2459, 2018.

CHARACTERIZATION OF AIRBORNE *ASPERGILLUS* SERIES *VERSICOLORES* COLLECTED IN FRENCH BIOAEROSOLS

ANTOINE GÉRY[1], ELIZABETH CHOSSON[1], VIRGINIE SÉGUIN[1],
JEAN-PHILIPPE RIOULT[1], JULIE BONHOMME[1,2] & DAVID GARON[1]
[1]ToxEMAC-ABTE, Centre F. Baclesse, Unicaen and Unirouen, Normandie University, France
[2]Service de Microbiologie, Centre Hospitalier Universitaire de Caen, France

ABSTRACT

Indoor air quality and exposure to fungal bioaerosols is a major concern in Europe where people spend most of their time indoors. 30–50% of European homes have moisture problems facilitating mold growth, with significant health and economic consequences. *Aspergilli* series *Versicolores* are molds recurrently found in these homes which are known to cause allergies, aggravate asthma and produce sterigmatocystin, a potential human carcinogen. Recent phylogenetic studies highlighted new *Versicolores* species which led us to characterize fungal isolates of this section to better understand their repartition in French bioaerosols. Air samples were taken with a cyclonic biocollector indoor (hospital and mold-damaged homes) and outdoor (agricultural environment), temperature and relative humidity were recorded during sampling. Collection liquid was cultured on malt extract agar (MEA) medium supplemented with chloramphenicol (0.05%, w/v). Each colony was isolated and then purified on MEA. A total of 93 isolates belonging to the series Versicolores were characterized both macroscopically, microscopically and were also identified using molecular approach. We identified eight different species from bioaerosols: *Aspergillus creber* (n = 40), *A. jensenii* (n = 37), *A. protuberus* (n = 6), *A. puulaauensis* (n = 4), *A. sydowii* (n = 2), *A. tabacinus* (n = 2), *A. fructus* (n = 1), and *A. amoenus* (n = 1). The phylogenetic tree performed from the *BenA* sequences shows homologies between isolates of the same species recovered from bioaerosols collected in the same environment which confirms the interest of Bt2a/b primers set for the identification of species of the *Versicolores* series. This work constitutes the first characterization of *Aspergilli* of the series *Versicolores* from French bioaerosols.

Keywords: Aspergillus, series Versicolores, bioaerosols.

1 INTRODUCTION

In France, as in most European countries, air quality has become a major public health issue [1]. Indeed, it is estimated that air pollution causes 48,000 deaths each year in France, which represents nearly 8% of total annual mortality between 2016 and 2019 [2]. This air quality can be altered by different factors from natural or anthropic origin [3]. Among the factors of natural origin, molds are the most frequent in indoor environments [4] where they proliferate easily in the presence of humidity, which affects 30 to 50% of European homes [5], explaining the visible presence of molds belonging mainly to the genera *Aspergillus*, *Penicillium* and *Cladosporium* [6] in 14–20% of French homes [7]. These are mainly responsible for allergies [8] and sick building syndrome [9], can aggravate asthma [10] and are likely to cause infections in immunocompromised patients [11], [12]. The frequency and intensity of these symptoms are related to the level of contamination of the habitat and the duration of exposure of the population [13], which can reach up to 22 hours spent indoors per day in industrialized countries [14], [15]. However, as the relations between dampness, microbial exposure and health effects cannot be quantified precisely, no quantitative health-based guideline values or thresholds can be recommended for acceptable levels of contamination with microorganisms [5]. Some molds are also capable of producing cytotoxic

WIT Transactions on Ecology and the Environment, Vol 252, © 2021 WIT Press
www.witpress.com, ISSN 1743-3541 (on-line)
doi:10.2495/AIR210061

[16], [17] and/or genotoxic mycotoxins [18], [19]. This explains the important cost related to medical and renovation expenses in France which amount to 19 billion euros [20].

The molds of the genus *Aspergillus* belonging to the series *Versicolores* section *Nidulantes* are cosmopolitan. They have been found in the soil [21], in various foodstuffs for humans and animals [22], in hypersaline water [23] and in indoor environments [24]. On average, they are found in 70% of bioaerosols [6]. In addition to causing, allergies [25], aggravate asthma [26] and being linked to sick building syndrome [9], they can occasionally be responsible for onychomycosis [27] or pulmonary aspergillosis [28]. Their presence has also been shown to be an explanatory variable for various symptoms experienced by the inhabitants of dwellings damaged by moulds, such as fever, itching, headaches, dizziness or abnormally abundant expectorations [6]. *Aspergilli* of the series *Versicolores* except *A. sydowii* produce sterigmatocystin [29], a mycotoxin involved in the biosynthetic pathway of aflatoxins [30] and recognized as potentially carcinogenic to humans by IARC (group 2B) [31]. *Aspergillus versicolor* was considered until a few years ago as the most abundant species of the series *Versicolores* in bioaerosols. However, several taxonomic revisions have revealed a total of 12 new species within the series *Versicolores* through different phylogenetic studies. Among these, *Aspergillus creber* from the Latin word "creber" meaning "numerous" or "frequent" is now considered to be the most frequent species and was misidentified as *Aspergillus versicolor* [32], [33]. Studies on the diversity and ecology of the species of the series *Versicolores* confirm this hypothesis [34]. Furthermore, the different species of the series *Versicolores* produce different metabolites from each other and do not all have the same pathogenicity [35]–[37].

This work focuses on the molecular characterization of 93 environmental isolates of *Aspergillus* series *Versicolores* isolated from French bioaerosols in order to evaluate their diversity and to contribute to the assessment of fungal exposure to these molds.

2 MATERIAL AND METHODS

2.1 Bioaerosols collection

Bioaerosols were collected in a hospital (Centre François Baclesse, Caen, France) (n = 24) [38], in *Serpula lacrymans*- and mold-damaged homes (n = 65) [13], [39] and outdoor in agricultural environments (silage and hay) (n = 4) [40], [41].

Samples were cultured on malt extract agar (MEA) medium with 0.02% chloramphenicol (Cooper, Melun, France). Plates were incubated at 25°C and checked daily. Each fungal colony was isolated and purified on MEA. All *Aspergillus* isolates belonging to the series *Versicolores* (n = 93) were stored on slant agar at 4°C and in a cryoprotective agent composed of sterile water (Fresenius Kabi AG, Bad Homburg, Germany) and 10% glycerol (Carlo Erba, Val-de-Reuil, France) at −80°C before molecular characterization.

2.2 Molecular characterization

DNA extraction was performed using a modified protocol of the Nucleospin™ Plant II kit (Macherey-Nagel, Duren, Germany). Fungal colony was introduced in a 2 mL microtube with glass beads. The microtube underwent two incubation cycles of 15 min at 80°C and then 15 min at −80°C. It was placed into a Qiagen Tissue Lyser with 400 µL of lysis buffer PL1 (Macherey-Nagel, Duren, Germany) for 15 min at 20 Hz, and incubated with 10 µL of RNAse (Macherey-Nagel, Duren, Germany) and 20 µL of proteinase K at 10 mg/mL (Sigma-Aldrich, St. Louis, MO, USA) at 65°C for 15 min. Then, 400 µL of chloroform (Sigma-Aldrich, St.

Louis, MO, USA) was added to the mixture. The aqueous phase was recovered and extracted using the precipitation kit and washing buffer according to the protocol described by the supplier.

DNA was purified using the NucleoSpin gDNA Clean-up kit (Macherey-Nagel, Duren, Germany) following the instructions of the manufacturer. For each isolate, quantification and quality of purified DNA were realized using a NanoDrop 2000 spectrophotometer (Thermo Fisher Scientific, Waltham, MA, USA) [42].

In this study molecular characterization was performed by amplification of the beta-tubulin gene (*BenA*) using Bt2a/Bt2b (5′-GGTAACCAAATCGGTGCTGCTTTC-3′/5′-ACCCTCAGTGTAGTGACCCTTGGC-3′) primers [32]. Sequences obtained were compared using BLAST (Basic Local Alignment Search Tool, NCBI) to the reference sequences of the 17 *Aspergillus* species of the series *Versicolores* publicly available by Houbraken et al. [43]. Identification was considered reliable only by having a query cover ≥ 98% and a % ID ≥ 99%.

Phylogenetic tree was generated using CLC Workbench 21.0.4. Tree construction was based on neighbor-joining method. Branch supports of the best maximum likelihood tree were estimated by 500 bootstrap replicates.

3 RESULTS

3.1 Relative abundance in French bioaerosols

Among all mold species recovered from bioaerosols, *Aspergilli* series *Versicolores* were the fifth most common species in bioaerosols from the hospital, the first in *Serpula lacrymans*- and mold-damaged homes, and the third most common in agricultural environments, with concentrations ranging from 2.48 CFU/m^3 in agricultural environments to 3.44×10^5 CFU/m^3 in mold-damaged homes. *Aspergilli* series *Versicolores* were associated with 109, 100, and 28 other fungal species in bioaerosols collected from hospital, *Serpula lacrymans*- and mold-damaged homes, and agricultural environments, respectively. The most represented genera in all environments were *Aspergillus* (*A. fumigatus*, *A. niger*, *A. melleus*, *A. flavus*, *A. pseudoglaucus*), *Penicillium* (*P. chrysogenum*, *P. brevicompactum*, *P. crustosum*) and *Cladosporium* (*C. cladosporioides*, *C. herbarum*, *C. sphaerospermum*) [18], [32]–[34].

As shown in Fig. 1, among all the isolates collected in French bioaerosols (n = 93), we were able to molecularly identify eight species of the *Versicolores* series: *Aspergillus creber* (n = 40) comes first, followed by *A. jensenii* (n = 37), *A. protuberus* (n = 6), *A. puulaauensis* (n = 4), *A. sydowii* (n = 2), *A. tabacinus* (n = 2), *A. amoenus* (n = 1) and *A. fructus* (n = 1). Bioaerosols collected at the hospital (n = 24) contained more *A. jensenii* (n = 14), followed by *A. protuberus* (n = 6), *A. creber* (n = 3) and *A. puulaauensis* (n = 1). *Aspergillus protuberus* was only found in bioaerosols collected at the hospital. Bioaerosols collected from houses damaged by *Serpula lacrymans* and molds (n = 65) were mainly composed of *A. creber* (n = 37) and *A. jensenii* (n = 22). *A. puulaauensis* (n = 3), *A. sydowii* (n = 2), and *A. amoenus* (n = 1) were also present although much less abundant. Of the four isolates obtained from bioaerosols collected from agricultural environments, we identified two isolates of *A. tabacinus*, one isolate of *A. jensenii* and one isolate of *A. fructus*.

3.2 Phylogenetic study

As shown in Fig. 2, the *Sydowii* subseries is much more represented within the bioaerosols than the *Versicolores* subseries. Indeed, *Aspergillus creber*, *A. jensenii*, *A. puulaauensis* and

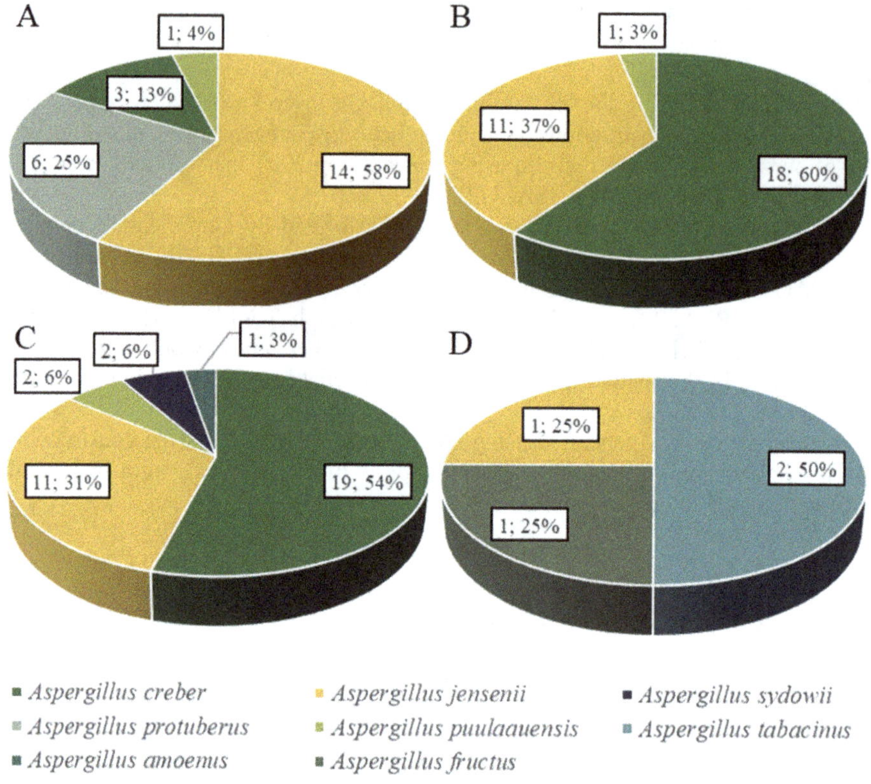

Figure 1: Relative abundances (number of isolates; percentage) of *Aspergillus* species belonging to the *Versicolores* series in bioaerosols from (A) hospital; (B) *Serpula lacrymans*-damaged homes; (C) mold-damaged homes; and (D) agricultural environments.

A. sydowii species represent 83 isolates (89.2%) of the isolates found in French bioaerosols. The species belonging to the *Versicolores* subseries (*Aspergillus protuberus*, *A. tabacinus*, *A. fructus* and *A. amoenus*) are much less present in bioaerosols. Indeed, these represent only ten isolates (10.8%) although *Aspergillus protuberus* is the third most frequently isolated species of the *Versicolores* series from bioaerosols. The use of the Bt2a/b primers set allowed the construction of a phylogenetic tree with a clear distinction of the eight species belonging to the *Versicolores* series found in French bioaerosols and to determine the specific richness for each environment where they were collected. In descending order, the highest species richness was found in bioaerosols from mold-damaged homes (n = 5), followed by those collected at the hospital (n = 4) and tied for third place *Serpula lacrymans*-damaged homes and in agricultural environments (n = 3). *Aspergillus tabacinus* and *A. fructus* were only found in bioaerosols collected from agricultural environments. *Aspergillus jensenii* was the only species found in all environments. We also observed a similar distribution of *Versicolores* species in contaminated homes regardless of whether they were contaminated by *Serpula lacrymans* or molds. The amplification of the *BenA* sequence has allowed the identification of nucleotide variations sufficient to discriminate between different strains within the same species and environment (especially for *A. jensenii*).

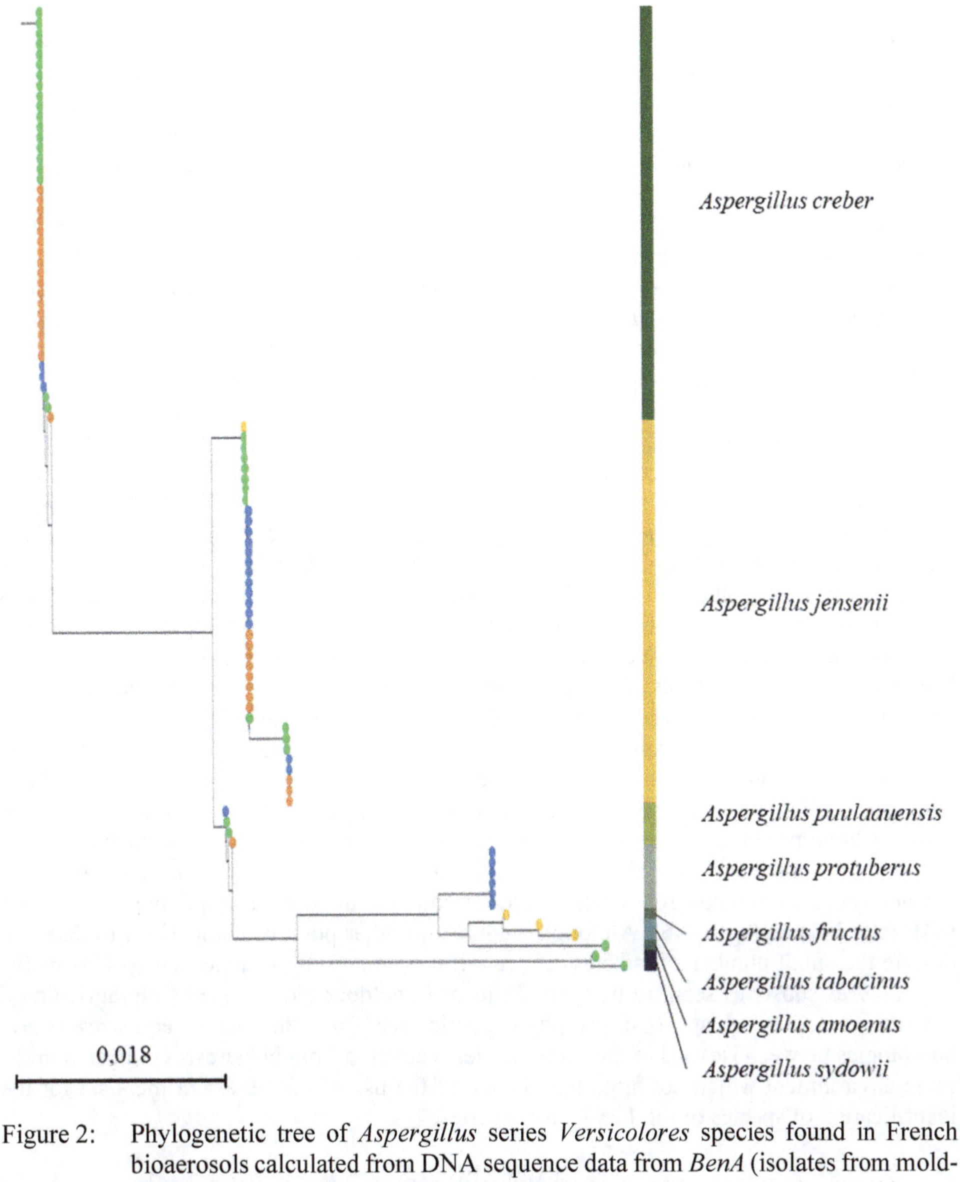

Figure 2: Phylogenetic tree of *Aspergillus* series *Versicolores* species found in French bioaerosols calculated from DNA sequence data from *BenA* (isolates from mold-damaged homes bioaerosols in green, from *Serpula lacrymans*-damaged homes bioaerosols in orange, from hospital bioaerosols in blue and from agricultural environments bioaerosols in yellow).

4 DISCUSSION

Ninety-three fungal isolates recovered from bioaerosols collected in a hospital, in contaminated homes and in agricultural environments were molecularly identified as belonging to species of the *Aspergillus Versicolores* series. This study constitutes the first report on the distribution of these species in French bioaerosols molecularly characterized by *benA* gene amplification. We were able to group these isolates under eight different species, which constitutes one of the highest specific richness reported to date. Indeed, other studies conducted on isolates collected from indoor environments (air and dust samples) identified by *CaM* gene amplification show that species richness within the *Versicolores* series varies from three (*Aspergillus creber*, *A. jensenii* and *A. protuberus*) to 11 species (*A. creber*, *A. jensenii*, *A. puulaauensis*, *A. tennesseensis*, *A. venenatus*, *A. amoenus*, *A. fructus*, *A. griseoaurantiacus*, *A. pepii*, *A. protuberus* and *A. sydowii*) [36], [44]. *Aspergillus creber* and *A. jensenii* were the most frequently isolated species from French bioaerosols (83% of isolates), which is consistent with studies on *Aspergillus* species belonging to the *Versicolores* series that always find these species, and especially with the study by Jakšić et al. where these two species represent 62% of isolates [36], [45]. However, we did not find *Aspergillus cvjetkovicii*, *A. griseoaurantiacus*, *A. pepii*, *A. tennesseensis*, or *A. venenatus*, which have been isolated from indoor air bioaerosols in the United States and Croatia [33], [36], [45] but are more frequently found in various foods and feeds and in soil [36], [46]. *A. protuberus* and *A. puulaauensis* remain quite recurrent in indoor air and dust, as reported by Micheluz et al. [44] and Jakšić et al. [36]. *Aspergillus sydowii* was only isolated from two bioaerosols collected from contaminated homes, confirming its presence at a rather low frequency in air compared to food matrices or human pathology [33], [37], [46]. *Aspergillus amoenus* was isolated by Jurjevic et al. [33] from different sources (mammary gland, cured meat, *Berberis* sp. fruit) but also from indoor dust in Japan [46], which may explain why we found so few among all the collected isolates. This is the first time that *Aspergillus tabacinus* has been isolated from bioaerosols, whereas it is more frequently isolated from tobacco, cured meat, or plants (corn) [33]. We also found an isolate of *Aspergillus fructus* that had never been identified in bioaerosols, although it had been previously recovered from dust of water-damaged and control houses in Croatia [24]. However, this species seems to be more frequently identified from fruits [33]. *Aspergillus versicolor* stricto sensu was not found in our bioaerosols, confirming the observations of Jurjevic et al. [33] and Kobayashi et al. [46], who reported its presence only in food products (dairy feeds, rice and noodles). Our work shows a large preponderance of species of the *Sydowii* subseries in indoor air bioaerosols which is in agreement with all studies that have investigated *Aspergillus* species of the *Versicolores* series in bioaerosols whether in Croatia [36], the United Kingdom [32], the USA [33], Italy [44] or Japan [46]. Although we have no other point of comparison to date and despite the small number of isolates recovered, it is interesting to note that species of the *Versicolores* subseries seem to be more frequent in outdoor air collected from agricultural environments. The creation of the phylogenetic tree from the *BenA* sequences shows homologies between isolates of the same species recovered from bioaerosols collected in the same environment which confirms the interest of the use of the Bt2a/b primers set for the identification of species of the *Versicolores* series.

REFERENCES

[1] Anses - Agence nationale de sécurité sanitaire de l'alimentation, de l'environnement et du travail, Enjeux autour de la qualité de l'air. https://www.anses.fr/fr/content/enjeux-autour-de-la-qualit%C3%A9-de-l%E2%80%99air. Accessed on: 31 Aug. 2021.

[2] Medina, S. et al., Impact de la polution de l'air ambiante sur la mortalité en France métropolitaine. Direction de la communication Santé Publique France: France. Apr. p. 63. Report No: 2609–2174, 2021.

[3] Jurewicz, J., Dziewirska, E., Radwan, M. & Hanke, W., Air pollution from natural and anthropic sources and male fertility. *Reprod. Biol. Endocrinol.*, **16**(1), p. 109, 2018.

[4] Andersen, B., Frisvad, J.C., Søndergaard, I., Rasmussen, I.S. & Larsen, L.S., Associations between fungal species and water-damaged building materials. *Appl. Environ. Microbiol.*, **77**(12), pp. 4180–4188, 2011.
[5] WHO Regional Office For Europe, *WHO Guidelines for Indoor Air Quality: Dampness and Mould*, World Health Organization: Geneva, 2009. http://www.ncbi.nlm.nih.gov/books/NBK143941/. Accessed on: 11 May 2021.
[6] Delanoë, A., Heutte, N., Gente, S., Séguin, V. & Garon, D., Relationships between exposure to bioaerosols, moldy surface and symptoms in French mold-damaged homes. *Atmosphere*. **11**(3), p. 223, 2020.
[7] ANSES, Moisissures dans le bâti. p. 374, Report No.: 979-10-286-0126-3, 2016.
[8] Rick, E,, Woolnough, K., Pashley, C. & Wardlaw, A., Allergic fungal airway disease. *J. Investig. Allergol. Clin. Immunol.*, **26**(6), pp. 344–354, 2016.
[9] Norbäck, D. et al., Rhinitis, ocular, throat and dermal symptoms, headache and tiredness among students in schools from Johor Bahru, Malaysia: Associations with fungal DNA and mycotoxins in classroom dust. *PLoS ONE*, **11**(2), e0147996, 2016.
[10] Tiotiu, A.I. et al., Impact of air pollution on asthma outcomes. *IJERPH*, **17**(17), p. 6212, 2020.
[11] Gletsou, E., Ioannou, M., Liakopoulos, V., Tsiambas, E., Ragos, V. & Stefanidis, I., *Aspergillosis* in immunocompromised patients with haematological malignancies. *J. BUON*, **23**(7), pp. 7-10, 2018.
[12] Spellberg, B., Edwards, J. & Ibrahim, A., Novel perspectives on mucormycosis: Pathophysiology, presentation, and management. *Clin. Microbiol. Rev.*, **18**(3), pp. 556-569, 2005.
[13] Delanoë, A. et al., Bioaerosols exposure assessment in mold-damaged houses in Normandy, France. *WIT Transactions on Ecology and the Environment*, vol. 230, pp. 313–320, 2018.
[14] Boulanger, G. et al., Socio-economic costs of indoor air pollution: A tentative estimation for some pollutants of health interest in France. *Environment International*, **104**, pp. 14–24, 2017.
[15] World Health Organization, Indoor air pollutants: Exposure and health effects. Report on a WHO meeting, Nördlingen, 8–11 June 1982. World Health Organization, Regional Office for Europe: Copenhagen, 1983. 42 pp.
[16] Bailey, B. et al., The mycotoxin deoxynivalenol significantly alters the function and metabolism of bovine kidney epithelial cells in vitro. *Toxins*, **11**(10), p. 554, 2019.
[17] Schmutz, C., Cenk, E. & Marko, D., The *Alternaria* mycotoxin alternariol triggers the immune response of IL-1β-stimulated, differentiated caco-2 cells. *Mol. Nutr. Food Res.*, **63**(20), 1900341, 2019.
[18] Benkerroum, N., Chronic and acute toxicities of aflatoxins: Mechanisms of action. *IJERPH*, **17**(2), p. 423, 2020.
[19] Malir, F., Ostry, V., Pfohl-Leszkowicz, A., Malir, J. & Toman, J., Ochratoxin A: 50 years of research. *Toxins*, **8**(7), p. 191, 2016.
[20] Kopp, P. et al., Étude exploratoire du coût socio-économique des polluants de l'air intérieur. *ANSES*, Report No.: 978-2-11-138465-1, 2014.
[21] Domsch, K.H., Gams, W. & Anderson, T.H., *Compendium of Soil Fungi*. Academic Press: London and New York, 1980.
[22] Pitt, J.I. & Hocking, A.D., Fungi and food spoilage, 2009. DOI: 10.1007/978-0-387-92207-2.

[23] Kis-Papo, T., Kirzhner, V., Wasser, S.P. & Nevo, E., Evolution of genomic diversity and sex at extreme environments: Fungal life under hypersaline Dead Sea stress. *Proceedings of the National Academy of Sciences*, **100**(25), pp. 14970–14975, 2003.

[24] Jakšić Despot, D. & Šegvić Klarić, M., A year-round investigation of indoor airborne fungi in Croatia. *Archives of Industrial Hygiene and Toxicology*, **65**(2), pp. 209–218, 2014.

[25] Vincent, M. et al., Investigation of inflammatory and allergic responses to common mold species: Results from in vitro experiments, from a mouse model of asthma, and from a group of asthmatic patients. *Indoor Air*, **27**(5), pp. 933–945, 2017.

[26] Mintz-Cole, R.A., Gibson, A.M., Bass, S.A., Budelsky, A.L., Reponen, T. & Hershey, G.K.K., Dectin-1 and IL-17A suppress murine asthma induced by *Aspergillus versicolor* but not *Cladosporium cladosporioides* due to differences in β-glucan surface exposure. *JI*, **189**(7), pp. 3609–3617, 2012.

[27] Torres-Rodríguez, J.M., Madrenys-Brunet, N., Siddat, M., López-Jodra, O. & Jimenez, T., *Aspergillus versicolor* as cause of onychomycosis: Report of 12 cases and susceptibility testing to antifungal drugs. *J. Eur. Acad. Dermatol. Venereol.*, **11**(1), pp. 25-31, 1998.

[28] Charles, M.P., Invasive pulmonary aspergillosis caused by *Aspergillus versicolor* in a patient on mechanical ventilation. *AMJ*, pp. 632-634, 2011.

[29] Kubosaki, A. et al., A new protocol for the detection of sterigmatocystin-producing *aspergillus* section *versicolores* using a high discrimination polymerase. *Biocontrol. Sci.*, **25**(2), pp. 113–118, 2020.

[30] Schwelm, A. & Bradshaw, R.E., Genetics of dothistromin biosynthesis of dothistroma septosporum: An update. *Toxins*, **2**(11), pp. 2680–2698, 2010.

[31] IARC IA for R on C. Agents classified by the IARC Monographs, 56, 2012.

[32] Visagie, C.M. et al., *Aspergillus*, *Penicillium* and *Talaromyces* isolated from house dust samples collected around the world. *Studies in Mycology*, **78**, pp. 63–139, 2014.

[33] Jurjevic, Z., Peterson, S.W. & Horn, B.W., *Aspergillus* section *Versicolores*: Nine new species and multilocus DNA sequence based phylogeny. *IMA Fungus*, **3**(1), pp. 59–79, 2012.

[34] Géry, A., Rioult, J.-P., Heutte, N., Séguin, V., Bonhomme, J. & Garon, D., First characterization and description of *aspergillus* series *versicolores* in French bioaerosols. *JoF*, **7**(8), p. 676, 2021.

[35] Kobayashi, N. et al., Microflora of mycotoxigenic fungi in rice grains in Kyushu region of Japan and their changes during storage under non-controlled conditions. *Biocontrol. Sci.*, **24**(3), pp. 161–166, 2019.

[36] Jakšić, D., et al., Fungi and their secondary metabolites in water-damaged indoors after a major flood event in eastern Croatia. *Indoor Air*, **31**(3), pp. 730–744, 2021.

[37] Siqueira, J.P.Z. et al., Species diversity of *Aspergillus* section *Versicolores* in clinical samples and antifungal susceptibility. *Fungal Biology*, **120**(11), pp. 1458–1467, 2016.

[38] Heutte, N. et al., Suivi de la qualité de l'air dans un centre de lutte contre le cancer: Evaluation et caractérisation de l'exposition aux contaminants fongiques (Biohospitalair). *Hygiènes*, 2016. https://www.hygienes.net/boutique/hygienes-2/suivi-de-qualite-de-lair-centre-de-lutte-contre-cancer-evaluation-caracterisation-de-lexposition-aux-contaminants-fongiques-biohospitalair/. Accessed on: 24 Jun. 2021.

[39] Pottier, D. et al., Airborne molds and mycotoxins in *Serpula lacrymans*-damaged homes. *Atmospheric Pollution Research*, **5**(2), pp. 325–334, 2014.

[40] Lanier, C., Richard, E., Heutte, N., Picquet, R., Bouchart, V. & Garon, D., Airborne molds and mycotoxins associated with handling of corn silage and oilseed cakes in agricultural environment. *Atmospheric Environment*, **44**(16), pp. 1980–1986, 2010.

[41] Séguin, V. et al., Effect of agricultural and environmental factors on the hay characteristics involved in equine respiratory disease. *Agriculture, Ecosystems and Environment*, **135**(3), pp. 206–215, 2010.

[42] Desjardins, P. & Conklin, D., NanoDrop microvolume quantitation of nucleic acids. *JoVE*, **1**, p. 2565, 2010.

[43] Houbraken, J. et al., Classification of *Aspergillus*, *Penicillium*, *Talaromyces* and related genera (*Eurotiales*): An overview of families, genera, subgenera, sections, series and species. *Studies in Mycology*, **95**, pp. 5–169, 2020.

[44] Micheluz, A. et al., Detection of volatile metabolites of moulds isolated from a contaminated library. *Journal of Microbiological Methods*, **128**, pp. 34–41, 2016.

[45] Jakšić, D. et al., Species diversity and cytotoxic potency of airborne sterigmatocystin-producing *Aspergilli* from the section *Versicolores*. *Sci. Total Environ.*, **562**, pp. 296–304, 2016.

[46] Kobayashi, N. et al., Distribution of sterigmatocystin-producing *Aspergilli* in Japan. *Food Saf (Tokyo)*, **6**(2), pp. 67–73, 2018.

INDOOR VOC CONCENTRATIONS AT NURSERY AND PRIMARY SCHOOLS: IMPACT OF COVID-19 PREVENTIVE MEASURES

JULIANA P. SÁ, PEDRO T. B. S. BRANCO, MARIA C. M. ALVIM-FERRAZ,
FERNANDO G. MARTINS & SOFIA I. V. SOUSA
Laboratory for Process Engineering, Environment, Biotechnology and Energy (LEPABE),
Faculty of Engineering, University of Porto, Portugal

ABSTRACT

The importance of evaluating indoor air pollutants, such as volatile organic compounds (VOC), became a topic of utmost interest, especially during COVID-19 pandemic due to the increasing cleaning and disinfection of hands, surfaces and spaces, the most common measures to prevent the spread of COVID-19. VOC presence in indoor environments can impair human health, particularly above threshold limit values. Thus, this study aimed to quantify the differences between VOC concentrations before and during COVID-19 pandemic in indoor air of one nursery and one primary school in Porto, Portugal. This study was carried out in early 2020 (before COVID-19 pandemic) and early 2021 (during COVID-19 pandemic) in two classrooms – I07_A (nursery school) and S07_B (primary school). Both classrooms presented a similar school timetable between the two periods, but the personal and environmental hygiene throughout the day using alcohol-based sanitisers were hugely increased in 2021. Total VOC (TVOC) were monitored continuously with research-grade instruments for a minimum period of two consecutive weekdays. Two average periods were considered: (i) an average day period (hourly means of all weekdays); and (ii) an average occupancy period (hourly means during occupancy periods considering the school timetable). Descriptive statistical analysis, as well as normality (Shapiro–Wilk Test) and significance (Wilcoxon Signed Rank Test) tests were performed using the R software version 4.0.5. The statistical significance level considered was set to 0.05. TVOC concentrations exceeded the limit value in the Portuguese legislation (1,200 μg/m^3) during occupancy period in both classrooms during COVID-19 pandemic, although they never exceeded before the pandemic. Moreover, a statistically significant increase (p-value < 0.05) on TVOC concentrations from 2020 to 2021 were observed in the two studied classrooms for both average day (mean difference: 647 μg/m^3 and 1,170 μg/m^3 for I07_A and S07_B, respectively) and average occupancy periods (mean difference: 521 μg/m^3 and 2,730 μg/m^3 for I07_A and S07_B, respectively). Therefore, it was possible to conclude that the continued use of alcohol-based products, as a result of COVID-19 prevention measures, could increase TVOC concentrations to unsafe levels in schools.

Keywords: indoor air quality, school, VOC, COVID-19 pandemic.

1 INTRODUCTION

Children constitute a sensitive population to environmental contaminant's exposure, such as indoor air pollutants, since their lungs and immune system are still under development and they have a relatively higher amount of air inhalation. Moreover, schools are the most important indoor environment for children, apart from home, where they spent a significant part of their time [1]. Effectively, evidence has been made that indoor air pollution (IAP) impacts children's respiratory health [2]. Particularly, the presence of volatile organic compounds (VOC) in indoor environments can impair human health with both acute and chronic effects, especially above threshold limit values [3].

With the outbreak of COVID-19 pandemic declared in March 2020, followed by a general lockdown and closing of schools until September 2020, the Portuguese authorities, following the World Health Organization (WHO) recommendations, adopted a contingency plan for safe reopening of schools with some measures to prevent the spread of COVID-19. The most

WIT Transactions on Ecology and the Environment, Vol 252, © 2021 WIT Press
www.witpress.com, ISSN 1743-3541 (on-line)
doi:10.2495/AIR210071

common measures included the frequent hand hygiene using alcohol 70% sanitisers, as well as the maintenance of a clean environment by disinfection of surfaces and shared objects using sodium hypochlorite solution with a concentration of 0.05% or alcohol products at 70% (mostly for surfaces that do not support sodium hypochlorite) as well as improving ventilation [4], [5]. Those products are potentially relevant sources of VOC in indoor environments [6]. Therefore, the COVID-19 pandemic, and particularly the application of preventive measures, reinforced the importance of evaluating indoor air pollutants, such as VOC, making it a topic of utmost interest.

Quantifying the impact of COVID-19 preventive measures on indoor air quality (IAQ) in schools is still quite a recent topic, thus information is still scarce, and particularly the impact on VOC concentrations has not been studied. Thus, this study mainly aimed to quantify the differences between VOC concentrations before and during COVID-19 pandemic in one nursery and one primary school in Porto, Portugal.

2 METHODOLOGY

This study was carried out in one classroom from a nursery school (I07_A) and one classroom from a primary school (S07_B), both part of the SENSINAIR project. Both schools are located in urban context in Porto district, North of Portugal (41°N, 8°W), next to a road with high traffic density (front) and a residential area (backyard).

Two IAQ monitoring campaigns were conducted in both studied classrooms, one in early 2020 (before COVID-19 pandemic) and the other in early 2021 (during COVID-19 pandemic). Thus, information on characteristics of nursery and primary schools and respective studied classrooms, as well as information on relevant activities, occupation periods, number of occupants, and potential sources of pollution were collected in each campaign. Accordingly, Table 1 summarises the main characteristics of each studied classroom, including children's age, floor, area, average occupancy, occupant density and occupancy periods.

Table 1: Summary of the main characteristics of each studied classroom.

Classroom		Children's age (years)	Floor	Area (m^2)	Average occupancy	Occupant density ($occupants/m^2$)	Occupancy period
I07_A	2020	2	Ground floor	40	20	0.50	9–12 h; 13 h 30–17 h 15
	2021				16	0.40	
S07_B	2020	9–10	2nd floor	40	18	0.45	9–18h
	2021	8–9			18	0.45	

Both classrooms (I07_A and S07_B) presented similar school timetables between the two studied periods. Moreover, S07_B kept the same number of occupants between the two campaigns, while the occupant density decreased from 0.50 to 0.40 $occupants/m^2$ in the second campaign (2021) for I07_A. Furthermore, none of the studied classrooms had mechanical ventilation, whereby natural ventilation was adopted through opening the available windows and/or doors, with a notorious increase in the COVID-19 pandemic campaign. Cleaning activities were usually carried out by cleaning staff more than once a day (during and at the end of the occupancy period) in both studied classrooms. However, the COVID-19 preventive measures implemented in 2021 promoted personal and environmental hygiene throughout the day, through the regular use of hand sanitiser and the cleaning of surfaces and shared objects/toys in the classrooms using alcohol-based disinfectants.

In both indoor air quality campaigns (before and during COVID-19 pandemic), Total VOC (TVOC) continuous monitoring (1 minute log) was performed using two research-grade instruments – HazScanner Model IEMS and GrayWolf AndvanceSence BE – for a minimum period of four consecutive days (two weekdays followed by the weekend). The instruments were deployed precisely in the same place in both campaigns on a table or a shelf, usually near a wall or in the centre of the studied classrooms at about the height of the children's breathing.

After each IAQ monitoring campaign, data were collected at the two studied classrooms for analysis. Continuous measurements logged each minute allowed to plot the TVOC time series using 1 minute values and perform the TVOC hourly mean daily profile, the comparison with reference limit values, and the descriptive statistical analysis using 1 hour mean. In addition, two average periods were considered for the analysis, namely: (i) an average all day period (hourly means of all weekdays); and (ii) an average occupancy period (hourly means during occupancy periods considering the school timetable, which corresponded to children's exposure period). Normality was assessed using Shapiro–Wilk Test, and the significance of the differences between TVOC concentrations in the two campaigns was assessed using Wilcoxon Signed Rank Test.

TVOC concentrations were compared with the Portuguese legislated limit value [7] considering as exceedance if the daily maximum of 8 hours running means was above 1,200 $\mu g/m^3$ (600 $\mu g/m^3$, plus 100% of margin of tolerance in the absence of a mechanical ventilation system working in the classrooms).

All statistical analyses were performed using the R software version 4.0.5, and the statistical significance level was set at 0.05.

3 RESULTS AND DISCUSSION

Fig. 1 presents the TVOC time series profiles for S07_B and I07_A before and during COVID-19 pandemic.

As can be seen in Fig. 1, the highest TVOC concentrations were found on weekdays during occupancy (more evident during the COVID-19 pandemic period), while non-occupancy and weekends showed almost constant baseline concentrations (below the detection limit). In addition, those profiles are in accordance with other studies carried out in the same type of classrooms and schools of this study [8]–[10].

Thus, the daily mean scenario was considered for further analyses, and the TVOC hourly mean daily profile, including both classrooms and both periods (before and during COVID-19 pandemic) were plotted in Fig. 2.

Table 2 summarises the main statistical parameters (minimum, maximum, mean, median and standard deviation) of the TVOC hourly mean data for each studied classroom (I07_A and S07_B) before and during COVID-19 pandemic, considering the entire period of weekdays (24 h) and the occupancy period. Also, the difference in TVOC concentrations from 2020 to 2021 in both classrooms were quantified and presented in Table 2.

The reference Portuguese legislated limit value for TVOC (1,200 $\mu g/m^3$) was exceeded during the occupancy period in the two studied classrooms (S07_B and I07_A) during COVID-19 pandemic. However, the TVOC reference limit value in both classrooms was never exceeded before the pandemic, although the highest concentrations were also found during the occupancy. The levels found may cause acute and/or chronic health effects for classroom occupants, especially children, thus a health risk assessment should be performed to evaluate both long-term and short-term exposure to TVOC high concentrations [8], [11].

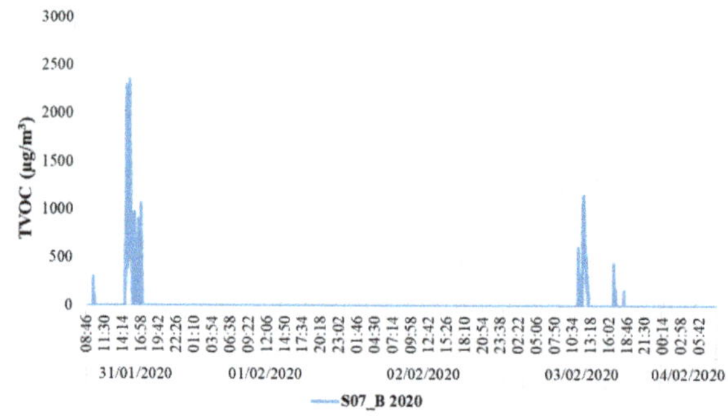

(a)

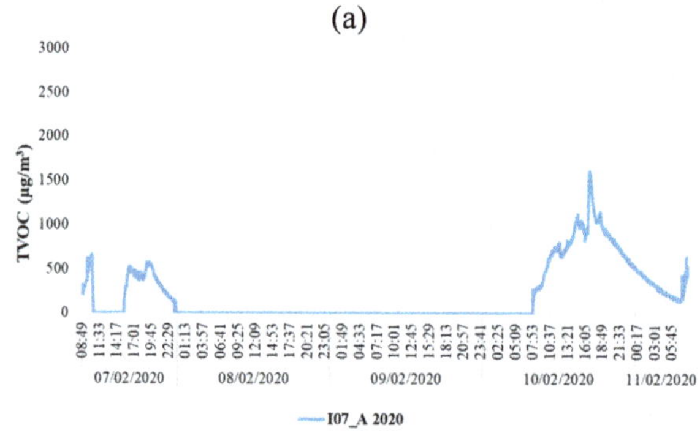

(b)

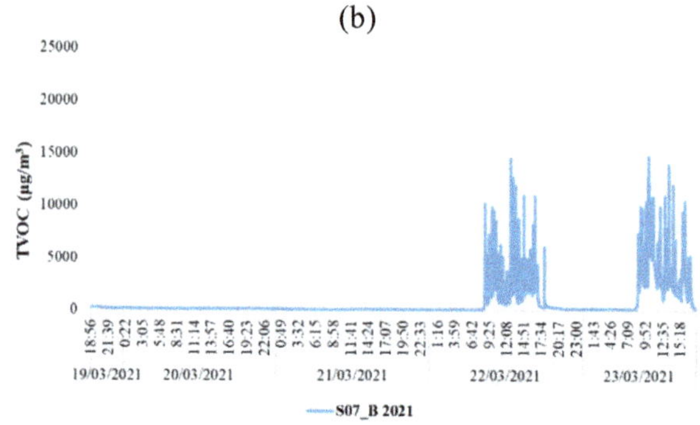

(c)

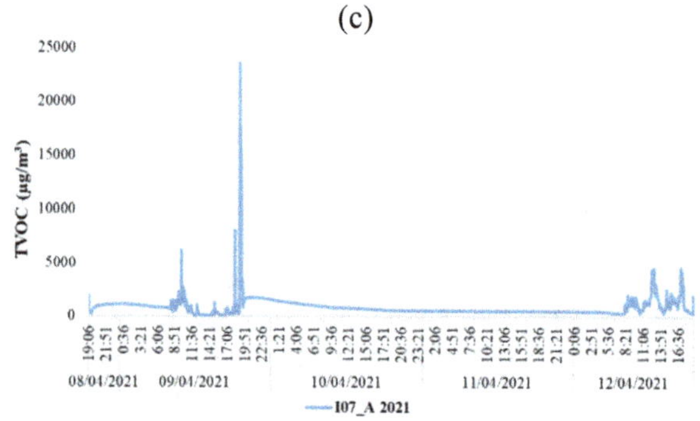

(d)

Figure 1: TVOC time series profiles considering 1 min mean for: (a) S07_B in 2020; (b) I07_A in 2020; (c) S07_B in 2021; and (d) I07_A in 2021.

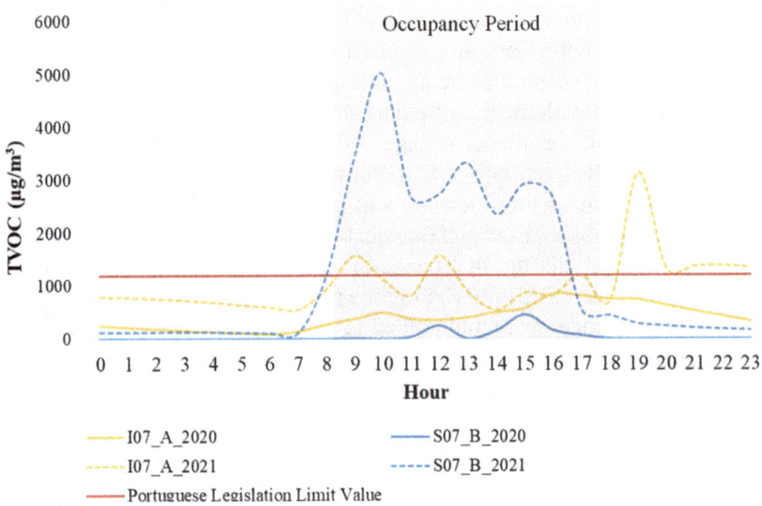

Figure 2: TVOC hourly mean daily profile for S07_B and I07_A before (2020) and during COVID-19 pandemic (2021) considering the Portuguese legislation reference limit value.

Table 2: Descriptive statistics of TVOC concentrations in the two studied classrooms (I07_A and S07_B) before and during COVID-19 pandemic, as well as the differences in the classrooms from 2020 to 2021, considering two periods (weekday and occupancy).

	Weekday (24 h)	**Occupancy period**
Classroom I07_A in 2020		
Mean (SD)	397 (230)	518 (185)
Median [Min, Max]	369 [82.1, 836]	484 [348, 836]
Classroom I07_A in 2021		
Mean (SD)	1040 (550)	1040 (354)
Median [Min, Max]	839 [537, 3130]	878 [537, 1570]
Difference in classroom I07_A		
Mean (SD)	647 (495) *	521 (441) *
Median [Min, Max]	563 [–27.4, 2400]	445 [–19.9, 1220]
Classroom S07_B in 2020		
Mean (SD)	46.1 (106)	123 (147)
Median [Min, Max]	0 [0, 443]	58.5 [0.962, 443]
Classroom S07_B in 2021		
Mean (SD)	1220 (1480)	2850 (1170)
Median [Min, Max]	256 [105, 5010]	2720 [537, 5010]
Difference in classroom S07_B		
Mean (SD)	1170 (1440) *	2730 (1200) *
Median [Min, Max]	256 [105, 5010]	2520 [479, 5010]

Difference in classroom I07_A and difference in classroom S07_B correspond to the difference in TVOC concentrations from 2020 to 2021 for I07_A and S07_B respectively; * statistically significant (p-value < 0.05).

As stated above, the presence of high TVOC concentrations occurred mainly during occupancy periods either before or during COVID-19 pandemic. In the first case (before the pandemic), its occurrence seemed to be a result of typical children activities carried out in the classrooms associated with the use of paints and glues (mostly in S07_B), as well as with the children's hygiene, such as diaper change, where some products that emitted VOC may have been used (exclusively in I07_A). In turn, the constant application of COVID-19 preventive measures, such as the intensive and periodic cleaning of surfaces, objects and hands using alcohol-based sanitisers and disinfectants, seemed to be the main explanation for the concentrations found during the pandemic period. Besides, the typical children's activities carried out within classrooms before the pandemic, and which appear to be the main cause of the TVOC presence, were maintained in 2021, so they could also have contributed to TVOC concentrations. Moreover, a peak between 18:00 and 20:00 in I07_A during COVID-19 pandemic was detected (Figs 1(d) and 2), which seemed to be caused by a reinforced cleaning of all materials and toys using an alcohol-based solution, usually performed before the weekend.

According to the results shown in Table 2, a statistically significant increase (p-value < 0.05) on TVOC concentrations from 2020 to 2021 were observed in the two studied classrooms for both average all day (mean difference: 647 μg/m^3 and 1,170 μg/m^3 for I07_A and S07_B, respectively) and average occupancy periods (mean difference: 521 μg/m^3 and 2,730 μg/m^3 for I07_A and S07_B, respectively). It is possible to observe a higher mean difference between the two periods for S07_B, which could be explained by the TVOC baseline concentration during the non-occupancy period being below the limit of detection before COVID-19 pandemic.

From a global perspective, one of the most common COVID-19 preventive measures – the continued use of sanitisers and disinfectants – increased the TVOC concentrations in the two studied classrooms to unsafe levels, which were not achieved before the pandemic. Moreover, the increase on natural ventilation during the COVID-19 pandemic, was insufficient to disperse VOC to the levels registered in 2020.

Therefore, it should be considered the reduction of alcohol-based products for the disinfection of surfaces and materials since surface transmission is not thought to be a significant risk (although possible) [12]. At the same time, the increase of ventilation should be even more encouraged to reduce the risk of transmission of diseases such as COVID-19 through aerosols [13]. The promotion of cross ventilation when using VOC emitting products should be considered since it will allow not only to prevent the accumulation of VOC in the indoor air but also to reduce CO_2 concentrations, which are usually above recommended thresholds in this type of classrooms, through a higher air change rate [14], [15].

4 CONCLUSIONS

As far as the author's knowledge goes, this was the first study quantifying the impact of COVID-19 preventive measures on TVOC concentration before and during COVID-19 pandemic in one nursery and one primary school.

The present study concluded that the more intense use of VOC-emitting compounds for personal (hands) and environmental (surfaces) disinfection, suggested by the COVID-19 preventive measures for reopening schools, has caused a statistically significant increase (p-value < 0.05) in the TVOC concentrations in the studied classrooms, from both nursery and primary schools, when compared to a period before the pandemic. That increment was more evident during occupancy periods, often exceeding the reference limit value in the Portuguese legislation, which may lead occupants (especially children) to be exposed to harmful levels of air pollution that may cause acute and/or chronic health effects.

Therefore, disinfection measures should be balanced with proper ventilation, namely cross ventilation, which allows not only to reduce VOC but also to prevent the accumulation of exhaled air and, thus, reduce the risk of COVID-19 infection by aerosols.

As future work, it is recommended to perform an individual VOC identification and extend this analysis to more schools, classrooms, and pollutants. Also, it should be considered the performance of a health risk assessment to evaluate both long-term and short-term exposure to TVOC high concentrations. In addition, a balance in the spread of COVID-19 preventive measures are crucial, so it is also important to quantify their impact in other indoor air pollutants (such as CO_2 and PM).

ACKNOWLEDGEMENTS

This work was financially supported by: Base Funding – UIDB/00511/2020 of the LEPABE – funded by national funds through FCT/MCTES (PIDDAC); Project PTDC/EAM-AMB/32391/2017, funded by FEDER funds through COMPETE2020 – POCI and by national funds (PIDDAC) through FCT/MCTES.

REFERENCES

[1] Morawska, L. et al., Airborne particles in indoor environment of homes, schools, offices and aged care facilities: The main routes of exposure. *Environment International*, **108**, pp. 75–83, 2017.

[2] Annesi-Maesano, I. et al., Indoor air quality and sources in schools and related health effects. *Journal of Toxicology and Environmental Health, Part B*, **16**(8), pp. 491–550, 2013.

[3] Paterson, C.A. et al., Indoor $PM_{2.5}$, VOCs and asthma outcomes: A systematic review in adults and their home environments. *Environmental Research*, **202**, 111631, 2021.

[4] DGS, Referencial Escolas: Controlo da transmissão de COVID-19 em contexto escolar. Direção Geral da Saúde: Lisbon, Portugal, 2020.

[5] WHO, Considerations for school-related public health measures in the context of COVID-19: Annex to considerations in adjusting public health and social measures in the context of COVID-19, 14 September 2020. World Health Organization, United Nations Children's Fund (UNICEF) & United Nations Educational Scientific and Cultural Organization, 2020. https://apps.who.int/iris/handle/10665/334294.

[6] EPA, Technical overview of volatile organic compounds. United States Environmental Protection Agency, 2017. https://www.epa.gov/indoor-air-quality-iaq/technical-overview-volatile-organic-compounds#3. Accessed on: 7 Aug. 2021.

[7] Portaria no. 353-A/2013, Ministérios do Ambiente, Odenamento do Território e Energia, da Saúde e da Solidariedade, Emprego e Segurança Social – Diário da República – 1ª Série, No. **253**(9), pp. 6644(2)-6644, 2013.

[8] Branco, P.T.B.S. et al., Children's exposure to indoor air in urban nurseries: Part II: Gaseous pollutants' assessment. *Environmental Research*, **142**, pp. 662–670, 2015.

[9] Nunes, R.A.O. et al., Gaseous pollutants on rural and urban nursery schools in Northern Portugal. *Environmental Pollution*, **208**, pp. 2–15, 2016.

[10] Sá, J. et al., Evaluation of low-cost mitigation measures implemented to improve air quality in nursery and primary schools. *International Journal of Environmental Research and Public Health*, **14**(6), p. 585, 2017.

[11] Sexton, K. et al., Children's exposure to volatile organic compounds as determined by longitudinal measurements in blood. *Environmental Health Perspectives*, **113**(3), pp. 342–349, 2005.

[12] Zhang, N. et al., Evidence for lack of transmission by close contact and surface touch in a restaurant outbreak of COVID-19. *Journal of Infection*, **83**(2), pp. 207–216, 2021.
[13] Fermo, P. et al., Improving indoor air quality through an air purifier able to reduce aerosol particulate matter (PM) and volatile organic compounds (VOCs): Experimental results. *Environmental Research*, **197**, 111131, 2021.
[14] Alonso, A. et al., Effects of the COVID-19 pandemic on indoor air quality and thermal comfort of primary schools in winter in a Mediterranean climate. *Sustainability*, **13**(5), p. 2699, 2011.
[15] Jones, E. et al., Healthy schools: risk reduction strategies for reopening schools. Harvard TH Chan School of Public Health Healthy Buildings Program, 2020.

ASSESSMENT OF INDOOR AND OUTDOOR EXPOSURE TO AIRBORNE PARTICLES IN SUBWAY METRO STATIONS AND THEIR ASSOCIATED HEALTH RISKS

AMIT PASSI[1], SARAGUR M. SHIVA NAGENDRA[1] & M. PRAKESH MAIYA[2]
[1]Department of Civil Engineering, Indian Institute of Technology Madras, India
[2]Department of Mechanical Engineering, Indian Institute of Technology Madras, India

ABSTRACT

This investigative study examines the exposure of metro commuters to inhalable airborne particulate matter (PM_{10}, $PM_{2.5}$, and PM_1) in and around the subway premises, and their associated health risk due to regular exposure to high levels of PM. PM monitoring was carried out in five metro stations (four underground and one elevated) in Chennai, India. PM measurements were taken for 3 days a week (two weekdays and one weekend) in each metro station. The adverse impacts on health and the risk of disease were assessed using AIRQ+, a tool developed by World Health Organization (WHO). Exposure to PM was found to be up to twofold higher than the WHO stipulated safe limits. Exposure to $PM_{2.5}$ was noted to be dominant (59%–70% of total PM mass). PM accumulation was found to be high in stations having a large number of commuters and those located in the busiest areas of the city. Infiltration of outdoor road emissions was found to have a significant influence on the deterioration of subway air quality. Underground metro stations were found to have the least dispersion and high accumulation of air pollutants whereas elevated metro stations were noted to have high dispersion and least accumulation. Health risk assessment indicated the non-carcinogenic risk to passengers (Hazard quotient = 1.18 and Cancer risk < 10^{-4}), who were exposed to PM regularly for a period of 1 year. Continuous, long-term exposure up to 8 years will lead to the development of carcinogenic risk (Hazard quotient > 10 and Cancer risk > 10^{-4}). Further, the health impact evaluation indicated high incidence per 100,000 population; asthma (816.92), unhealthy days (424.7), and lost workdays (416.57) due to acute exposure and ischemic heart disease (60.11), stroke (50), acute lower respiratory infections (30.52), and chronic obstructive pulmonary diseases (21.17) due to chronic exposure.

Keywords: subway stations, airborne particles, indoor exposure, outdoor exposure, health risk.

1 INTRODUCTION

In urban cities, residents spend a significant amount of time commuting between their home and workplace regularly. During their commute, they are inadvertently exposed to a variety of air pollutants on the roads, public, and private transportation modes they use. Subway metro is one of the fastest, cost-effective, convenient, reliable, and comfortable (air-conditioned) public transport systems. Regional metro rail projects are developing at a fast pace and gaining a lot of popularity in recent years. The subway system has helped to substantially reduce the quantum of road traffic and their resultant noxious emissions. However, the intrinsic compact design, underground location, and airtight construction features of the subway system have posed new challenges in terms of poor indoor air quality (IAQ). Several studies conducted in the past have indicated that airborne particulate matter (PM) concentrations were usually high in indoor transport microenvironments [1]–[3]. Regular exposure to short-duration high-level PM concentrations results in adverse acute and chronic illnesses in passengers. Epidemiological and toxicological research investigations have shown a strong interconnection between PM exposure and severe health impacts [4], [5].

Several studies conducted in the past have reported PM exposure levels that were up to 10 times higher in subway stations than in the outdoor environment [6], [7]. The PM

WIT Transactions on Ecology and the Environment, Vol 252, © 2021 WIT Press
www.witpress.com, ISSN 1743-3541 (on-line)
doi:10.2495/AIR210081

concentrations studied in the subway systems were associated with both indoor and outdoor environments. The underground and confined space of the metro favors infiltration of outdoor road emissions in addition to particles generated inside the metro [7]. The indoor sources of particle generation include abrasion of wheels and rails, sparking of electric wires, wear of brake pads, catenaries, and pantographs. Resuspension of particles due to the continuous to and fro activities of passengers and trains is another one of the main causes of sustained high levels of air pollutant concentrations. Several factors that influence particle concentrations in subway metro include the age of metro, design of tunnel and station, passenger density, frequency of trains, depth of station, wheel or rail track material, train speed, air-conditioning /ventilation system, frequency of station cleaning, etc. [8]. Due to the various operations of the metro, exposure to several $PM_{2.5}$ metals such as Fe, Mn, Mo, Cu, Ni, Cr, Ca, Na, Zn, and Pb was observed in metro stations [9], [10]. Thus, exposure of passengers to harmful airborne particles prevalent in the metro environment is a cause for concern that cannot be neglected. Inhalation of high concentrations of fine particles prevalent in the subway air results in their direct deposition in the lower respiratory system. Consequent to such exposure regularly, the deposited fine particles lead to the development of a variety of health symptoms. The different sizes of inhaled PM have the potential to impact passengers' health in different ways with varying degrees of severity [11]. The diameter of the PM is inversely proportional to the severity of health risk [12]. Exposure to airborne particles can have diverse short to long-term health issues such as respiratory infections, cardiovascular diseases, asthma, heart diseases, lung cancer, and mortality (WHO).

A comprehensive health impact assessment (HIA) procedure was developed by US EPA (Environmental Protection Agency). It defines the probability of occurrence of a particular health disease depending upon the duration of exposure to a pollutant. Further, WHO has developed a tool (AIRQ+) to calculate the health impacts of exposure to the contaminant of concern [13]. The present study aims to assess the exposure of metro passengers to PM in and around the subway premises, evaluate the health risks of exposure to high-level concentrations, calculate acute and chronic health impacts, and estimate the probability of disease.

2 METHODS

2.1 Study area

The air quality study was conducted in selected Chennai metro stations in India. Chennai is a bustling metropolitan city with an occupied area of 426 km^2 and population of 11.23 million (2021). The metro system in Chennai became operational in July 2015. Chennai Metro is the third largest metro network in India with a current operational length of 54.15 km. Chennai Metro has 19 elevated and 21 underground level metro stations carrying 1.25 lac passengers per day. Currently, two metro lines operate in the city; partially elevated and partially underground. Both metro lines serve to connect the urban areas of Chennai city. Line 1 connects the local areas of the city to the Airport and Line 2 connects to the Chennai Bus Depot and Central Railway Station. Fig. 1 presents the details of Chennai metro stations selected for the study.

2.2 Monitoring and analysis of exposure to PM

PM exposure concentrations were measured inside subway metro stations and around the premises of metro stations using Grimm spectrometer 1.109. Grimm spectrometer is a highly

www.witpress.com, ISSN 1743-3541 (on-line)

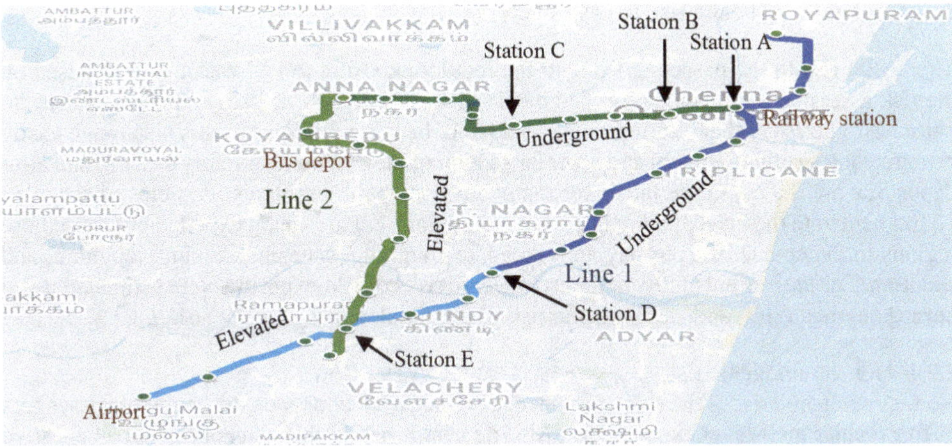

Figure 1: Chennai metro system and selected metro stations.

efficient and portable instrument used for collecting data on mass and number distribution of particles. In this study, the parameters considered were PM_{10}, $PM_{2.5}$, and PM_1. Also, data was taken from the continuous ambient air quality monitoring station located in Alandur, which was installed by CPCB (cpcb.nic.in) for outdoor exposure assessment. The particle exposure monitoring was conducted in four underground metro stations selected from both lines and 1 elevated metro station. PM measurements were taken for 3 days a week (two weekdays and one weekend) in each metro station during March 2019. Each one-minute data was collected continuously for all 24 hours of the day. Data collected inside the metro stations and around the premises of metro stations were analyzed using descriptive statistics. The data is presented in the form of line graphs showing the hourly variation of data including standard deviation for both weekdays and weekends. Fig. 2 presents the location of measurements taken in and around subway premises. Inside metro stations, the measurements are representative of platform area and outside subway stations, measurements are representative of roadside areas around subway premises.

Figure 2: Location of measurements taken in and around subway premises. (a) Inside subway station; and (b) Outside subway station.

2.3 Assessment of health risk to passengers

The health risk to the passengers due to the regular exposure to PM was analyzed based on the risk assessment guidance provided by US EPA. For health risk assessment, exposure to particles of size less than 2.5 μm was considered, because particles in this size range easily penetrate the terminal bronchi and alveolar section of the human respiratory system, and after deposition are the causative factor for many adverse health problems. Particles of size 2.5–10 μm penetrate the head and tracheobronchial regions but are easily expelled out from these regions in the course of everyday activities like coughing, sneezing, talking, laughing, and shouting. For the current study, non-carcinogenic risks of PM exposure were estimated using hazard quotient (HQ) and carcinogenic risk was estimated using cancer risk (CR).

2.3.1 Non-carcinogenic risk

Non-carcinogenic risk is the risk of occurrence of negative impact on the health of passengers due to regular high-level exposure to PM. The health risk in this category is not a causative factor for cancer. It is estimated as:

$$HQ = \frac{CDI}{Rfd}, \tag{1}$$

where, *CDI* is chronic daily inhalation (mg/kg/day) and *Rfd* is the reference dose of pollutant exposure (mg/kg/day).

$$CDI = \frac{PM*IR*ED*Ef*LE}{BW*ATL*NDY} *PPF, \tag{2}$$

where, *PM* is particle mass (mg/m^3), *IR* is particle inhalation rate (m^3/hours), *ED* is exposure duration (hours/week), *Ef* is exposure frequency (weeks/year), *LE* is the length of exposure (years), *BW* is bodyweight of passengers, *ATL* is the average time of human life (years), *NDY* is the number of days per year (days/year), and *PPF* is particle penetration factor.

$$Rfd = \frac{Rfc*IR}{BW}, \tag{3}$$

where *Rfc* is the reference dose of exposed air pollutant (mg/m^3).

As per US EPA, (HQ < 1) indicates no risk; (HQ > 1) indicates non-carcinogenic risk; and (HQ > 10) indicates high chronic risk.

2.3.2 Carcinogenic risk

Carcinogenic risk is the probability of occurrence of serious health impacts on passengers due to regular high-level exposure to PM that can cause cancer.

$$CR = \frac{CDI}{CSF}, \tag{4}$$

where, *CSF* is cancer slope factor (kg day/mg).

$$CSF = IUR* \frac{BW}{IR} *10^3, \tag{5}$$

where, *IUR* is inhalation unit risk of suspended particles (m^3/mg).

As per US EPA, CR < 10^{-6} is considered acceptable for individual compounds, and CR < 10^{-4} is considered acceptable as cumulative risk.

2.4 Evaluation of the impact of exposure to PM on the health of subway passengers

The short-term and long-term health impacts on metro passengers were evaluated by deploying AIRQ+ software developed by WHO, using data collected from in and around the premises of subway stations. In the current study, the impact of continued exposure to PM and the consequent burden of disease have been quantified. The AIRQ+ software works with the epidemiological data of detrimental health impacts associated with prolonged exposure to air pollution. To evaluate the health impacts for present-day Indian conditions, the incidence rates of various acute health impacts such as respiratory diseases [14], cardiovascular diseases [14], asthma [15], and overall mortality [16], as well as chronic health impacts such as mortality due to ALRI [17], COPD [18], LC [19], IHD [20], and stroke [21] were taken from epidemiological research studies conducted in India. To calculate the estimated attributable incidences, reference values were taken as per WHO guidelines of PM exposure for an average of 24 hours; $PM_{10} = 50$ μg/m^3 and $PM_{2.5} = 25$ μg/m^3.

3 RESULTS AND DISCUSSION

3.1 Exposure of passengers to PM

Table 1 presents the PM concentrations that the passengers are exposed to inside and around the premises of subway metro stations. The data presented in Table 1 is the average of 3 days (24/7) for each metro station.

Table 1: Exposure of passengers to airborne PM present inside and around the subway metro environment.

Stations	Variables	Inside subway stations			Around subway premises		
		PM_{10}	$PM_{2.5}$	PM_1	PM_{10}	$PM_{2.5}$	PM_1
Station A	Mean	80.98	54.38	40.24	103.00	61.47	48.48
	SD	24.46	12.40	7.81	41.00	19.09	13.36
Station B	Mean	32.27	25.12	20.83	74.98	45.58	37.83
	SD	5.36	3.69	3.46	12.04	1.72	1.43
Station C	Mean	23.18	17.36	14.07	49.11	29.66	23.73
	SD	8.47	6.35	5.31	21.87	10.76	8.61
Station D	Mean	32.12	21.29	16.61	35.95	19.95	15.17
	SD	8.90	5.67	4.90	12.58	6.06	4.60
Station E	Mean	47.32	29.90	24.58	90.21	53.89	42.03
	SD	8.69	4.01	3.18	33.07	14.72	11.48

It can be observed from (Table 1) that the concentrations of PM were high around the premises of subway stations than inside metro stations. The highest concentrations of air pollutants were observed at Station A. Metro Station A is located in the central hub of Chennai. It is one of the busiest areas in the city. The overall ratios of outdoor/indoor PM concentrations were 1.75, 1.48, and 1.47 for PM_{10}, $PM_{2.5}$, and PM_1, respectively (Table 2). The considerably high outdoor air pollutant concentrations indicate that road traffic emissions outside the subway could influence the indoor environment of the subway system.

Around the subway premises, the average $PM_{2.5}/PM_{10}$ ratio was 0.59, and $PM_1/PM_{2.5}$ ratio was 0.79. These ratios indicate that around subway premises, the total amount of PM

Table 2: PM ratios; outdoor/indoor, $PM_{2.5}/PM_{10}$, and $PM_1/PM_{2.5}$.

Stations	Outdoor/Indoor			Inside subway stations		Around subway premises	
	PM_{10}	$PM_{2.5}$	PM_1	$PM_{2.5}/PM_{10}$	$PM_1/PM_{2.5}$	$PM_{2.5}/PM_{10}$	$PM_1/PM_{2.5}$
Station A	1.27	1.13	1.20	0.67	0.74	0.60	0.79
Station B	2.32	1.81	1.82	0.78	0.83	0.61	0.83
Station C	2.12	1.71	1.69	0.75	0.81	0.60	0.80
Station D	1.12	0.94	0.91	0.66	0.78	0.55	0.76
Station E	1.91	1.80	1.71	0.63	0.82	0.60	0.78
Average	1.75	1.48	1.47	0.70	0.80	0.59	0.79
SD	0.53	0.41	0.39	0.06	0.04	0.02	0.03

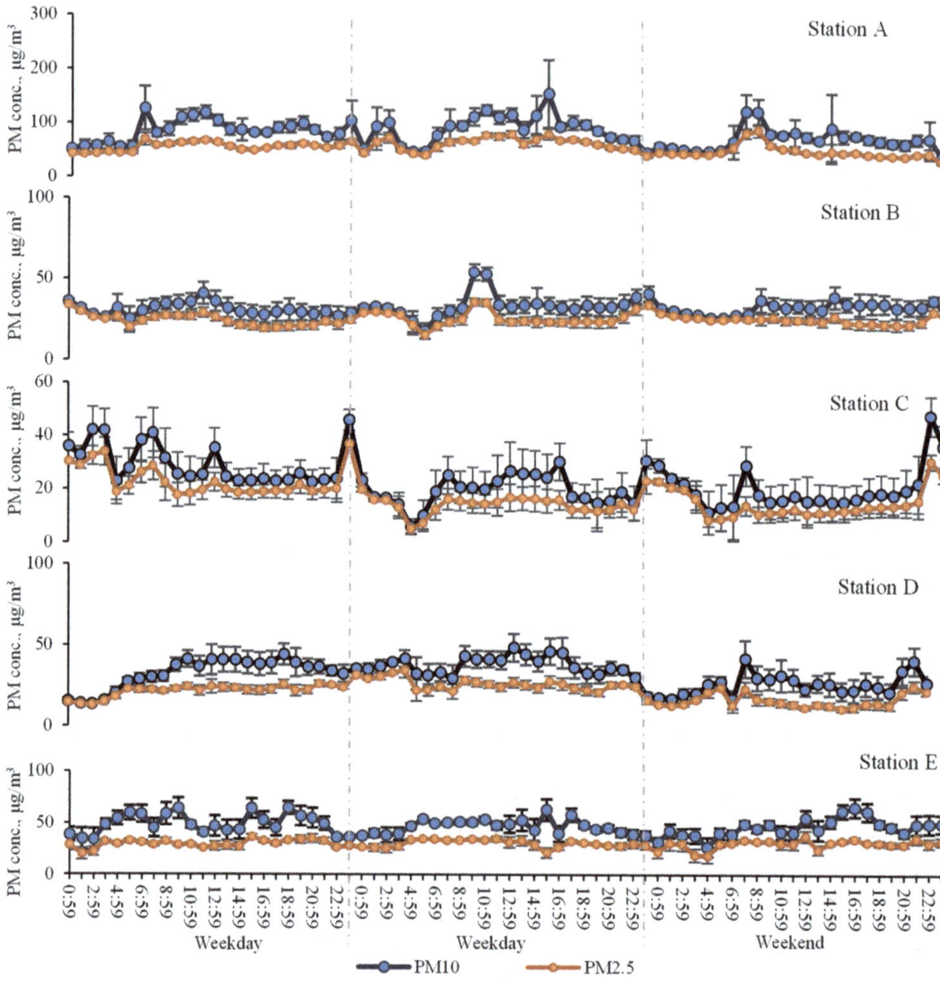

Figure 3: Hourly variations of particulate matter concentrations (conc.) inside studied subway metro stations during weekdays and weekends.

comprised of > 59% particles of < 2.5 μm size. The high composition of fine particles indicates the influence of road traffic vehicular exhaust emissions. Metro stations in Chennai city are situated on busy urban roads with substantially high daily average vehicular traffic.

Inside subway metro stations, the average $PM_{2.5}/PM_{10}$ ratio was 0.70, and $PM_1/PM_{2.5}$ was 0.80. These ratios indicate that inside metro areas, the total amount of PM comprised of > 70% particles of the fine size range (< 2.5μm). There are various sources of fine particles inside metro stations such as from tunnels (wear of rails, wheels, catenaries, and pantographs; service of metro lines; sparking of electric wires; wear of brake pads), resuspension due to the movement of trains and passengers, train piston force, and infiltration of outdoor fine particles suspended in the air [8]. Fig. 3 presents the hourly variations of PM inside subway metro stations during weekdays and weekends.

The average concentrations of PM (PM_{10}, $PM_{2.5}$) during weekdays and weekends were (86.42, 57.71 and 70.09, 47.72), (32.32, 25.11 and 32.18, 25.12), (24.64, 18.68 and 20.27, 14.73), (35.16, 23.86 and 26.03, 16.14), and (47.85, 29.83 and 46.28, 30.05) for Stations A–E, respectively. A notable difference of mean values was observed during weekdays and weekends in the stations having high daily average passenger travel > 5,000 (Stations A and D) while the pollutant concentrations were almost similar in stations having low daily average passenger travel < 1,500 (Stations B and C). Also, PM concentrations in the elevated metro station (Station E) were found to be almost similar during weekdays and weekends. One of the main reasons for the low and similar pollutant concentrations in elevated metro stations is direct interaction with the ambient environment and significant space for the dispersion of air pollutants. On the other hand, the highly closed and compact features of underground metro stations result in the least space for dispersion and a high accumulation of air pollutants.

Chennai metro operates from 5:27–23:05 hours on weekdays and 6:54–20:10 hours on weekends. Fig. 3 shows that the variation in the PM level commenced around 5:00 AM and continued till 23:00 hours after which a dip in concentration was observed till the metro operation resumed the next day. There was no notable difference observed for the variation of PM during peak (8:00–11:00 hours and 16:00–20:00 hours) and non-peak (11:00–16:00 hours) hours. The concentration variations were almost similar. This might be due to the suspension of fine particles for a longer duration in the air inside the metro stations because of prevailing environmental conditions. The concentration peaks in the figures indicate the activities of passengers at the metro stations.

Fig. 4 presents the hourly variation of PM around the premises of subway metro stations during weekdays and weekends. It can be observed from Fig. 4 that PM concentration variations around the premises of subway stations were comparatively high. The average concentrations of PM (PM_{10}, $PM_{2.5}$) during weekdays and weekends were (110.42, 64.76 and 88.16, 54.88), (75.67, 45.67 and 73.61, 45.42), (54.37, 32.55 and 38.58, 23.89), (34.17, 20.53 and 30.51, 18.80), and (88.98, 52.53 and 92.65, 56.59) for Stations A–E, respectively. A notable difference in concentration values was observed for stations A and C. Station A is situated in the busiest area of the city and Station C is located in the commercial area (educational institute). Thus, due to the high activity during weekdays in comparison to weekends, notable variations were seen.

Thus, it is evident that high outdoor road $PM_{2.5}$ emissions have a considerable influence on the indoor air quality of subway metro stations. Underground metro stations have the least dispersion and high accumulation of air pollutants whereas elevated metro stations have high dispersion and least accumulation. Metro stations featured with high daily average passenger travel and located in the busiest areas of the city were highly impacted by road emissions,

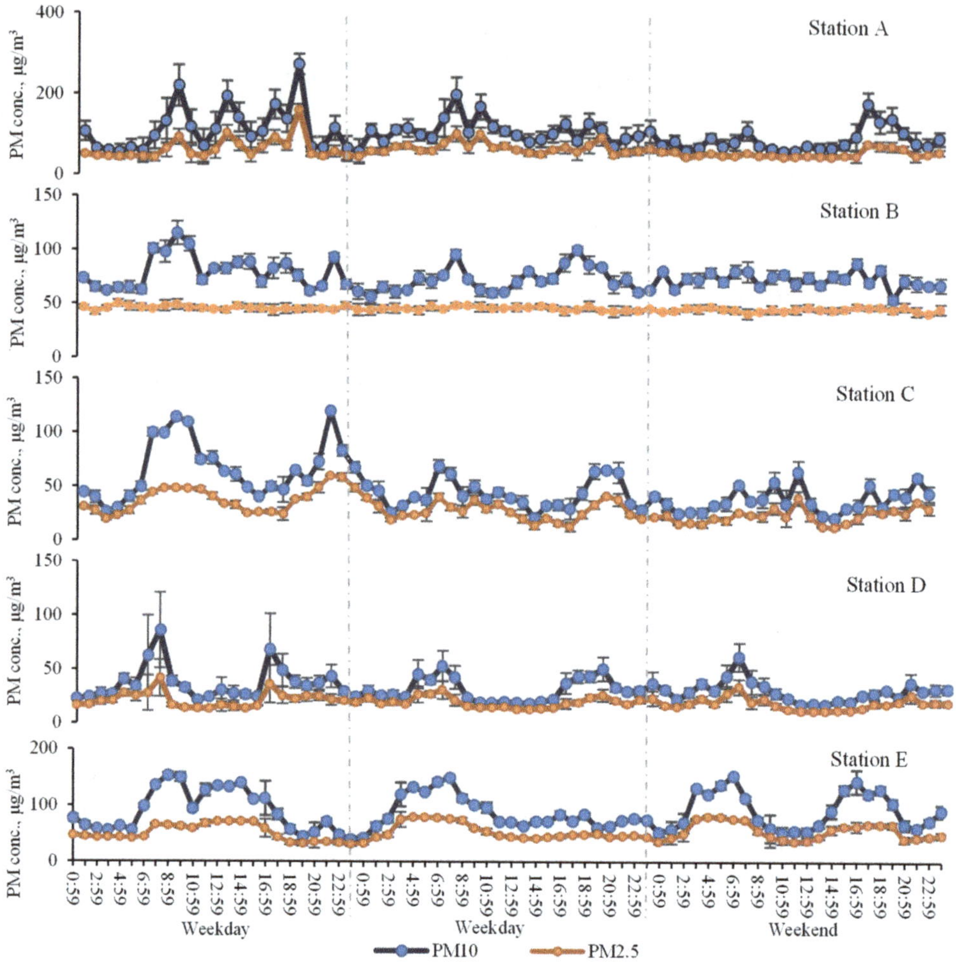

Figure 4: Hourly variation of particulate matter concentrations (conc.) around the premises of subway metro stations during weekdays and weekends.

passenger movement, and air pollutant accumulation. The least influence of passenger movement was observed in terms of peak and non-peak hours for metro stations with low to medium density of passenger flow.

3.2 Health risk assessment of passengers

3.2.1 Health risk to passengers

Regular exposure to substantially high air pollutant concentrations in and around the premises of subways can severely impact the health of passengers. For the calculation of health risk, PM mass was taken as per actual calculation (Table 1), IR was taken for light intensity work (sitting + walking operation) = 0.72 (EPA), ED = 10 hours/week (Metro stations in Chennai city are mostly used by daily office workers. So, taken as 5 days * 2 hours per day travel), Ef = 50 weeks/year, LE = 1 year (calculated for 1-year regular exposure),

BW = 60 kg (as per average body weight of adult in India – ICMR-NIN), ATLT = 70 years, NDY = 70 years, PPF = 0.082 [22], Rfc = 0.025 (WHO), and IUH = 0.01 [23]. Table 3 presents the health risk assessment of passengers exposed to high air pollutant concentrations in and around the subway premises.

Table 3: Health risk to passengers exposed to PM in the subway.

Stations	Inside subway stations			Around subway premises		
	CDI	HQ	CR	CDI	HQ	CR
Station A	1.04716E-06	2.18	3.63598E-05	1.18369E-06	2.46	4.11003E-05
Station B	4.8372E-07	1	1.67958E-05	8.77705E-07	1.82	3.04759E-05
Station C	3.3429E-07	0.69	1.16066E-05	5.71144E-07	1.19	1.98301E-05
Station D	4.09968E-07	0.85	1.42341E-05	3.84164E-07	0.80	1.33382E-05
Station E	5.75765E-07	1.20	1.99906E-05	1.03773E-06	2.16	3.60298E-05
Average	5.70181E-07	1.18	1.97974E-05	8.10885E-07	1.68	2.81549E-05

From Table 3, it can be seen that the average HQ is > 1 and < 10. It indicates the non-carcinogenic health risk of exposure to PM. As CR < 10^{-4}, it indicates that cancer risk is within the acceptable limit (CR < 10^{-4} is considered because calculations are for $PM_{2.5}$ as a whole; the composition includes multiple elements such as Fe, Mn, Cu, Ni, Cr, Ca, Pb, etc.). It should be noted that here exposure risk is calculated for regular 1 year of exposure. The HQ and CR will increase with the increase in exposure duration. The overall average calculations showed that a continuous increase in the exposure duration up to 8.5 years resulted in HQ > 10 and CR > 10^{-4}. This may have severe long-term impacts on the health of passengers.

3.2.2 Evaluation of the impact of exposure to PM on the health of metro passengers

Table 4 presents the summary of acute health impacts that were calculated using AIRQ+ software for highly polluted metro station A. The PM data obtained for the areas in and around the subway metro premises surpassed the WHO stipulated guidelines for safe exposure levels in the air.

In Table 4, EAP (estimated attributable proportions), ENAC (estimated number of attributable cases), and ENAC-R (estimated number of attributable cases per 100,000 population at risk) are presented. The risk of asthma was found to be highest followed by restricted activity days and workdays lost (due to illness or deterioration in health).

3.2.3 Burden of disease

Prolonged exposure to high air pollutant concentrations can lead to chronic health impacts such as acute lower respiratory infections in children (ALRI), chronic obstructive pulmonary diseases (COPD), lung cancer (LC), ischemic heart disease (IHD), and mortality due to stroke. Table 5 presents the negative health impacts due to long-term PM exposure (burden of disease).

Observations from Table 5 indicate the highest prevalence of IHD followed by stroke mortality. There are also chances for children below age 5 to develop ALRI. LC has the lowest prevalence rate.

4 CONCLUSIONS

The rapid growth of the subway metro network at a global level in recent years has resulted in a marked increase in the number of passengers choosing this travel option. It has gained

Table 4: Evaluation of health impact on exposure to PM in the premises of subway metro stations.

Health impacts	Incidence rate	RR*	EAP (%)	ENAC	ENAC-R (per 100,000)
Total mortality	939.73	1.0123 (1.0045–1.0201)	4.36 (1.62–7)	51 (19–82)	40.98 (15.26–65.79)
Hospital admissions, respiratory diseases	932	1.019 (1–1.0402)	6.63 (0–13.39)	77 (0–156)	61.83 (0–124.78)
Hospital admissions, cardiovascular diseases	882	1.0091 (1.0017–1.0166)	3.25 (0.62–5.83)	36 (7–64)	28.66 (5.45–51.40)
Restricted activity days	2753.73	1.047 (1.042–1.053)	15.42 (13.93–17.17)	531 (480–591)	424.70 (383.68–472.73)
Work days lost	2753.73	1.046 (1.039–1.053)	15.13 (13.02–17.17)	521 (448–591)	416.57 (358.63–472.73)
Incidence of asthma symptoms (children)	6000	1.028 (1.006–1.051)	13.62 (3.12–23.17)	1021 (234–1738)	816.92 (187.25–1390.47)

*RR = relative risk with 95% confidence interval.

Table 5: Assessment of the burden of disease in subway metro passengers.

Health impacts	Incidence rate	RR	EAP (%)	ENAC	ENAC-R (per 100,000)
Mortality due to ALRI (0–5 years)	85.5	1.555	35.69 (27.99–43)	38 (30–46)	30.52 (23.93–36.76)
Mortality due to COPD (adults)	64.7	1.486	32.73 (19.89–45.17)	26 (16–37)	21.17 (12.87–29.23)
Mortality due to LC (adults)	6.9	1.359	26.42 (17.43–34.86)	2 (2–3)	1.82 (1.20–2.41)
Mortality due to IHD (adults)	144	1.716	41.74 (24.81–58.21)	75 (45–105)	60.11 (35.73–83.82)
Mortality due to stroke (adults)	145	1.526	34.49 (18.11–47.63)	63 (33–86)	50 (26.26–69.06)

popularity as the preferred travel mode choice due to its advantageous features such as being fast, convenient, and comfortable. Based on the air quality study conducted in and around the premises of subway stations, it was identified that concentrations of PM were high around the premises of the subway in comparison to inside metro stations. Concentrations of PM surpassed the WHO recommendations of 50 μg/m^3 and 25 μg/m^3 for PM_{10} and $PM_{2.5}$, respectively, in both environments. Exposure to fine particles was dominant both inside and around the metro premises (> 59% of total PM mass). Air pollutant accumulation was found to be highest in stations with a high density of passengers and stations located in the busiest areas of the city. Infiltration of outdoor road emissions was found to have a considerable influence on IAQ deterioration. Underground metro stations were found to have the least dispersion and high accumulation of air pollutants whereas elevated metro stations were seen to have high dispersion and least accumulation. A notable difference in air pollutant concentrations was observed during weekdays and weekends in stations having a high daily average number of passengers travelling by metro. The least difference in air pollutant concentrations was noted during peak and non-peak hours because fine particles were dominant; they tend to remain suspended in the air for a longer duration. Thus, continuous operations of metro trains and the constant movement of passengers cause an elevated level of pollutant concentrations in the air throughout the day.

Health risk assessment of passengers due to regular high-level exposure to PM for 1 year indicated the HQ = 1.18, i.e., non-carcinogenic risk of exposure and CR was found to fall within the WHO limit of 10^{-4}. Health impact evaluation of PM indicated the highest incidences (per 100,000) of asthma (816.92) followed by restricted activity days (424.7) and workdays lost (416.57). The least incidences were reported due to cardiovascular diseases and mortality (< 50). Further, the evaluation carried out to determine the burden of disease indicated the highest mortality (per 100,000 population) in adult passengers due to IHD (60.11) and stroke (50) followed by ALRI in children and COPD in adults, and least mortality due to LC (< 2). It was also identified that regular exposure of metro passengers to high concentrations of airborne PM for a period of 8.5 years will result in carcinogenic risks (HQ>10 and CR < 10^{-4}).

ACKNOWLEDGEMENT

The authors would like to thank CMRL (Chennai Metro Rail Limited) for permitting us to conduct IAQ measurements inside subway metro stations.

REFERENCES

[1] Goel, R., Gani, S., Guttikunda, S.K., Wilson, D. & Tiwari G., On-road $PM_{2.5}$ pollution exposure in multiple transport microenvironments in Delhi. *Atmospheric Environment*, **123**, pp. 129–138, 2015. https://doi.org/10.1016/j.atmosenv.2015.10.037.

[2] Yang, F. et al., Heterogeneity of passenger exposure to air pollutants in public transport microenvironments. *Atmospheric Environment*, **109**, pp. 42–51, 2015. https://doi.org/10.1016/j.atmosenv.2015.03.009.

[3] Chan, L.Y., Lau, W.L., Lee, S.C. & Chan, C.Y., Commuter exposure to particulate matter in public transportation modes in Hong Kong. *Atmospheric Environment*, **36**, pp. 3363–3373, 2002.

[4] Valavanidis, A., Fiotakis, K. & Vlachogianni, T., Airborne particulate matter and human health: Toxicological assessment and importance of size and composition of particles for oxidative damage and carcinogenic mechanisms. *Journal of Environmental Science and Health – Part C*, **26**(4), pp. 339–362, 2008. https://doi.org/10.1080/10590500802494538.

[5] Dominici, F. et al., Fine particulate air pollution and hospital admission for cardiovascular and respiratory diseases. *The Journal of the American Medical Association*, **295**(10), pp. 1127–1134, 2006.

[6] Barmparesos, N., Assimakopoulos, V.D., Assimakopoulos, M.N. & Tsairidi, E., Particulate matter levels and comfort conditions in the trains and platforms of the Athens underground metro. *AIMS Environmental Science*, **3**(2), pp. 199–219, 2016. https://doi.org/10.3934/environsci.2016.2.199.

[7] Martins, V. et al., Exposure to airborne particulate matter in the subway system. *Science of the Total Environment*, **511**, pp. 711–722, 2015. https://doi.org/10.1016/j.scitotenv.2014.12.013.

[8] Passi, A., Nagendra, S.M.S. & Maiya, M.P., Characteristics of indoor air quality in underground metro stations: A critical review. *Building and Environment*, **198**, 107907, 2021. https://doi.org/10.1016/j.buildenv.2021.107907.

[9] Ryswyk, K.V. et al., Metro commuter exposures to particulate air pollution and $PM_{2.5}$-associated elements in three Canadian cities: the urban transportation exposure study. *Environmental Science and Technology*, **51**(10), pp. 5713–5720, 2017. https://doi.org/10.1021/acs.est.6b05775.

[10] Aarnio, P. et al., The concentrations and composition of and exposure to fine particles ($PM_{2.5}$) in the Helsinki subway system. *Atmospheric Environment*, **39**, pp. 5059–5066, 2005. https://doi.org/10.1016/j.atmosenv.2005.05.012.

[11] Khamraev, K., Cheriyan, D. & Choi, J.-H., A review on health risk assessment of PM in the construction industry: Current situation and future directions. *Science of the Total Environment*, **758**, 143716, 2021. https://doi.org/10.1016/j.scitotenv.2020.143716.

[12] Tian, G. et al., Indirect effect of PM_1 on endothelial cells via inducing the release of respiratory inflammatory cytokines. *Toxicology in Vitro*, **57**, pp. 203–210, 2019. https://doi.org/10.1016/j.tiv.2019.03.013.

[13] Miri, M. et al., Human health impact assessment of exposure to particulate matter: An AirQ+ software modeling. *Enviro. Sci. Pollut. Res.*, **24**, pp. 16513–16519, 2017. https://doi.org/10.1007/s11356-017-9189-9.

[14] Pascal, M. et al., Assessing the public health impacts of urban air pollution in 25 European cities: Results of the Aphekom project. *Science of the Total Environment*, **449**, pp. 390–400, 2013. https://doi.org/10.1016/j.scitotenv.2013.01.077.

[15] Global Asthma Network, Global asthma report, 2018. www.globalasthmanetwork.org.

[16] World Health Organization, *Health Impact Assessment of Air Pollution: Introductory Manual to AirQ+*, Regional Office for Europe, 2020.

[17] Krishnan, A. et al., Epidemiology of viral acute lower respiratory infections in a community-based cohort of rural north Indian children. *Journal of Global Health*, **9**(1), 010433, 2019. https://doi.org/10.7189/jogh.09.010433.

[18] Koul, P.A., Chronic obstructive pulmonary disease: Indian guidelines and the road ahead. *Lung India: Official Organ of Indian Chest Society*, **30**(3), pp. 175–177, 2013. https://doi.org/10.4103/0970-2113.116233.

[19] Malhotra, R.K., Manoharan, N., Nair, O., Deo, S. & Rath, G.K., Trends in lung cancer incidences in Delhi, India 1988–2012: Age-period-cohort and joinpoint analyses. *Asian Pacific Journal of Cancer Prevention*, **19**(6), pp. 1647–1654, 2018. https/doi.org/10.22034/APJCP.2018.19.6.1647.

[20] Prabhakaran, D., Jeemon, P. & Roy, A., Cardiovascular diseases in India: Current epidemiology and future directions. *American Heart Association Inc.*, pp. 1605–1620, 2016. https://doi.org/10.1161/CIRCULATIONAHA.114.008729.

[21] Pandian, J.D. & Sudhan, P., Stroke epidemiology and stroke care service in India. *Journal of Stroke*, **15**(3), pp. 128–134, 2013. https://doi.org/10.5853/jos.2013.15.3.128.

[22] Asgharian, B., Hofmann, W. & Bergmann, R., Particle deposition in a multiple-path model of the human lung. *Aerosol Science and Technology*, **34**, pp. 332–339, 2001.

[23] Greene, N.A. & Morris, V.R., Assessment of public health risk associated with atmospheric exposure to $PM_{2.5}$ in Washington, DC, USA. *International Journal of Environmental Research and Public Health*, **3**(1), pp. 86–97, 2006.

FUNCTIONAL GROUPS CHARACTERISATION OF INDOOR PARTICULATE MATTER IN SCHOOLS

JOANA V. BARBOSA, JULIANA P. SÁ, MARIA C. M. ALVIM-FERRAZ,
FERNANDO G. MARTINS & SOFIA I. V. SOUSA
Laboratory for Process Engineering, Environment, Biotechnology and Energy (LEPABE),
Faculty of Engineering, University of Porto, Portugal

ABSTRACT

School-age children spend most of the day at school. During this period, and regarding indoor activities, they are exposed to pollutants, namely particulate matter (PM), which can originate from dust re-suspension, furniture, building materials, cleaning activities and be influenced by ventilation rates. Identifying the chemical composition of PM is essential to minimise its effects on human health. Thus, the main aim of this study was to chemically characterise the composition of indoor PM in a pre-school classroom in the PM_{10} fraction. Indoor concentrations of PM_1, $PM_{2.5}$ and PM_{10} were continuously measured in a pre-school classroom for approximately 2.5 days. Indoor hourly mean concentrations were compared with WHO guidelines. Simultaneously, particles were gravimetrically collected for approximately 7 days. FTIR was used to chemically characterise the PM functional groups. The results showed an increase in PM concentration at the beginning of morning activities, which alternates with decreases and increases, according to children's occupation period, as expected. WHO guidelines were exceeded more than 61% of the time for $PM_{2.5}$ and 17% for PM_{10}. FTIR spectra allowed to identify the chemical composition of PM, namely the bands assigned to ammonium, aliphatic carbon, carbonyl group, nitrates, and sulfates. The presence of sulfates (probably ammonium sulfate and more acidic sulfate species) and nitrates (ammonium nitrate), which are formed in the atmosphere, suggests the influence of outdoor air.

Keywords: indoor air quality, particulate matter (PM), FTIR characterization, pre-school classroom, children.

1 INTRODUCTION

Indoor air pollution has been associated with the decline in lung function, being more evident in children, as their organism is still in the development stage [1]. School-age children spend most of the day at school, and during this period, they are exposed to many pollutants, such as particulate matter (PM), carbon monoxide, nitrogen dioxide, ozone, volatile organic compounds and formaldehyde. Many factors can contribute to poor indoor air quality namely, emissions from building materials, furniture, equipment, ventilation system, wear or textile products, consumer products, cleaning and cooking activities [2]–[4].

The study of PM has attracted high interest because of their associated health impacts. Their reduced dimensions allow penetration into the respiratory system. Thus, identifying the chemical composition of PM is essential to minimise its effects on human health.

Typically, PM is composed of nitrates, chlorides, sulfates, metals, ammonium, minerals and carbonaceous material, polycyclic aromatic hydrocarbons (PAH) and metals [5]. Exposure to these substances can cause adverse effects on respiratory health, such as asthma, allergic reactions and lung cancer [6], [7].

Over the last years, many studies have been conducted assessing children exposure to PM in schools, and as expected, most of them showed that indoor PM concentrations depend on children activities, classroom occupancy and outdoor PM concentrations [2], [3], [8], [9]. On the other hand, the few studies that collected particles inside the classrooms and carried out their chemical characterisation attributed their presence to both human (skin flakes) and mineral origin (building materials, chalk and soil material brought by children's shoes) [10].

WIT Transactions on Ecology and the Environment, Vol 252, © 2021 WIT Press
www.witpress.com, ISSN 1743-3541 (on-line)
doi:10.2495/AIR210091

Although FTIR spectroscopy has been used to characterise the chemical composition of PM both from indoor and outdoor environments [11]–[15], to the best of the authors knowledge, only one study characterised dust settled in different indoor micro-environments of an academic institution in India [16]. Thus, the main aim of this study was to chemically characterise the composition of indoor PM in a Portuguese pre-school classroom in the PM_{10} fraction.

2 MATERIAL AND METHODS

2.1 Sites description

This study was carried out in a pre-school classroom, part of the SENSINAIR project, located in an urban area in Porto district, North of Portugal (41°N, 8°W), next to a road with high traffic density (front) and a residential area (backyard). The main characteristics of the classroom studied are summarised in Table 1.

Table 1: Pre-school classroom description.

Children's age (years)	Floor	Area (m^2)	Average occupancy	Occupant density (occupants /m^2)	Occupancy period	Dust cleaning	Ventilation	Board
5–6	Ground floor	40	18	0.45	9–12 h; 14–18 h	Sweep, vacuum, wash	Natural (keeping the doors and windows open)	Chalkboard

Although the main cleaning activities were performed during and at the end of the occupancy period, alcohol-based disinfectants use, and surfaces cleaning were constant throughout the day because of preventive measures implementation due to COVID-19.

2.2 Sampling and analysis

Indoor PM_1, $PM_{2.5}$ and PM_{10} concentrations were continuously measured using a TSI DustTrak™ DRX 8533 Aerosol Monitor (TSI, USA), with light-scattering, operating at 3 L/min, during a sampling period of approximately 2.5 days. Indoor hourly mean concentrations were compared with WHO guidelines for PM_{10} (50 $\mu g/m^3$) and $PM_{2.5}$ (25 $\mu g/m^3$) and with Portuguese legislation limits whose values are the same. Simultaneously, particles were gravimetrically collected using internal polytetrafluoroethylene (PTFE) membranes filters with a 2.00 μm pore size and 37 mm diameter, during a sampling period of approximately 7 days. The device was equipped with a size-selective impactor for PM_{10} and was previously calibrated by the manufacturer. Filter weight was measured using an analytical balance (Mettler Toledo AG245 with a reading accuracy of 0.1 mg). Filters were appropriately stored in Petri dishes for consequent chemical analysis.

Infrared (IR) spectra were recorded with a Vertex 70 spectrometer (Bruker), equipped with an ATR cell with a A225/Q PLATINUM ATR Diamond crystal with single reflection accessory. Each spectrum was taken with an average of 64 scans, with a 4 cm^{-1} resolution in the 4000–500 cm^{-1} range.

3 RESULTS AND DISCUSSION

Table 2 summarises the minimum, maximum and mean concentrations of PM_1, $PM_{2.5}$ and PM_{10} measured in the pre-school classroom, and Fig. 1 shows their mean daily profiles.

Table 2: Parameters of the PM hourly mean data.

	Particulate matter hourly mean values (µg/m³)		
	PM_1	**$PM_{2.5}$**	**PM_{10}**
Minimum	11.00	11.32	12.67
Maximum	57.38	58.27	104.37
Mean	28.45	29.32	36.71

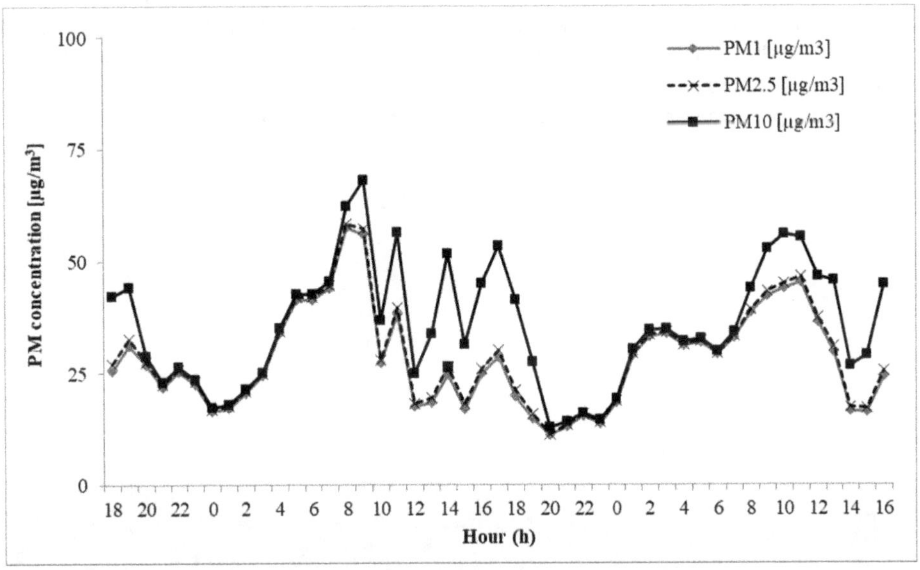

Figure 1: Mean PM_1, $PM_{2.5}$ and PM_{10} daily profile.

As expected, PM concentration profiles were similar among the different fractions, showing an increase in PM concentration at the beginning of morning activities, which alternates with decreases and increases, according to children's occupation period. Furthermore, concentrations decreased during the night after the end of activities and a last increase, possibly related to cleaning activities. Other authors also reported an increase in PM concentration during the beginning of children room occupancy alternated with decreases during the lunch period, when children went to the lunchroom, followed by a new increase until the end of the activities [3], [17].

Usually, the activities performed by the younger children (pre-school) inside the classroom and their greater physical activity combined with a large number of students per room and poor ventilation to maintain thermal comfort contribute to a constant re-suspension of sedimented particles [18]. Furthermore, sweeping during cleaning activities and the presence of a chalkboard clearly increased the PM levels.

To avoid contact between children from different classrooms (due to the SARS-CoV-2 pandemic), the children included in this study had all their meals inside the classroom, hampering the observation of such evident differences in the PM concentrations profiles between the occupancy and non-occupancy periods. This unusual occupation associated with an elevated re-suspension caused by occupants' activities, may have contributed to the fact that both Portuguese legislation limit and WHO guidelines were exceeded around 61% of the time for $PM_{2.5}$ and 17% for PM_{10}.

PM functional groups were chemically characterised by FTIR (Fig. 2). As expected, the typical PTFE bands were observed at 1205, 1149, 639, 554 and 509 cm^{-1} [19]. Excluding these bands, the FTIR spectra allowed to identify the chemical composition of PM, namely the bands at 3286 (NH_4^+, antisymmetric stretching) and 1410 cm^{-1} corresponding to ammonium and two bands at 2918 and 2850 cm^{-1} assigned to aliphatic carbon (CH stretching). The presence of amides can be attributed to the rubber of sneakers (shoe sole), as it is one of its main constituents [16], [20], while the presence of CH compounds, namely, alkanes and phthalate esters, may be attributed to cooking activities and floor polishes [14]. Furthermore, Sahu et al. [16] reported that the particles found in dust were soot, minerals, gypsum and the functional groups identified by FTIR (mainly alkyl halide and alkane), confirmed the presence of VOC (volatile organic carbon) in indoor dust.

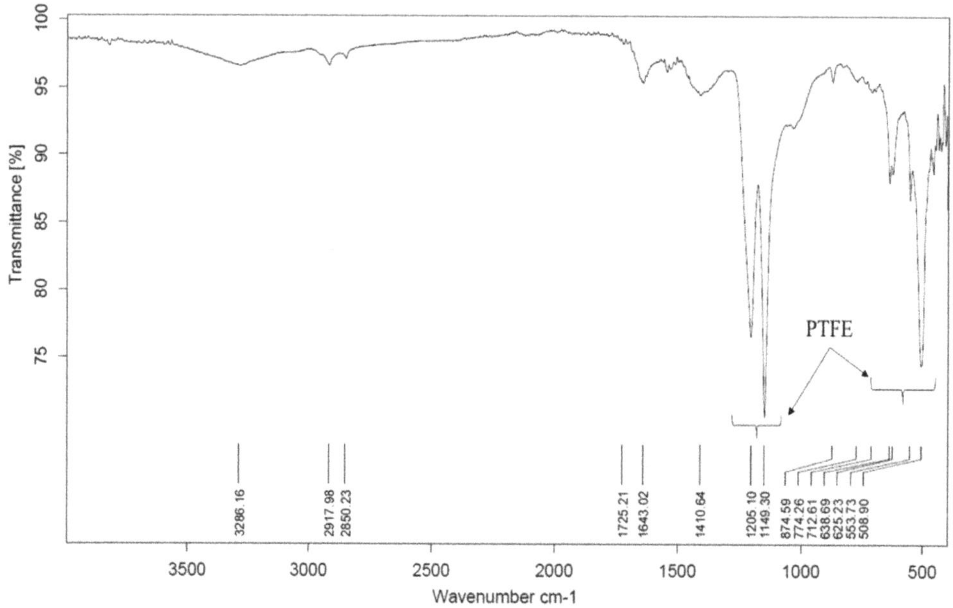

Figure 2: FTIR spectrum of the collected particles.

The carbonyl group band (C=O stretching) at 1725 cm^{-1}, typical of aliphatic aldehydes, ketones and carboxylic acids, was also observed. Finally, the organic nitrate at 1643 cm^{-1} (N-H, bending), the inorganic nitrate at 874 cm^{-1} and the shoulder at 1100 cm^{-1} assigned to sulphates (S=0, stretching) were also identified. The presence of sulfates (probably ammonium sulfate and more acidic sulfate species), and nitrates (ammonium nitrate), which are formed in the atmosphere from the photochemical oxidation of NO_x and SO_2, suggests the influence of outdoor air [12], [14]. These findings are in accordance with Reff et al. [14] and Estokova et al. [11].

4 CONCLUSIONS

This study allowed to chemically characterise the composition of indoor PM in a Portuguese urban pre-school classroom in the PM_{10} fraction, using FTIR spectra.

The main bands identified were assigned to ammonium, aliphatic carbon, carbonyl group, sulphates, organic and inorganic nitrate. The presence in indoor samples of sulfates (probably ammonium sulfate and more acidic sulfate species), and nitrates (ammonium nitrate), which are formed in the atmosphere, suggests that outdoor sources influenced the indoor air.

Globally, PM concentration profiles were similar among the different fractions and WHO guidelines were exceeded around 61% of the time for $PM_{2.5}$, and 17% for PM_{10}. These exceedances may be attributed to the presence of many students associated with a poor ventilation system. In addition, the use of a chalkboard, their classroom walking activities and having meals inside the classroom may have contributed to increasing PM accumulation, leading to an elevated re-suspension.

ACKNOWLEDGEMENTS

This work was financially supported by Base Funding – UIDB/00511/2020 of the Laboratory for Process Engineering, Environment, Biotechnology and Energy – LEPABE – funded by national funds through the FCT/MCTES (PIDDAC); PTDC/EAM-AMB/32391/2017, funded by FCT/MCTES. Sofia I.V. Sousa thanks the Portuguese Foundation for Science and Technology (FCT) for the financial support of her work contract through the Scientific Employment Stimulus – Individual Call – CEECIND/02477/2017.

REFERENCES

[1] Sram, R.J. et al., Health impact of air pollution to children. *Int. J. Hyg. Environ. Health*, **216**(5), pp. 533–540, 2013.

[2] Guo, H., Morawska, L., He, C., Zhang, Y.L., Ayoko, G. & Cao, M., Characterization of particle number concentrations and $PM_{2.5}$ in a school: Influence of outdoor air pollution on indoor air. *Environ. Sci. Pollut. Res.*, **17**(6), pp. 1268–1278, 2010.

[3] Nunes, R.A.O., Branco, P.T.B.S., Martins, F.G. & Sousa, S.I.V., Particulate matter in rural and urban nursery schools in Portugal. *Environ. Pollut.*, **202**, pp. 7–16, 2015.

[4] Perrino, C., Tofful, L. & Canepari, S., Chemical characterization of indoor and outdoor fine particulate matter in an occupied apartment in Rome, Italy. *Indoor Air*, **26**(4), pp. 558–570, 2016.

[5] Kelly, F.J. & Fussell, J.C., Size, source and chemical composition as determinants of toxicity attributable to ambient particulate matter. *Atmos. Environ.*, **60**, pp. 504–526, 2012.

[6] Zhang, L. et al., Personal exposure measurements of school-children to fine particulate matter ($PM_{2.5}$) in winter of 2013, Shanghai, China. *PLoS One*, **13**(4), pp. 1–16, 2018.

[7] Xu, D. et al., Acute effects of ambient $PM_{2.5}$ on lung function among schoolchildren. *Sci. Rep.*, **10**(1), pp. 1–8, 2020.

[8] Vilcekova, S., Meciarova, L., Burdova, E.K., Katunska, J., Kosicanova, D. & Doroudiani, S., Indoor environmental quality of classrooms and occupants' comfort in a special education school in Slovak Republic. *Build. Environ.*, **120**, pp. 29–40, 2017.

[9] Branco, P.T.B.S., Alvim-Ferraz, M.C.M., Martins, F.G., Ferraz, C., Vaz, L.G. & Sousa, S.I.V., Impact of indoor air pollution in nursery and primary schools on childhood asthma. *Sci. Total Environ.*, **745**, 2020.

[10] Rovelli, S. et al., Airborne particulate matter in school classrooms of northern Italy. *Int. J. Environ. Res. Public Health*, **11**(2), pp. 1398–1421, 2014.

[11] Estokova, A. & Stevulova, N., Investigation of suspended and settled particulate matter in indoor air chapter. *Intech*, pp. 455–480, 2012.

[12] Sreejith, M.V., Thomas, J.R., Aravindakumar, C.T. & Aravind, U.K., Characterisation of atmospheric particulate matter over a site in southern Kerala, India: Using ATR-FTIR and confocal micro-raman spectroscopy. *Mater. Today Proc.*, **33**, pp. 1410–1414, 2019.

[13] Reff, A. et al., Functional group characterization of indoor, outdoor, and personal $PM_{2.5}$: Results from RIOPA. *Indoor Air*, **15**(1), pp. 53–61, 2005.

[14] Reff, A. et al., A functional group characterization of organic $PM_{2.5}$ exposure: Results from the RIOPA study. *Atmos. Environ.*, **41**(22), pp. 4585–4598, 2007.

[15] Radulescu, C., Stihi, C., Iordache, S., Dunea, D. & Dulama, I.D., Characterization of urban atmospheric $PM_{2.5}$ by ATR-FTIR, ICP-MS and SEM-EDS techniques. *Rev. Chim.*, **68**(4), pp. 805–810, 2017.

[16] Sahu, V., Elumalai, S.P., Gautam, S., Singh, N.K. & Singh, P., Characterization of indoor settled dust and investigation of indoor air quality in different micro-environments. *Int. J. Environ. Health Res.*, **28**(4), pp. 419–431, 2018.

[17] Juliana, P.S., Branco, P.T.B.S., Alvim-Ferraz, M.C.M., Martins, F.G. & Sousa, S.I.V., Evaluation of low-cost mitigation measures implemented to improve air quality in nursery and primary schools. *Environ. Res. Public Heal.*, pp. 14–17, 2017.

[18] Branco, P., Alvim-Ferraz, M., Martins, F. & Sousa, S., Indoor air quality in urban nurseries at Porto city: Particulate matter assessment. *Atmos. Environ.*, **84**, pp. 133–143, 2014.

[19] Fazullin, D.D., Mavrin, G.V., Sokolov, M.P. & Shaikhiev, I.G., Infrared spectroscopic studies of the PTFE and nylon membranes modified polyaniline. *Mod. Appl. Sci.*, **9**(1), pp. 242–249, 2015.

[20] Martín-Martínez, J.M., Shoe industry. *Handbook of Adhesion Technology*, eds L.F.M. da Silva, A. Öchsner & R.D. Adams, Springer, Berlin, Heidelberg, 2011.

SECTION 3
EMISSION STUDIES

ESTIMATION OF EMISSIONS BY ROAD VEHICLES IN THE METROPOLITAN REGION OF SALVADOR, BRAZIL

KATTY SANTOS DA SILVA MARQUES, ANDERSON DA SILVA PALMEIRA,
DAVIDSON MARTINS MOREIRA & ERICK GIOVANI SPERANDIO NASCIMENTO
Centro Universitário SENAI CIMATEC, Brazil

ABSTRACT

This study focuses on atmospheric emissions from road vehicles in the Metropolitan Region of Salvador (RMS), northeaster Brazil. To investigate and analyse the contributions to air pollution, the methodology proposed and developed by the Ministry of the Environment (MMA) is used in the 1st National Inventory of Road Vehicle Emissions in 2011, the application of which has been published annually by the State Environmental Company of São Paulo (CETESB), from the 1st Vehicle Emissions Report, of which electric, hybrid and CNG vehicles are not considered. Usually, in the preparation of inventories, the emission rates are estimated according to the circulating fleet categorized by tonnage (light/heavy and passenger/cargo), type of fuel used, intensity of use and emission factors. Emissions are due to exhaust, fuel evaporation rate and tire, brakes, and track wear. Although, the emission inventories present uncertainties associated with the calculation of emissions according to the characteristics of the fuel, the circulating fleet and the technology on board, these inventories are powerful tools in this type of research. In this sense, the objective of this work was to elaborate an inventory of emissions by road vehicles in the RMS, referring to the period from 2013 to 2017, to subsidize the management of air quality in the RMS. The results show that RMS owns 36.48% of the circulating fleet in the State of Bahia, comprising approximately 65% of automobiles and 7.0% of heavy vehicles (trucks and buses) and heavy vehicles emit the highest levels of NO_x and PM, with approximately 84% and 70%, respectively.

Keywords: metropolitan region of Salvador, road vehicles, NO_x, MP.

1 INTRODUCTION

The growing volume of motor vehicles circulating on the highways is a worrying trend in air quality management. This trend is evident in the fact that most vehicles that make up the circulating fleet have outdated technology and use fossil fuels, contributing significantly to air pollution.

As a result, there is an increase in atmospheric emissions, where some of the compounds emitted are carbon monoxide, nitrogen oxides, particulate matter, etc. The complex interactions of compounds of different nature associated with the microclimate potentiate the greenhouse effect, triggering effects such as acid rain, heat islands, thermal inversion, etc. According to the World Health Organization (WHO) in 2018 it was estimated that 7 million people die annually from air pollution [1].

The RMS is located between the Recôncavo of Bahia and the Northeast of Bahia (North Coast), being the eighth largest pole of national wealth, where the municipalities of Camaçari, Candeias, Dias d'Ávila, Itaparica, Lauro de Freitas, Madre de Deus are located, Mata de São João, Pojuca, Salvador, São Francisco do Conde, São Sebastião do Passé, Simões Filho and Vera Cruz, as shown in Fig. 1.

Fig. 1 represents the RMS as an urban-industrial region where 1,446,336 vehicles are located, that is, 36.48%, of the State's fleet [2]. Since air pollution is an environmental, social, and economic challenge. Therefore, in this perspective, this work proposed to investigate road emissions in the RMS, and for this, an inventory of road atmospheric

WIT Transactions on Ecology and the Environment, Vol 252, © 2021 WIT Press
www.witpress.com, ISSN 1743-3541 (on-line)
doi:10.2495/AIR210101

Figure 1: Geographic location of the study area containing the location of the main mobile, fixed sources, and air quality monitoring stations.

emissions was prepared, considering the period from 2013 to 2017, representing the state of the art in this sense.

2 LITERATURE REVISION

The emission inventories elaborate for road vehicles makes it possible to investigate the contribution of these sources. Analysing the RMS, authors such as Lyra [3] and De Aguiar Filho and de Albuquerque [4], developed studies in this direction.

In the study by Lyra [3], a survey of vehicle and industrial emissions and some one-off emissions aimed at generating necessary information to propose an integrated model of air quality management and found that despite industrial concentration vehicle emissions are on average 75% of all air emissions in the region.

In the study by De Aguiar Filho and de Albuquerque [4], they proposed to investigate the vehicle emissions where the emissions of carbon monoxide (CO), total hydrocarbons (THC), nitrogen oxides (NO_x), sulfur oxides (SO_x) and particulate matter (PM) were estimated and the authors found high co emissions that are responsible for cardiovascular complications and changes in the nervous system, increased NO_x emissions, and THC that contribute to the production of tropospheric ozone and the increase of pm emissions that the smaller the particle size are called inhalable particles (PM_{10}) and fine inhalable particles ($PM_{2.5}$) that penetrate deeply into the lungs and the cardiovascular system.

Even though rms has one of the largest air quality monitoring networks in Brazil, totalling 24 stations (10 stations in the city of Salvador and the other 14 in those installed in other cities), the region still lacks continuous studies in this sense, aiming at investigations using models of air quality enabling a more realistic scenario of air quality as has been done for other metropolitan regions [5].

The southeast region of Brazil concentrates the most developed metropolitan regions in the continuous elaboration of studies focused on air quality and dispersion of pollutants. In the greater São Paulo metropolitan Region (RMSP) [6], several studies have been carried out in this regard, among them the UEDA study, which investigated vehicle emissions in 2011 and found that CO, THC, and NO_x emissions represent more than 75% of total emissions.

The Greater Vitória Metropolitan Region (RMGV) has developed in this direction and studies such as that by Loriato et al. [7] has been elaborated in the region where using the local inventory elaborated by the State Institute of the Environment (IEMA) for the year of 2010. From this, an atmospheric modelling was done realized an atmospheric modelling using the integration of models as the meteorological model Weather Research and Forecasting Model (WRF) [6], the model Sparse Matrix Operator Kernel Emissions (SMOKE) which processes the inventory to the required format in air quality models, and the Community Multiscale Air Quality Model (CMAQ) chemical air quality model that processes wrf and smoke information [6] and generates local chemical concentrations hourly [8].

Among the results obtained, the authors found that vehicle emissions contribute about 74.63% of PM_{10}, 73.42% of $PM_{2.5}$, 33.38% NO_x, 49.61% of CO. Demonstrating the relevance of these studies as strategic tools for local air quality management [6], [9], [10].

3 METHODOLOGY

Emission inventories are relevant tools in studies focusing on air quality, being used as input data in air quality models. In general, based on these data, emissions are estimated in annual periods. The approach described consists in the elaboration of emission inventories with the objective of estimating atmospheric emissions from road vehicles and something of this type has not yet been done in the region.

In these quantifications, vehicles are considered by categories: automobiles, light commercial vehicles, trucks, buses, and motorcycles [11], [12]. According to the fuel consumed, vehicles are categorized into: vehicles powered by gasoline, ethanol, flex (four-stroke internal combustion engine that has the capacity to be refuelled and work with more than one type of fuel, mixed in the same tank and burned in the combustion chamber simultaneously) and diesel [11], [12]. Hybrid, electric and GNV vehicles are not analysed. And we consider that the emissions occurring by exhaustion (discharge pipe), evaporation (supply, braking, acceleration, deceleration, and rest) and by the friction between the tires, brakes, and tracks [11], [12].

Fig. 2 summarizes the methodology developed by the Ministry of the Environment for road vehicles in a summarized way, being the most widespread method today [12].

In Fig. 2, we see the input data highlighted in blue and the outputs that are obtained by applying the estimation equations to these data in red.

The circulating fleet (*Fc*) is a particularly important data for this type of evaluation, being obtained through the sales history of new vehicles, applying survival rates where the probability of scrapping of vehicles is considered according to the time of use of the vehicles (from the year of manufacture), see eqn (1):

$$Fc = V \cdot S \tag{1}$$

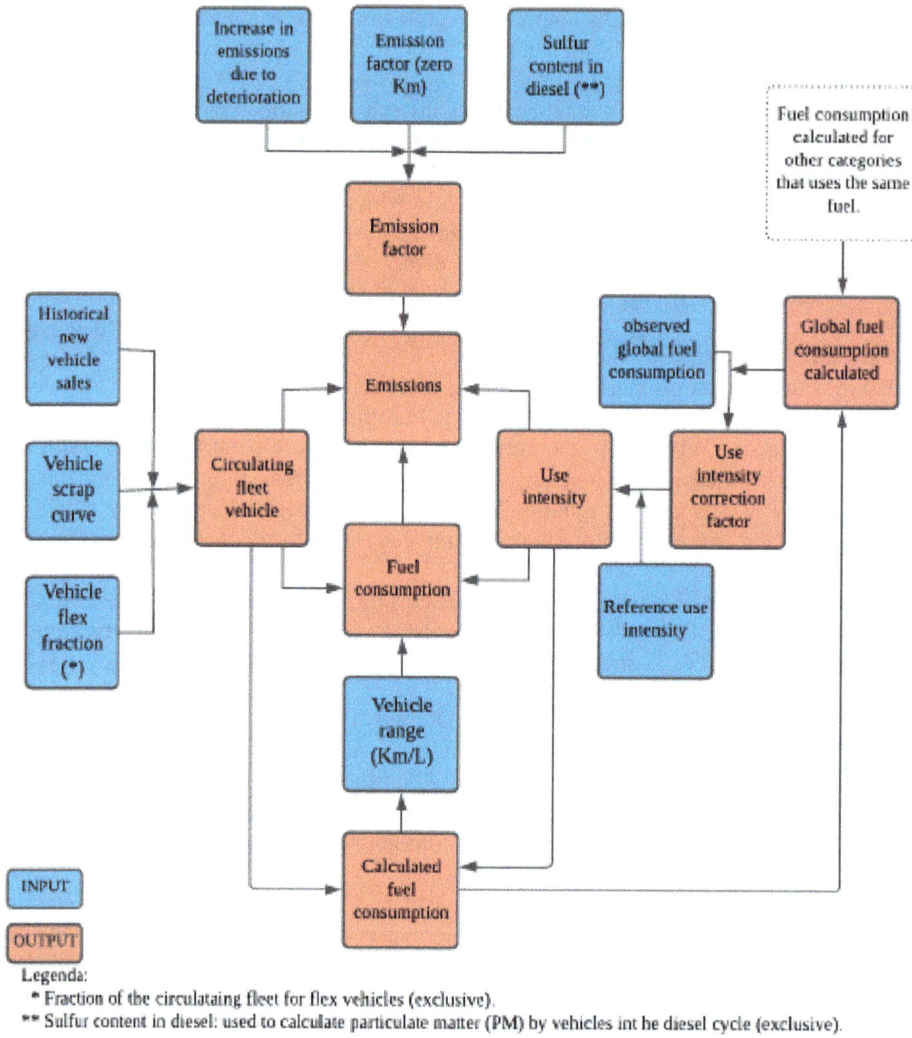

Figure 2: Summary of the steps of the inventory elaboration method.

where is the circulating vehicle fleet, V is the number of vehicles that came into circulation and S is the vehicle survival rate (or scrapping function of automotive vehicles) according (model-year) k in the (calendar-year) i. To determine a function that better describes the scrapping process for light vehicles, three functions were tested: Logistics, Gompertz and Weibull, the latter being used in international literature. However, the Gompertz function was chosen [13], [14] for better adapting to the data under analysis [3]–[8], [10]. In Otto cycle engines, see eqn (2), and in diesel cycle engines and motorcycles, see eqn (3), the survival curves are obtained through a Gompertz function:

$$S(t) = 1 - e^{-e(a+b.t)}, \tag{2}$$

where $S(t)$ is the survival rate, that is, the fraction of remaining vehicles not yet scrapped at age t, and a and b are variable parameters according to each type of vehicle.

$$S(t)=\frac{1}{1+e^{a(t-t_0)}}+\frac{1}{1+e^{a(t+t_0)}}, \tag{3}$$

In eqn (3), "a" and "t_0" are variable parameters according to each type of vehicle. To estimate the fraction of flex-fuel vehicles, that is, those that use gasoline and alcohol as fuel, an analysis was made based on the price of these fuels where if on average the price of ethanol is less than 70% of the price of gasoline the owner supplies with ethanol and if the price of ethanol is greater than 70% of the price of gasoline, the owner supplies with gasoline. From this, a list of these prices was made [15] and it was observed that in the period evaluated for the region, the owners of flex vehicles supplied primarily with gasoline.

For the calculation of emissions, the circulating fleet is used (*Fc*), data on the intensity of use of each category and the emission factors, see Fig. 2. Intensity of use is the average distance travelled for each vehicle category in a unit of time (year). For Brazil, the average values estimated in the first national inventory for road vehicles are used [11], [12] and adjusts to the local reality, as eqn (4), adapting based on the fuel consumption of each vehicle category according to actual consumption, eqn (4) [11], [12]. This relation is given by:

$$Ce_{C,T,A}=\frac{\sum_{T=1}^{t}\sum_{A=1}^{a}(Fc_{T,A}\cdot Iur_{T,A})}{Ql_{T,A}}, \tag{4}$$

where *Ce* is the estimated total annual fuel consumption for the vehicle categories and their respective model-year; is the reference usage intensity for each vehicle category and their respective model year; is the range that represents the mileage travelled per litter of fuel consumed by each vehicle category and their respective model year; *T*, *C*, and *A* are respectively the type of fuel, the category of vehicle and the model year.

From the estimated consumption (*Ce*), the actual consumption (*Cr*) and the intensity of reference use (*Iur*) the adjusted intensity of use is obtained (*Iua*) by vehicle category in their respective model-year as eqn (5):

$$Iua_{C,T,A}=Iur_{T,A}\cdot\frac{Cr}{Ce_{C,T,A}}, \tag{5}$$

where, is the intensity of use adjusted for each vehicle category and their respective model year; *Cr* is the actual consumption of each type of fuel; *Ce* is the total annual consumption of a type of fuel estimated for all vehicle categories and their respective model year.

Emission factors are representative values that estimate the amount of a given pollutant released as a function of an activity associated with the release of that pollutant (see USEPA) [16]. In Brazil, there is still insufficient data on road vehicle emission factors under real conditions of use. With this, data referring to homologation tests for new vehicles are used and for this reason an increase in the rounded mileage was made to reduce the uncertainties associated with the maintenance conditions and the driving patterns of the vehicle. The increment values were determined for the pollutants CO, NO_x, NMHC and

RCHO and were added to the emission factors every 80,000 km, except for vehicles of the diesel cycle and motorcycles due to the absence of this information.

Through the methodology adopted [11], [12] road vehicle emissions from exhaust, evaporation and wear of tires, brakes and tracks were estimated. Exhaust emissions refer to the operation of the vehicle's exhaust system and are estimated as eqn (6):

$$E_{P,T,C} = \frac{Fc_{T,C,A} \cdot Iua_{T,C,A} \cdot Fe_{P,T,C,A}}{10^6}, \tag{6}$$

where E is the emission rate (t/year); Fe is the emission factor expressed in terms of the mass of pollutant emitted per km travelled (g/km) and P is the pollutant evaluated [11], [12]. Evaporative emissions from road vehicles can occur through three processes. Daytime emissions that refer to emissions by fuel evaporation as a function of local temperature, emissions from the engine cooling process after use and emissions from fuel loss while the vehicle is in motion. For all cases, specific temperature ranges are considered (10°C to 15°C, 10°C to 25°C and 20°C to 35°C), to estimate the number of days in percentages, an analysis of the average daily temperatures of the season was made. conventional weather forecast in Salvador available at the National Meteorological Institute (INMET) and the same behaviour is considered for the entire region.

Daytime emissions were obtained from eqn (7):

$$E_{NMHC_{d,T,C,A}} = \frac{Fc_{T,C,A} \cdot ed_{T,C,A,F} \cdot Df}{10^6}, \tag{7}$$

where, E_{NMHCd} is the NMHC emission rate by evaporation in the daytime phase (t/year); ed is the emission factor for NMHC evaporation in the daytime (g/day) given in terms of vehicle category, fuel type, year of manufacture and temperature range; Df is the number of days to be considered for choosing the emission factor in each temperature range, and d represents the evaporative emissions in the daytime phase [11], [12].

In relation to the emissions from the engine cooling process after use and the emissions from fuel loss with the vehicle in progress, the emission factors es and er are given in g/trip so for these a conversion to g/km was made by use of the average distance traveled per trip in individual transport in the RMS (km/trip). Emissions from the engine cooling process after use were obtained using eqn (8):

$$E_{NMHC_{s,T,C,A}} = \frac{Fc_{T,C,A} \cdot Iua_{T,C,A} \cdot \frac{es_{T,C,A,F}}{9,48} \cdot Df\,(\%)}{10^6}, \tag{8}$$

where, E_{NMHCs} is the NMHC emission rate in the engine cooling process after use (t/year); es is the emission factor for NMHC evaporation in the engine cooling process (g/day) given in terms of vehicle category, fuel type, year of manufacture and temperature range [10], [11].

Emissions from fuel loss with the vehicle in progress were obtained using eqn (9):

$$E_{NMHC_{r,T,C,A}} = \frac{Fc_{T,C,A} \cdot Iua_{T,C,A} \cdot \frac{er_{T,C,A,F}}{9,48} \cdot Df\,(\%)}{10^6}, \tag{9}$$

where E_{NMHCr} is the NMHC emission rate of the fuel loss with the vehicle in progress (t/year); *er* is the emission factor for NMHC evaporation in the engine cooling process (g/day) given in terms of vehicle category, fuel type, year of manufacture and temperature range.

Emissions due to wear of brakes, tires and tracks are generated from the wear of vehicle brakes as a function of use, friction between tires and tracks and the resuspension of material deposited on the roads. To estimate these emissions, emission factors were used based on the TIER 1 methodology [16], [17]. And they were obtained from eqn (10):

$$E_{P,T} = \sum_{C=1}^{c}\sum_{A=1}^{a}\frac{(Fc_{T,C,A}\cdot Iua_{T,C,A})\cdot Fe_{P,T}}{10^6}, \tag{10}$$

where $E_{P,T}$ is the emission rate (t/year); *P* is the pollutant evaluated.

4 RESULTS AND DISCUSSION

The circulating fleet is one of the most important data in estimating and evaluating pollutant emissions by road vehicles. It is based on the analysis of the composition of the circulating fleet that we obtain the perspective not only of the total pollutant emissions, but also of the relative contributions of each category of vehicles, favoring the performance of evaluations of specific scenarios in a more appropriate way.

Thus, the results obtained bring automobiles, motorcycles, and light commercial vehicles with a greater relative participation in the composition of the circulating fleet of RMS, and these vehicles have more advanced embedded technology and as a result, with lower emissions of pollutants. At the same time, in this circulating fleet, there are vehicles with high levels of intensity of use, causing an increase in the level of aging of the fleet and with inadequate maintenance generating higher levels of emissions, when compared to trucks and buses with more current technologies.

Besides, in an expressive and growing way, we see that automobiles represent the largest category of vehicles in circulation, as shown in Fig. 3.

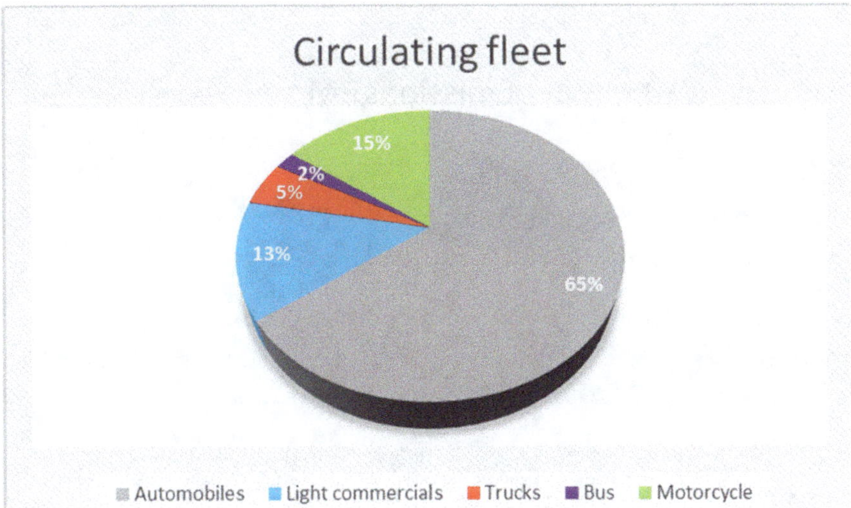

Figure 3: Relative composition of the circulating fleet.

Fig. 3 shows the participation of each category of vehicles in the circulating fleet of RMS, and we see that buses have the lowest participation followed by trucks. Motorcycles represent 15% of the total circulating fleet, while light commercial vehicles account for 13%.

However, in Fig. 4 we see that the total emissions of nitrogen oxides (NO_x) come from trucks representing 58% of emissions, despite representing 5.28% of the circulating fleet.

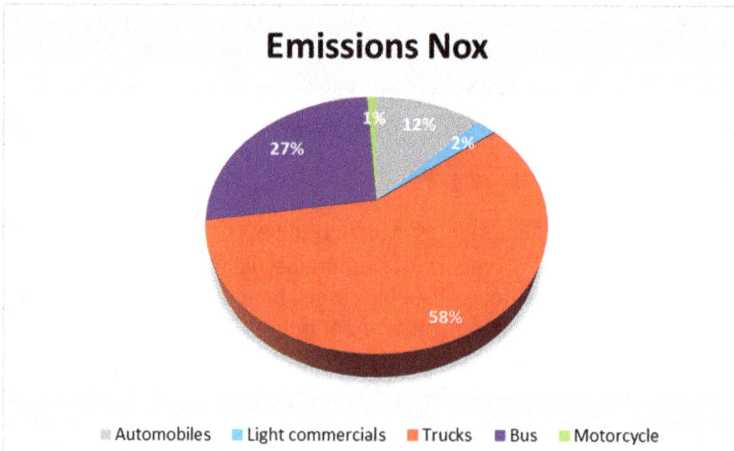

Figure 4: Relative contributions of vehicle categories to NO_x emissions.

Although automobiles represent the largest category of road vehicles (65%), they contribute about 12% of NO_x emissions while buses contribute 27% of these emissions representing the smallest category of road vehicles with about 2%.

The total emissions of particulate matter (PM) through the exhaust system are shown in Fig. 5.

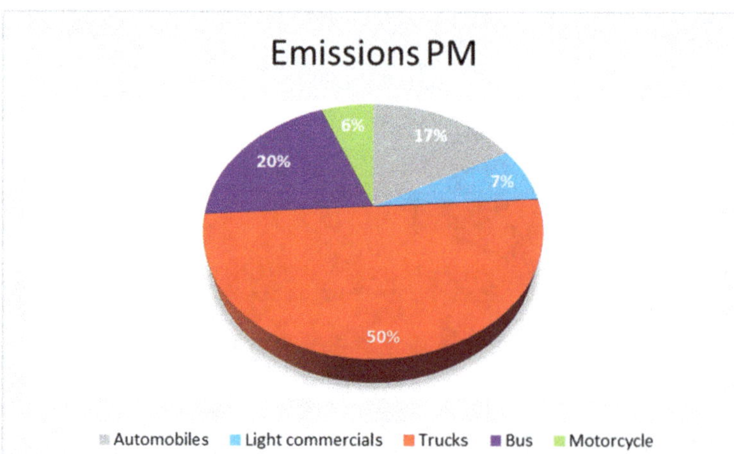

Figure 5: Relative contributions of vehicle categories to PM emissions.

Fig. 5 demonstrates that even though buses and trucks represent the smallest participation in the circulating fleet, they emit the highest rates of PM in relation to the other categories, such as, for example, the category of automobiles.

However, the results obtained are relevant and similar to the results found for other metropolitan regions, this is because it occurs how to obtain from automobiles, motorcycles and light commercial vehicles despite having a greater participation in the circulating fleet, they had less information about the pollutants obtained because these vehicles an embedded technology is more advanced, fuel is cleaner even taking into account fleet aging and inadequate maintenance when compared to trucks and buses.

5 CONCLUSION

The main objective of this work was to investigate and analyze the contributions to air pollution in the RMS, generating more current information to promote continuous updates in this sense. For this purpose, the methodology proposed and developed by the Ministry of the Environment (MMA) and the Environmental Company was used from the State of São Paulo (CETESB), which quantifies emissions from vehicles powered by fossil fuels, from the Otto cycle and the diesel cycle, excluding electric and hybrid vehicles.

Based on this approach, it was estimated that RMS owns 36.48% of the circulating fleet in the State of Bahia, comprising approximately 65% automobiles and 7.0% heavy vehicles (trucks and buses). And even though heavy vehicles emit the highest levels of NO_x and PM, contributing approximately 84% and 70%, respectively, of the pollution of these substances that contaminate the environment and affect human health. These results are pertinent and represent the contributions of motor vehicles to atmospheric pollution in large urban centers.

Thus, it is evident that the emission inventories are fundamental tools for the development of studies aimed at the analysis of air quality, as well as, in photochemical modeling. Although, there are still many uncertainties associated with its elaboration, such as, for example, the calculation of emissions, the characteristics of the fuels used, the circulating fleet of each region, as well as the technology on board.

For future work, it is suggested to draw up inventories covering other sources with a polluting potential to also know their contribution. And the study of these sources in models of air quality promoting a more realistic scenario of local air quality.

ACKNOWLEDGEMENTS

We are grateful to the Bahia Research Support Foundation (FAPESB) and to the HPC of the Centro Universitário SENAI CIMATEC.

REFERENCES

[1] Nueve de cada diez personas de todo el mundo respiran aire contaminado, 2018. Organización mundial de la salud (OMS), avenue appia 20, 1211 ginebra 27, suiza. https://www.who.int/es/news/item/. Accessed on: 26 Apr. 2021.

[2] Instituto Brasileiro de Geografia e Estatistica (IBGE), Banco de Dados – Cidades, 2018 (in Portuguese). http://www.cidades.ibge.gov.br/v3/cidades/home-cidades. Accessed: 20 Jan. 2019.

[3] Lyra, D.G.P.d., Modelo integrado de gestão da qualidade do ar da região metropolitana de Salvador/Diógenes Ganghis Pimentel de Lyra. – Campinas, SP: 2008 (in Portuguese).

[4] De Aguiar Filho, A.M. & de Albuquerque, É.L., Inventário de Emissões Veiculares na Região Metropolitana de Salvador para o ano de 2010, 2012 (in Portuguese).

[5] Loureiro, L.N., Panorâmica Sobre Emissões Atmosféricas Estudo De Caso: Avaliação Do Inventário Emissões Atmosféricas da Região Metropolitana Do Rio De Janeiro Para Fontes Móveis. Tese Universidade federal do Rio de Janeiro, 2005 (in Portuguese).

[6] De Almeida Albuquerque, T.T. et al., WRF-Smoke-CMAQ modeling system for air quality evaluation in São Paulo megacity with a 2008 experimental campaign data. *Environ Sci Pollut Res*, **25**, pp. 36555–36569, 2018. DOI: 10.1007/s11356-018-3583-9.

[7] Loriato, A.G., et al., High resolution emissions inventory using WRF-SMOKE-CMAQ integrated modeling for Great Vitoria region. *Revista Brasileira de Meteorologia*, **33**(3), pp. 521–536, 2018.

[8] Araujo, I.B.d., Quantificação, Espacialização e especiação de emissões atmosféricas de origem veicular na Região Metropolitana da Grande Vitória/Igor Baptista de Araújo. – Vitoria, ES: 2016, 2005 (in Portuguese).

[9] Kitagawa, Y.K.L., Nascimento, E.G.S., Guarieiro, L.L.N., Albuquerque, T.T.D.A. & Moreira, D.M., Assessment of primary air pollutants in a tropical metropolitan region by combining local and global emissions inventories. *WIT Transactions on Ecology and the Environment*, vol. 236, WIT Press: Southampton and Boston, pp. 99–110, 2019.

[10] Santos, F.S.d., Diagnóstico das emissões atmosféricas em Minas Gerais: Um estudo para fontes fixas e veiculares. Escola de Engenharia/Dissertação de mestrado. Belo Horizonte – Universidade Federal de Minas Gerais, 2018 (in Portuguese).

[11] BRASIL, Ministério do Meio Ambiente. Inventário Nacional de Emissões Atmosféricas por Veículos Automotores Rodoviários – Ano - Base: 2012. Relatório Final. Brasília: Ministério do Meio Ambiente. 114 p., 2013 (in Portuguese).

[12] CETESB, Qualidade do ar no estado de São Paulo 2013. Publicação CETESB. Série Relatórios/CETESB, ISSN 0103-4103, São Paulo, 2013.

[13] Mattos, J.A.B. & Correia, E.L., Uma nova estimativa da frota de veículos automotives no brasil. *Anais do VII Congresso Brasileiro de Energia*, p. 1267, 1996.

[14] Schmitt, F.W., Metodologia de Avaliação de Eficiência Energética em Veículos Leves e Seus Impactos em Termos de Uso de Combustíveis/William Frederic Schmitt. – Rio de Janeiro: UFRJ/COPPE, 2010.

[15] Agência Nacional do Petróleo, Gás natural e Biocombustíveis. Anuário Estatístico Brasileiro do Petróleo, Gás Natural e Biocombustíveis (ANP), 2016 (in Portuguese). http://www.anp.gov.br. Rio de Janeiro: ANP. 265 p. Accessed 20 Jun. 2019.

[16] U.S. Environmental Protection Agency, Compilation of air pollutant emission factor, volume I: stationary point and area sources, AP42, fifth edition. *Chapter 13: Miscellaneous Sources. 13.2.1 Paved Roads*, Research Triangle Park, NC, 2011. https://www3.epa.gov/ttn/cheif/ap42/ch13/final/c13/s0201.pdf. Accessed 20 Jun. 2019.

[17] Ntziachristos, L. & Boulter, P., Road transport: automobile tyre and brake wear and road abrasion. *EMEP/EEA Air Pollutant Emission Inventory Guidebook 2016: Technical Guidance to Prepare National Emission Inventories*, Publications Office of the European Union: Luxembourg, pp. 1–32, 2016.

VARIABILITY OF EMISSION RATE DEFINITION IN REGULATORY ODOUR DISPERSION MODELLING FROM CIVIL WASTEWATER TREATMENT PLANTS

MARCO RAVINA, SALVATORE BRUZZESE, DEBORAH PANEPINTO & MARIACHIARA ZANETTI
Department of Environment, Land and Infrastructure Engineering, Politecnico di Torino Italy

ABSTRACT

The analysis of odour impacts in civil wastewater treatment plants (WWTPs) is a challenging task. Odour regulations still present a lack of standardization, that bring inherent levels of uncertainty to the analysis procedure. Dispersion models can provide support towards the characterization and reduction of odour nuisances. The application of dispersion models requires an adapt setting and a detailed characterization of the emission sources, in terms of emission rate. In this study odour dispersion of a large WWTP in northern Italy was considered. Simulations were carried out with the CALPUFF model. The study focused on the selection of the open field correction method for wind velocity used in the calculation of odour emission rates (OERs). Three different relationships were considered: the power law, the logarithmic law and the Deaves–Harris (D–H) law. The area underlying the 1 OU/m^3, 3 OU/m^3, and 5 OU/m^3 concentration isopleths was considered as indicator. The results showed that OERs and impact area varied depending on the selected method. Taking the power law as the reference, the average variability of the impact area was between –33% and –48% if the logarithmic law was applied, and –83% and –94% if the D–H law was applied. The present study provides knowledge towards a better alignment of the concept of the odour impact criteria.

Keywords: odour impact assessment, odour emission rate, dispersion modelling, wastewater treatment, odour impact criteria.

1 INTRODUCTION

Odour impact assessment is a complex and debated topic in the scientific research community. Wastewater treatment plants (WWTPs) are known sources of odours, because these facilities are often located in urban and territorial contexts. An olfactory impact evaluation strategy is required for WWTPs, to limit odour impacts in the surrounding area, allowing the correct process management. Odour impact assessment presents multiple aspects of complexity. It is recommended that an integrated multi-tool assessment strategy is applied. Qualitative and quantitative analyses must be supported by atmospheric dispersion modelling, odours measurement in the ambient air, participation of population, and control and monitoring activities [1]. Odour impact assessment is composed of the following phases: sampling, characterization, odour emission rates (OER) calculation, atmospheric dispersion modelling, and impact evaluation [2]. The definition of odour emission rates is significantly affected by the sampling conditions and instrumentation. The choice of dispersion model, as well as its parameterization and settings also affect the results. Finally, the interpretation of the results depends on different odour impact criteria, such as the peak-to-mean conversion of concentration values.

Recent research activity brought a significant contribution to the knowledge of odour sampling methods and tools. Odour impact assessment stages are studied worldwide, aiming at standardizing the representativeness of the results [3]. Scholar highlighted that among all impact assessment phases, odour sampling and characterization are the most critical [4].

Following the recommendations brought by the scientific community, institutions of all countries adopted guidelines and regulations to properly address the odour emissions

WIT Transactions on Ecology and the Environment, Vol 252,
www.witpress.com, ISSN 1743-3541 (on-line)
doi:10.2495/AIR210111

problem. Nevertheless, the theme complexity led to different approaches and instruments, resulting in a lack of homogeneity between regulations [5]. Besides, assessment procedures are often incomplete or lack of precise information, generating uncertainties in the results.

The objective of this research work was to evaluate the influence of the variability related to some individual stages of odour impact assessment in the application of the current regulatory criteria. The study focused on the open field correction method for wind velocity used in the calculation of OERs. A WWTP located in Northern Italy, whose odour emissions sources were measured in previous campaigns, was considered as a case study.

Odour dispersion modelling was carried out using the CALPUFF software.

This study was structured by developing a reference impact assessment and simulation according to the Lombardy Region guidelines [6], and subsequently running alternative simulations with a modified OER characterization.

2 METHODOLOGY

Odour dispersion of a large WWTP in northern Italy was considered in this study. The approach adopted for the assessment was the one reported in the local regulations of northern Italian regions, that follow international methodologies. In this approach, odour concentrations are evaluated by considering the following odour impact criteria (OIC) [5]: odour concentration thresholds, percentile compliance level, and the averaging time set in dispersion simulations.

This study was elaborated by comparing a reference simulation with other alternative simulations. The reference simulation was conducted according to the guidelines issued by Piedmont Region, Italy. Alternative simulations aimed at evaluating the influence of the open field correction method for wind velocity used in the calculation of OER. These were set up considering the same modelling domain, sources, and odour emission rates, but different alternative correction methods. The most applied relationships were considered, i.e., power law and logarithmic law. Additionally, the application of the D–H correlation was also tested and compared to other solutions.

The power law only depends on the stability class and the prevalent type of land use class of the surface, expressed through the Hellman's unitless α parameter (eqn (1)) [7]:

$$v_1 = v_2 \cdot \left(\frac{h_1}{h_2}\right)^{\alpha}, \tag{1}$$

where v_1 is the wind speed at the correct height (m s^{-1}), v_2 is the wind speed at the meteorological station height (m s^{-1}), h_1 is the correct height (m), and h_2 is the meteorological station height (m). The Power Law is considered valid in the 30–300 m range, although it cannot be applied at the upper and lower limits of the PBL. At heights with values close to the surface roughness length (1–10 m), it does not provide a valid estimate of the speed. The logarithmic law (eqn (2)) was observed to be more suitable for the evaluation of velocity profiles at a lower height. This relationship accounts for friction velocity u^* and surface roughness z_0. Unlike the power law, it is not an empirical expression as it is derived from the similarity theory, according to eqn (2):

$$v_1 = \frac{u^*}{K_v}\left[ln\frac{h_1}{z_0} + \psi\left(\frac{h_1}{L_m}\right)\right], \tag{2}$$

where u^* is the friction velocity (m s^{-1}), K_v is the Von Kármán's constant (–), z_0 is the surface roughness length (m), L_m is the Monin–Obukhov length (m), and ψ is a stability factor, related to the atmospheric stability class (–). The ψ function is a function of the atmospheric stability [8]. The application of a logarithmic law is often complex because of its constituent parameters. In addition to the most used relationships, Lucernoni [8] proposed the application of the Deaves–Harris (D–H) correlation [7] to odour emissions. The D–H correlation, also known as the logarithmic with parabolic defect model equation, is defined as eqn (3):

$$v_1 = \frac{u^*}{K_v}\left[ln\frac{h_1}{z_0} + 5.75\,ln\left(\frac{h_1}{H}\right) - 1.88ln\left(\frac{h_1}{H}\right)^2 - 1.33ln\left(\frac{h_1}{H}\right)^3 + 0.25ln\left(\frac{h_1}{H}\right)^4\right], \quad (3)$$

where, H is the equilibrium boundary layer height, equal to $\frac{u^*}{6f_c}$ (m), f_c is the Coriolis' parameter, equal to $2\Omega\sin(\varphi)$ (s^{-1}), Ω is the Earth rotation rate, equal to $7.2921\cdot10^{-5}$ (rad s^{-1}), and φ is the latitude (rad). The D–H law is an extension of the analysed previous laws since it includes both scale parameters u^* and z_0 (inherited from the logarithmic profile) and the PBL height parameter. This latter depends on wind speed and latitude. In principle, the D–H law can accurately describe the entire PBL, including at height values close to upper and lower boundary conditions [7]. However, the applicability of this correlation is complex, since it was calibrated only for a certain range of wind speeds (strong winds of about 20–30 m s^{-1} at 10 m height above ground). Cook [7] pointed out that in the wind speeds design range the correspondence between D–H and power law models is to be considered excellent. For low wind speeds, however, two laws deviate more. More research is still needed to understand which of them provides more realistic profiles in the analysis conditions.

The description of the reference and alternative simulations is reported in Table 1.

Table 1: Alternative simulations for the odour impact assessment of the WWTP case study.

Simulation	Correction method for wind speed
1 (reference simulation)	Power Law (eqn (1))
2	Logarithmic Law (eqn (2))
3	Deaves–Harris Law (eqn (3))

2.1 Study site

Odour dispersion modelling was applied to a WWTP located in Piedmont Region, Italy. Site orography is mainly flat and presents a slight slope in the south-east direction towards the river in which clarified water is discharged. Some reliefs reaching about 600 m of elevation above sea level are present in direction SW. The plant is surrounded by two towns, located NW and SE respectively. After the plant construction, urban expansion raised, so that the closest residential area is now located 1 km to the plant boundary, in direction NW. Except of urban settlements, the area is characterized by mainly agricultural and forest land use.

The two main treatment lines of the WWTP are represented by wastewater treatment and sludge treatment. The water line includes de-gritting facilities (grit and grease removal), primary sedimentation (circular open tanks), anoxic and aeration basins (for organics and nitrogen removal), secondary sedimentation, and final tertiary treatments. The sludge treatment line includes sludge pre-thickening, mesophilic anaerobic digestion, post-thickening (centrifuges), and final thermal dewatering.

Sampling and characterization of emission sources were carried out at the plant by mean of several campaigns in 2017 and 2019. This work used olfactometry analysis results to calculate the emission rates, as required by EN13725:2004 standard [9]. In these campaigns, air samples from diffused sources were collected employing a wind tunnel (WT). The following facilities were monitored: de-gritting tanks, primary sedimentation tanks, and the external area were treated sludge is temporarily stored). The characteristics of the WT sampling are reported in Table 2. WT sampling flow was 2.5 m^3/h, corresponding to an average velocity of 0.035 m/s. Odour concentrations were measured by an external laboratory according to standard EN 13725:2004. An ODOURNET TO8 olfactometer and a panel of evaluators was employed with this purpose The results of the olfactometry analysis, expressed in concentration of odour units (OU m^{-3}), are reported in Table 3. Odour concentrations were higher in correspondence of de-gritting tanks, primary settlers, and external storage areas for sands and sewage sludge. These concentration values were used to calculate odour emission rates to be introduced into CALPUFF.

Table 2: Characteristics of the WT used for sampling.

Parameter	Description	Value	Unit of measure
Q_{effl}	Air flow rate	2.5	$m^3 h^{-1}$
T	Temperature	23	°C
A_{base}	WT area	0.125	m^2
U_{WT}	Sweep air velocity	0.035	$m s^{-1}$

Table 3: Olfactometry analysis results.

Plant area	Average OU concentration (OU m^{-3})
De-gritting	180
Primary sedimentation tanks	540
Stabilized sludge external storage 1	1,100
Stabilized sludge external storage 2	3,600

In the reference simulation, odour emission rates (OER) were calculated according to the indications provided by the regional technical specifications, starting from the results of the olfactometry analysis. The OER values obtained are reported in Table 4. Each source was introduced in CALPUFF as a stationary source, i.e., OER was assumed to be constant in time.

Table 4: Emission sources characterization.

	Olfactometry results (OU m^{-3})	OER (OU s^{-1})
De-gritting	180	2,880.
Primary sedimentation tanks	540	12,742
Stabilized sludge external storage 1	1,100	5,767
Stabilized sludge external storage 2	3,600	63,360

2.2 The dispersion modelling Odour dispersion modelling and impact assessment

phase was carried out using CALPUFF [10]. CALPUFF is a multi-layer, multi-species, non-steady-state puff dispersion model that simulates the effects of time- and space-varying meteorological conditions on pollution transport, transformation, and removal. The modelling system consists of three main components and a set of pre-processing and post-processing programs. The main components of the modelling system are CALMET (a diagnostic three-dimensional meteorological model), CALPUFF (an air quality dispersion model), and CALPOST (a post-processing package). CALPUFF is a model that simulates puffs of material emitted from modelled sources, reproducing dispersion and transformation processes along the way. Temporal and spatial variations in the meteorological fields selected are explicitly incorporated in the resulting distribution of puffs throughout a simulation period. CALPUFF is widely employed for odour dispersion modelling activities.

Simulations were conducted on grid receptors, in a square domain of 16.2 × 16.2 km, with 10 vertical layers and a 200 m grid step. Meteorological data of the year 2018 were collected by a weather monitoring station installed onsite. The weather unit is composed of the following components:

- An ultra-sonic biaxial anemometer, installed at a height of 10 m above ground;
- A global class 2 radiometer;
- A temperature sensor PT100 1/3 DIN, with a non-vented anti-radiation shield;
- A hygrometer with a non-vented anti-radiation shield;
- A tilt-out tray pluviometer;
- A barometer;
- A Campbell CR800 data logging system.

Odour impact assessment was conducted based on current regulatory indications. Northern Italy technical regulations require the definition of odour concentration maps reporting peak hourly concentrations on an annual basis. The 98th percentile of hourly concentrations must be considered. Odour impact criteria (OIC) require the analysis of three different concentration isophlets:

- 1 OU m^{-3}, corresponding to the 50% of the population perceiving the smell;
- 3 OU m^{-3}, corresponding to the 85% of the population perceiving the smell;
- 5 OU m^{-3}, corresponding to the 90–95% of the population perceiving the smell.

In practical terms, a peak-to-mean factor of 2.3 must be applied to the hourly average odour concentration obtained from dispersion simulations.

3 RESULTS AND DISCUSSION

Wind distributions measured by the weather monitoring stations are reported in Fig. 1. This figure shows different wind distributions between night-time and daytime. Night-time is characterized by low wind (around 0.5–1 m/s) with main direction SE. Daytime is characterized by slightly higher wind (up to 5 m/s) with main direction NE. Wind speed is generally low. No significant seasonal variation was evidenced, although higher wind speed is generally observed during spring.

The peak hourly concentrations map of the 98th percentile related to odour dispersion modelling simulations, with reference to Table 1, is reported in Figs 2–4.

Results of Simulation 1 (Fig. 2) showed an extended odour impact area, mainly in the NE–SW direction, consistent with the wind distribution. The 1 OU m^{-3} isopleth reaches a distance of 6 km away from the plant boundary. 3 and 5 OU m^{-3} isopleths, which indicate a greater frequency of nuisance perception by the population, are extended beyond the plant borders and include the nearby residential area. The town located in direction NW is one of the most exposed receptors, together with the national road located north of the plant borders.

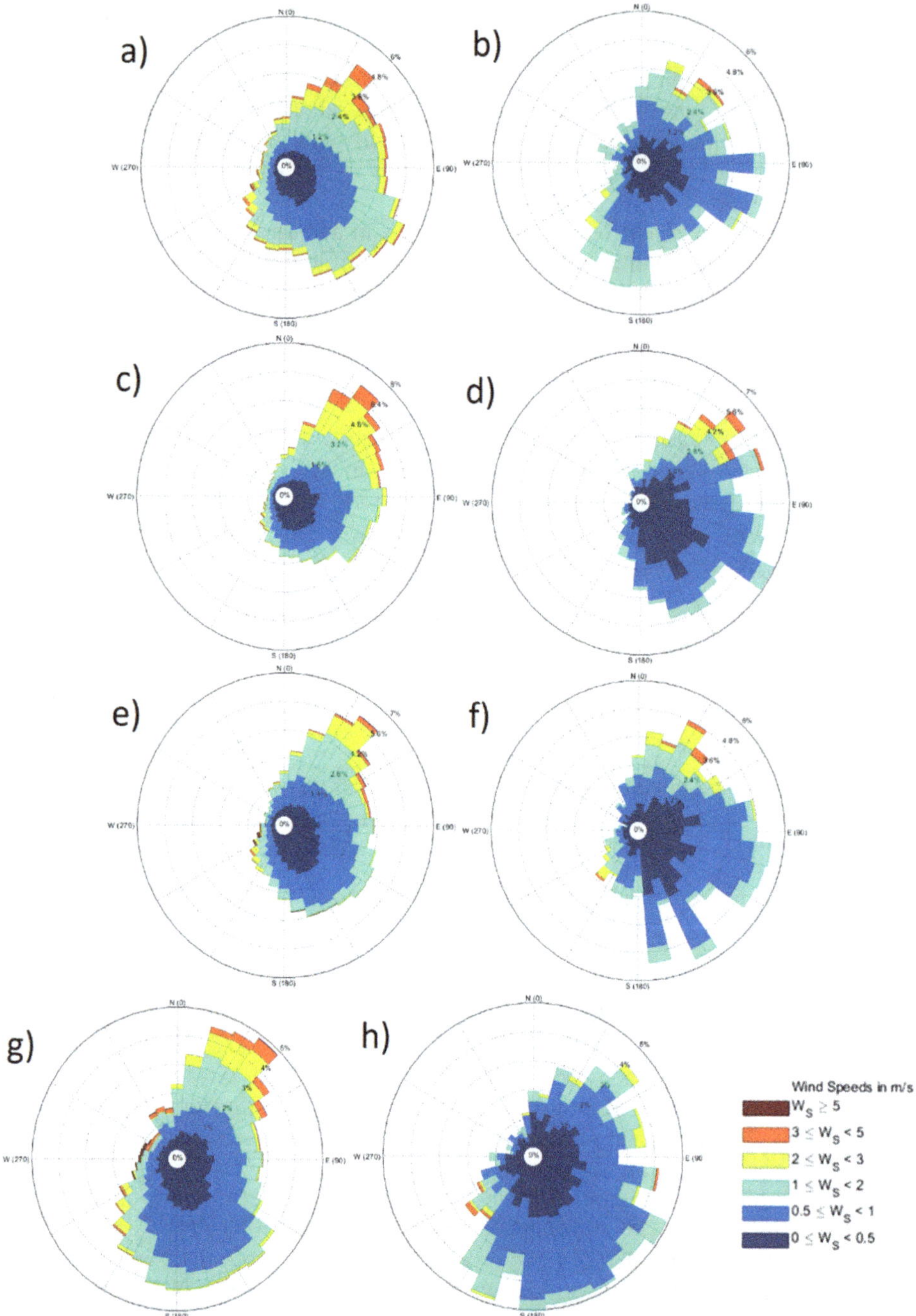

Figure 1: Wind distribution measured onsite. (a) Spring, daytime; (b) Spring, night-time; (c) Summer, daytime; (d) Summer, night-time; (e) Autumn, daytime; (f) Autumn, night-time; (g) Winter, daytime; and (h) Winter, night-time. Daytime distributions represent the average from 8:00 to 19:00. Night-time distributions represent the average from 19:00 to 8:00.

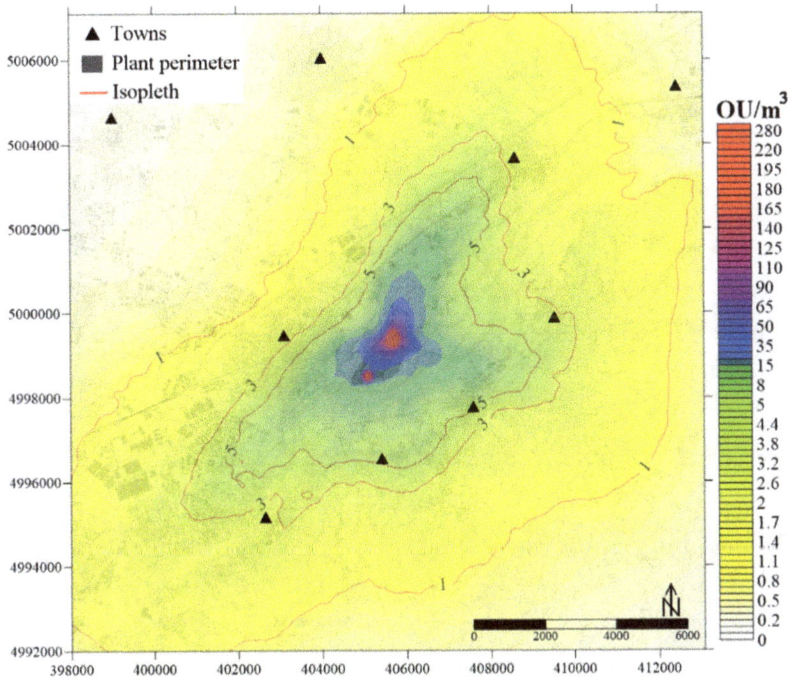

Figure 2: Simulation 1 – 98th percentile peak hourly odour concentration map.

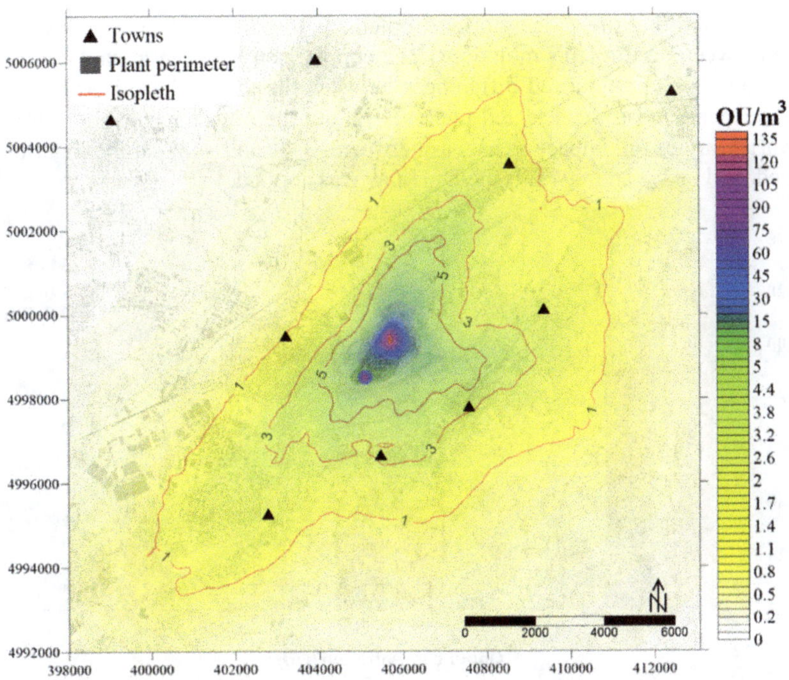

Figure 3: Simulation 2 – 98th percentile peak hourly odour concentration map.

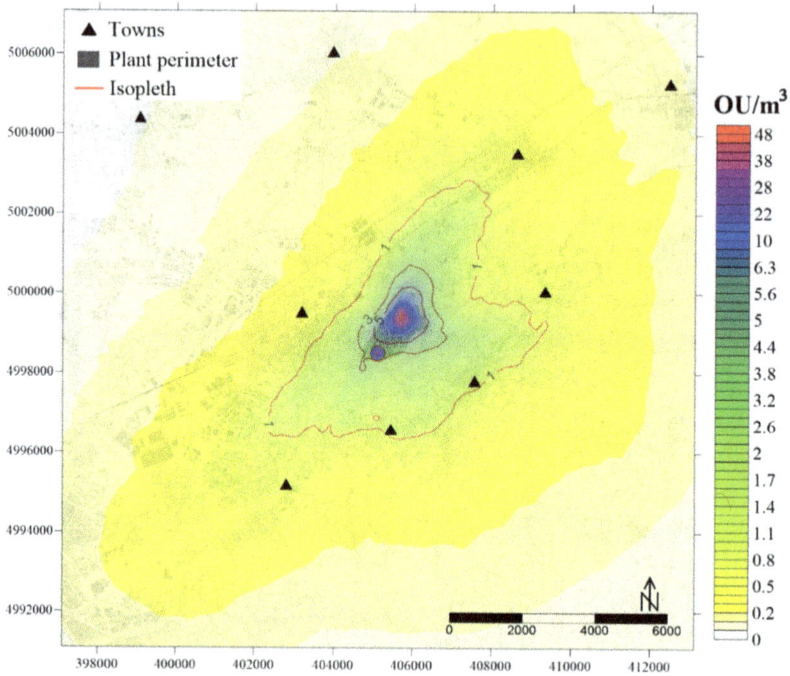

Figure 4: Simulation 3 – 98th percentile peak hourly odour concentration map.

The area underlying the concentration isopleths is the main indicator to investigate the difference between performed simulations. The comparison between impact areas – visible in Fig. 5 shows very pronounced differences between the three adopted laws. The impact area varies depending on the selected method. Taking the power law as the reference, the average variability of the impact areas was between –33% and –48% if the logarithmic law was applied, and –83% and –94% if the D–H law was applied.

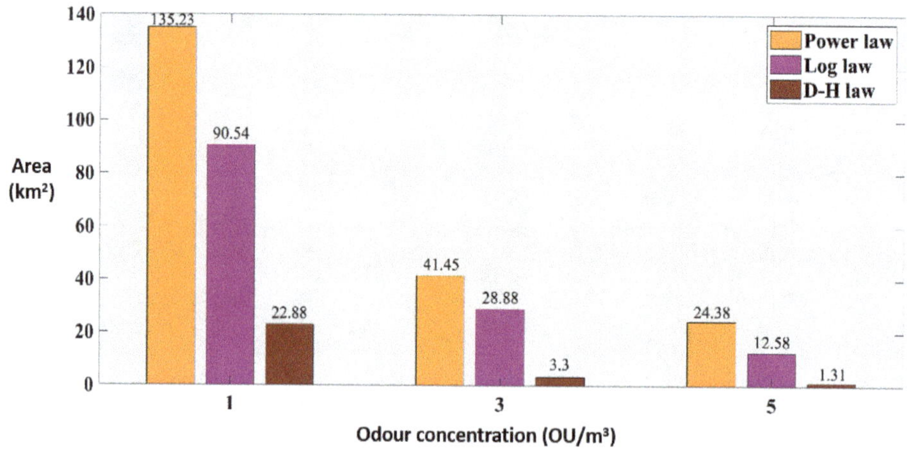

Figure 5: Simulations 1, 2, and 3 – Comparison between 1, 3, and 5 OU m^{-3} isopleths.

Simulations results showed large impact extension in the study domain. No odour field measurements were conducted in support of these results. However, it is likely that the OIC applied lead to an overestimation of the impact area. With regard to Simulation 1, the area covered by the 1 OU/m^3 isopleth reached an extension larger than the town surrounding the plant (Fig. 2). This would mean that 50% of the population living in that area would perceive the smell for 98% of the year. Such a situation is not evidenced by the information available, neither by reports from the monitoring agency, nor by surveys from living population. Such an overestimation of impacts connected to the application of OIC in northern Italian regions was already observed by other studies.

Besides this consideration, the results showed very pronounced differences (up to 94% in terms of impact area) between the three adopted laws. The adoption of the logarithmic law, which more accurately describes the speed profile for height close to z_0, corresponds to a significant reduction of the impact area. The application of the D–H law corresponds to a further reduction. There is not enough data available to define which simulation is the more realistic. The results obtained should be compared with field measurements, although the extensive application of dynamic olfactometry is cost and time expensive. Thus, from the results, it is not possible to define if a simulation is better than another. However, considered the general absence of complaints about odour nuisance by the living population, impact areas of Simulation 3 seem actually more realistic. Considering its theoretical formulation, the application of the D–H law can be considered interesting. This relationship describes the whole PBL, as it derives from combining a logarithmic function with a 4th order polynomial [11]. Nevertheless, its applicability must be deepened by future studies.

Uncertainty related to regulatory odour impact assessment is a debated topic. Since regulations provide no information about the methodology to be adopted, this study demonstrated how the selection of one or another solution could generate significant variability in the results. Most of the research work is currently done on finding the most effective correlation between emissive flows in the open field and within the dynamic hoods. Regarding the OER calculation methodology, there is still the need of efforts to reach an agreement on the law used to describe the wind speed profile. Lucernoni [8] studied the influence on the velocity law by comparing the same three relationship considered in this study. The results showed a variability of separation distances that is consistent with the results of this study. The same authors also highlighted that the logarithmic profile is more reliable at reduced heights from the ground (0–100 m). The same authors, in a subsequent literature work about the OER correction, also highlighted the possibility of adopting the D–H relationship [12].

4 CONCLUSIONS

The objective of the present study was to investigate the variability of emission rate definition in regulatory odour impact assessment. Odour dispersion modelling was applied to a case study – a wastewater treatment plant located in northern Italy. The representativeness of the analysis depends on the application of current regulatory provisions on odour emissions, which presently require more detailed and standardized methodologies. In this context, this study showed that a single phase of the analysis can bring inherent uncertainty in the overall procedure, as impact areas can vary up to 94% depending on the evaluator's choices. Existing regulatory tools must be checked and reviewed based on this and other studies.

The analysis of odour dispersion modelling showed a WWTP odorous impact on the study area that is probably overestimated, as no evidence is present when considering conditions of the population living in this area. Nor previous inspection and monitoring activities evidenced such olfactory nuisance. The adoption of the power law profile may have contributed to this overestimation: this is the simplest correlation and its representativeness is lower at low heights from the ground. The logarithmic law can be considered more representative, both from the literature studies analysed and based on the results of the analysis. The third proposed correlation, the Deaves–Harris law, is interesting but needs further analyses, as it still need a more detailed validation for this kind of applications. Impact maps need to be compared with the field reports results and a future survey on the population, including citizen science initiatives. Such an approach would lead to a better understanding of the relationships between model results and the actual perceived olfactory nuisance. Further studies are needed to optimize the evaluation method and the comparison between dispersion models results and odour impact criteria. This would allow institutions to translate the method into uniform regulatory references at the national level, limiting results variability and improving their representativeness.

REFERENCES

[1] Byliński, H., Kolasińska, P., Dymerski, T., Gębicki, J. & Namieśnik, J., Determination of odour concentration by TD-GC×GC–TOF-MS and field olfactometry techniques. *Monatshefte für Chemie*, **148**, pp. 1651–1659, 2017. DOI: 10.1007/s00706-017-2023-8.

[2] Belgiorno, V., Naddeo, V. & Zarra, T., *Odour Impact Assessment Handbook*. John Wiley & Sons, 2013.

[3] Ravina, M. et al., Integrated model for estimating odor emissions from civil wastewater treatment plants. *Environmental Science and Pollution Research*, **27**, pp. 3992–4007, 2020. DOI: 10.1007/s11356-019-06939-5.

[4] Ravina, M. et al., Characterization of odorous emissions from a civil wastewater treatment plant in Italy, Aveiro, Portugal, pp. 159–170, 2019.

[5] Brancher, M., Griffiths, K., Franco, D. & Lisboa, H., A review of odour impact criteria in selected countries around the world. *Chemosphere*, **168,** pp. 1531–1570, 2017. DOI: 10.1016/j.chemosphere.2016.11.160.

[6] Invernizzi, M., Bellini, A., Miola, R., Capelli, L., Busini, V. & Sironi, S., Assessment of the chemical–physical variables affecting the evaporation of organic compounds from aqueous solutions in a sampling wind tunnel. *Chemosphere*, **220**, pp. 353–361, 2019. DOI: 10.1016/j.chemosphere.2018.12.124.

[7] Cook, N.J., The Deaves and Harris ABL model applied to heterogeneous terrain. *Journal of Wind Engineering and Industrial Aerodynamics*, **66**(3), pp. 197–214, 1997. DOI: 10.1016/S0167-6105(97)00034-2.

[8] Lucernoni, F., The evaluation of the odour emission rate for passive area sources: A new approach. *Chemical Engineering Transactions*, **43**, 2015.

[9] EN 13725:2004, Air quality: Determination of odour concentration by dynamic olfactometry.

[10] Scire, J., Robe, F., Fernau, M. & Yamartino, R., A user's guide for the CALMET meteorological model (Version 5), 1998.

[11] Scire, J., Strimaitis, D. & Yamartino, R., A user's guide for the CALPUFF dispersion model. Earth Tech, Inc., p. 521, 2000.

[12] Lucernoni, F., Capelli, L., Busini, V. & Sironi, S., A model to relate wind tunnel measurements to open field odorant emissions from liquid area sources. *Atmospheric Environment*, **157**, pp. 10–17, 2017. DOI: 10.1016/j.biortech.2007.04.034.

EFFECT OF VEHICLE LIGHT ON THE NANOSTRUCTURE OF PARTICULATE MATTERS EMITTED FROM DIESEL AND GASOLINE VEHICLES

MEISAM AHMADI GHADIKOLAEI[1], PAK KIN WONG[1], SHOU HAO CHEN[1], KAR WEI NG[2] & JIN CHENG XU[2]
[1]Department of Electromechanical Engineering, University of Macau, Macau
[2]Institute of Applied Physics and Materials Engineering, University of Macau, Macau

ABSTRACT

Particulate matters (PMs) are currently one of the main threats to human health and the environment. The transportation sector, particularly diesel and gasoline vehicles, among others, has a significant contribution to the global PM level. While the PMs from vehicles have serious and dangerous negative impacts on human health due to their toxicity components along with their small diameters (penetrating more deeply into the human's body). Therefore, researchers are seeking some solutions to reduce the concentration of PMs from vehicles as well as investigate those parameters affecting the PM properties upon their release from the exhaust pipes. Vehicle light, among others, could be a factor for changing the PM properties, while no comprehensive study can be found in the literature in this field. This study is therefore presented to explore the impact of vehicle light emitted from halogen headlights (as an abundant one available on the market) on the nanostructure of both fresh and aged PMs from a diesel vehicle under an idle condition (fresh PMs), a diesel particulate filter (aged PMs), and a gasoline vehicle exhaust pipe (engine was off, aged PMs). The samples were exposed to the light for 6 hours at a distance of 0.1 m. Based on the results obtained from a scanning transmission electron microscope (STEM) and a scanning electron microscope (SEM), the vehicle light changes the nanostructure of PMs by reductions in fringe length (25.0%, 8.4%, and 7.1%) and fringe tortuosity (3.0%, 2.9%, and 1.8%), and enlargements in fringe separation distance (3.0%, 2.3%, and 1.9%), core size/shell size ratio (67.6%, 36.9%, and 62.1%), and core size/particle size ratio (38.4%, 25.0%, and 33.3%), resulting in enlargements in particles diameters from 21.2 to 30.4 nm, from 32.3 to 33.5 nm, and from 27.1 to 29.8 nm for all three samples tested, respectively.

Keywords: air pollution control, environmental issue, human health, properties of particulate matters, vehicle headlights, vehicular emission control.

1 INTRODUCTION

Nowadays, approximately nine out of ten people in the world breathe in contaminated air containing a high level of pollutants, particularly particulate matters with a diameter of 2.5 microns or smaller ($PM_{2.5}$), in both ambient (outdoor) and household (indoor) areas. It causes heart disease, stroke, chronic obstructive pulmonary disease, lung cancer, acute respiratory infections, and other diseases/problems, resulting in shortening the human's life on average by 20 months and approximately 7 million global deaths per year, and making the PMs as one of the serious leading risk factors for global mortality [1]–[4]. In addition, PMs have a negative impact on both the ground and underwater environments and even animals' lives due to their suspending potential in air and solubility potential in water [5].

It is known that, on the one hand, the negative impact of PMs on human health and the environment is mostly due to their toxicity components [6]–[11]. But on the other hand, the magnitude of these negative components on human health and the environment can/will be more or less, if the number, size, morphology, nanostructure, and mass of PMs be changed [12], [13]. For instance, if the PMs become smaller, they can penetrate more deeply into the human's and animal's bodies through the respiratory passageways, resulting in more damaging the lung and other tissues in comparison with the larger ones [12], [14]. Regarding

WIT Transactions on Ecology and the Environment, Vol 252, © 2021 WIT Press
www.witpress.com, ISSN 1743-3541 (on-line)
doi:10.2495/AIR210121

the environment, small PMs can be suspended longer in air, are more transportable, can be dissolved easily and faster in water and rain, have lower solar radiation absorption rate and less reacting potential with other types of atmospheric constituents, etc. [5], [15], [16]. Thus, there is a need for more concern in this field, particularly in the size and structure of PMs. Various parameters and conditions have impacts on the size and structure of PMs, including types and sources of PMs (from transportation, construction, and industrial sectors, etc.), ambient temperature, humidity, wind, light, etc., PMs age (fresh or mature) and suspension period, along with other parameters/conditions. The transportation fleet, among others, is one the main sources of PMs that are very close to the human's living areas. Therefore, there is a need for more attention on the PMs emitted from the transportation fleet and studying the possible parameters affecting their properties, particularly in the size and structure.

However, unfortunately, the existing literature regarding the properties of PMs emitted from the transportation fleet, particularly diesel and gasoline vehicles along with their particulate filters, is not rich enough as needed and further studies are still required (latest review papers in 2020–2021 [17]–[20]) to find some suitable and applicable solutions for reducing their negative impacts on human health and the environment. Therefore, this study is presented to enrich the literature by investigating the influence of light emitted from a vehicle headlight (Halogen type as an abundant one), on the nanostructure of fresh and mature PMs, including fresh PMs from a diesel vehicle and mature PMs deposited in a diesel particulate filter (DPF) and a gasoline vehicle exhaust pipe. Investigating the nanostructure of primary particles is one of the best methods for exploring and modelling the important parameters of PMs such as morphology and size. The content of this research is useful for getting information from the contaminated places containing a high level of PMs such as car parks (closed areas), tunnels, urban canyons, etc., when several vehicle headlights are switched on for a long period of time during the night and day and the PMs fly around them.

2 METHODOLOGY

Three PM sampling locations from one diesel vehicle operating under an idle condition, and inside of a diesel particulate filter and a gasoline vehicle exhaust pipe (engine was off) were selected in this research as shown in Fig. 1. Firstly, the PM samples were directly collected on the 3 mm copper transmission electron microscope (TEM) grids. In the next step, the samples were exposed to a vehicle headlight (Halogen type, 12V 65/55W) under the low beam condition for 6 hours at a distance of 0.1 m (light intensity of 438 W/m^2) as shown in Fig. 2. We selected halogen headlight because this type is one of the abundant types available on the market. A scanning transmission electron microscope (STEM) (model FEI Talos F200S; resolution of 0.16 nm) and a scanning electron microscope (SEM) (model ZEISS Sigma; resolution of 1.0 nm) were employed to analyse the samples before and after exposure to the vehicle light and an image processing software (Image-Pro Plus 6.0, Media Cybernetics) were used to process the results (images).

To obtain highly accurate results, 30 primary particles were chosen from the images for analysing their nanostructure. Fringe length, fringe separation distance, fringe tortuosity, and core size/shell size and core size/particle size ratios refer to the nanostructure of particles in this research, as illustrated in Fig. 3. Detailed information regarding the PM sampling and analysis procedure is available in Mao et al. [21] and Ghadikolaei et al. [22]. The experimental results were checked by the two-tailed student's T-test under a 95% significance level ($p < 0.05$). Uncertainty at a 95% confidence level was calculated according to Moffat [23].

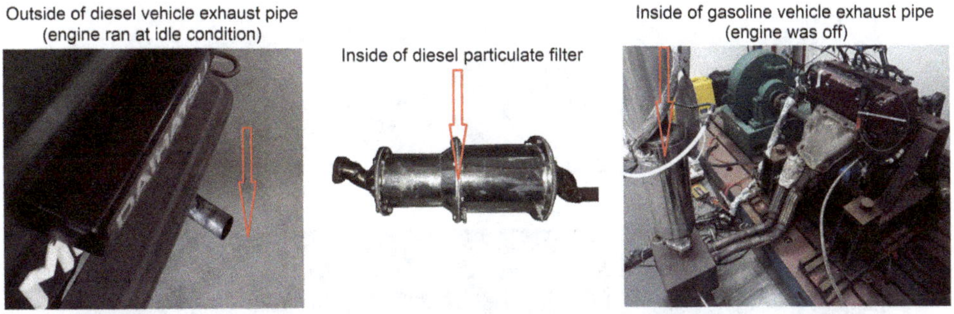

Figure 1: PM sampling locations for this research.

Figure 2: Exposure distance of TEM grids containing PMs to vehicle light.

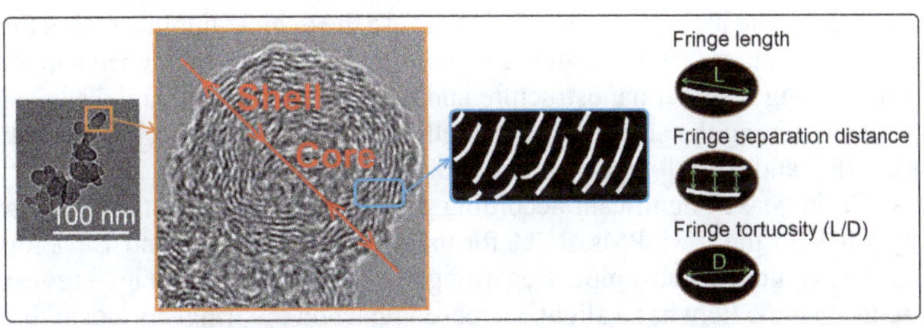

Figure 3: Nanostructure parameters of primary particles investigated in this study.

3 RESULTS AND DISCUSSION

The idea for exploring the impact of vehicle light on the properties of PMs has come from our observation on the changes in the shape, structure, and size of PMs due to exposure to the electron beam for a short period (few seconds to minutes, see Fig. 4) during operating the SEM. Therefore, the vehicle light for a longer period of use (several hours at the night and even in the daytime in contaminated places like car parks, tunnels, roadsides, etc.) may also change the PM properties. Thus, we performed several tests to investigate the effect of vehicle light on the nanostructure of fresh and mature PMs from diesel and gasoline vehicles

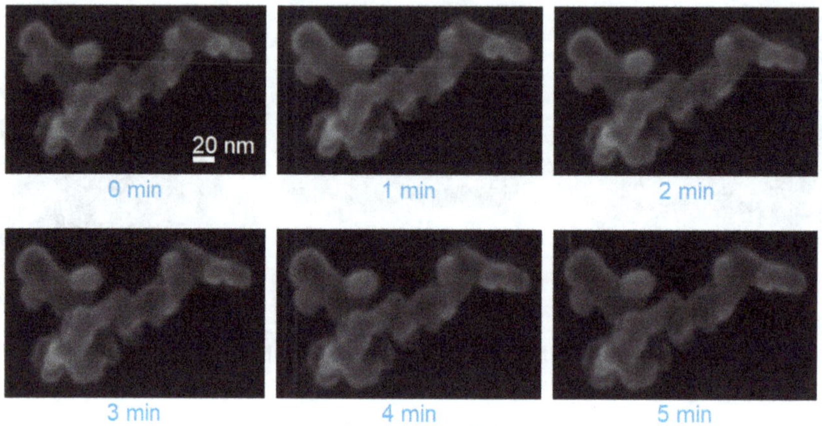

Figure 4: Changes in the shape, structure, and size of PMs emitted from the diesel vehicle at an idle condition, due to exposure to the electron beam of SEM from 0 min (original PMs) until 5 min under a low accelerating voltage of 5 kV.

for an exposure period of 6 hours at a distance of 0.1 m. The fresh PMs in this research were directly sampled from the engines right after their generation from the engine; while the mature PMs refer to those PMs that are produced from the engines but are deposited in some places (i.e., diesel particulate filter or exhaust pipe) for a long period of time, e.g., more than 5 years for the PM samples deposited in the diesel particulate filter and approximately 6 months for the gasoline vehicle exhaust pipe, for this research.

It can be seen from Fig. 5 that the vehicle light has the potential to change the morphology of the primary particles by creating edge defects for all three PM samples, while the fresh PM sample receives the highest changes. It is because the vehicle light increases the ambient temperature, among other factors, resulting in melting the volatile components in the primary particles and forming disorder nanostructure and defects, particularly at the edge sides (Fig. 5) along with shortening the size of fringe length by 25.0% for the fresh PMs, 8.4% for the aged PMs of DPF, and 7.1% for the aged PMs of gasoline vehicle exhaust pipe (Fig. 6). Also, there is a slight increase (significant according to T-test checking) in the fringe separation distance by 3.0% for the fresh PMs, 2.3% for the aged PMs of DPF, and 1.9% for the aged PMs of gasoline vehicle exhaust pipe. Regarding the fringe tortuosity, Fig. 6 reveals that the exposure to the vehicle light has a slight decrease impact on the fringe tortuosity by 3.0% for the fresh PMs, 2.9% for the aged PMs of DPF, and 1.8% for the aged PMs of gasoline vehicle exhaust pipe. Since the volatile components in the particles are melted and spread, the particles will have a shorter fringe length with larger carbon layers distance (fringe separation distance) and more straight carbon layers (smaller fringe tortuosity).

Vehicle light, also, causes increases in both core size/shell size and core size/particle size ratios by 67.6% and 38.4% for the fresh PMs, 36.9% and 25.0% for the aged PMs of DPF, and 62.1% and 33.3% for the aged PMs of gasoline vehicle exhaust pipe, respectively, as shown in Fig. 7. Since the core has a disordered structure, very small fringe length, and weak connections between the carbon layers, the exposure to the vehicle light can make it bigger compared with the shell side, resulting in having higher core size/shell size and core size/particle size ratios. It can be understood from the results that the fresh PMs, in

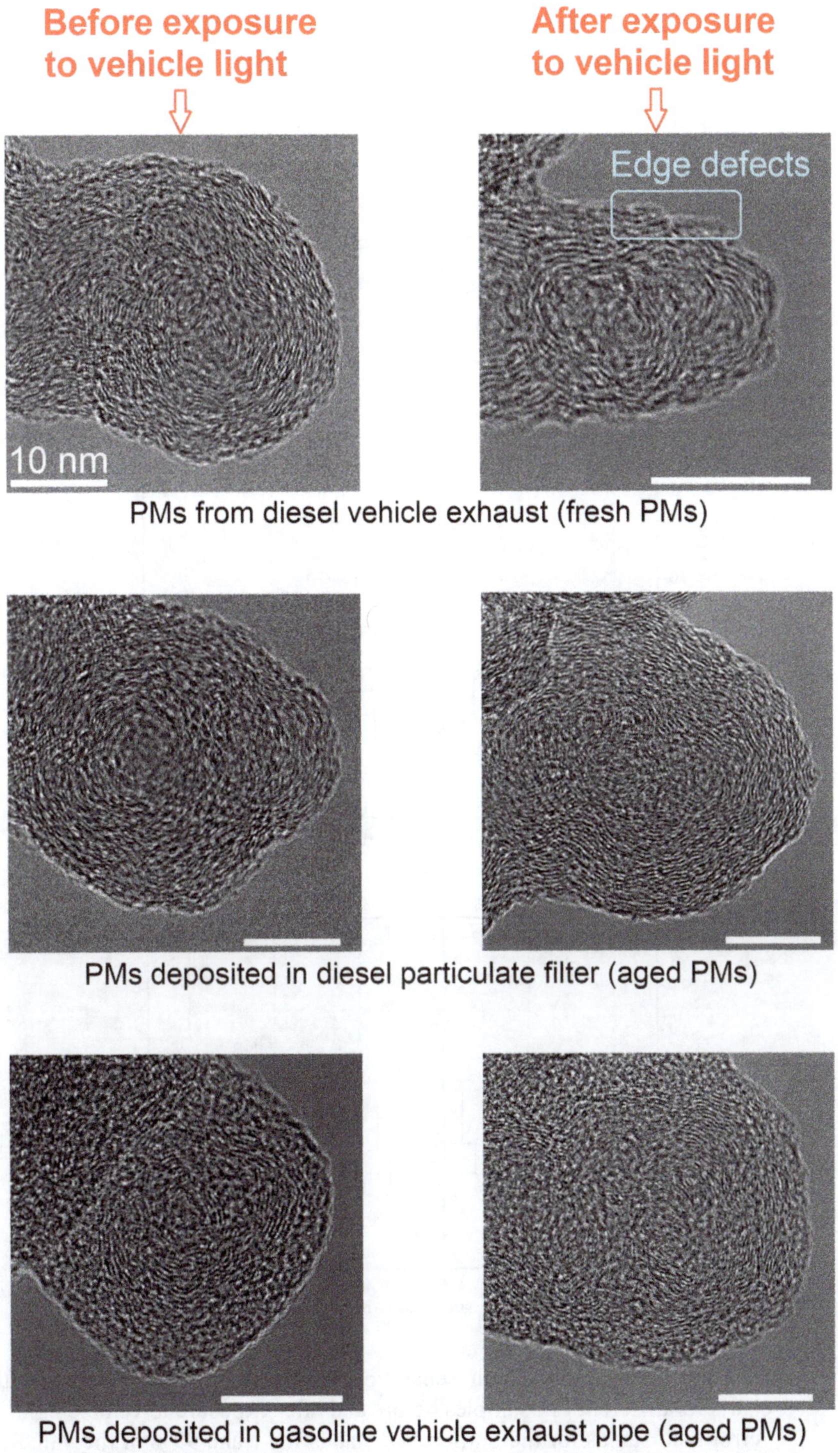

Figure 5: Effect of vehicle light on the morphology of primary particles.

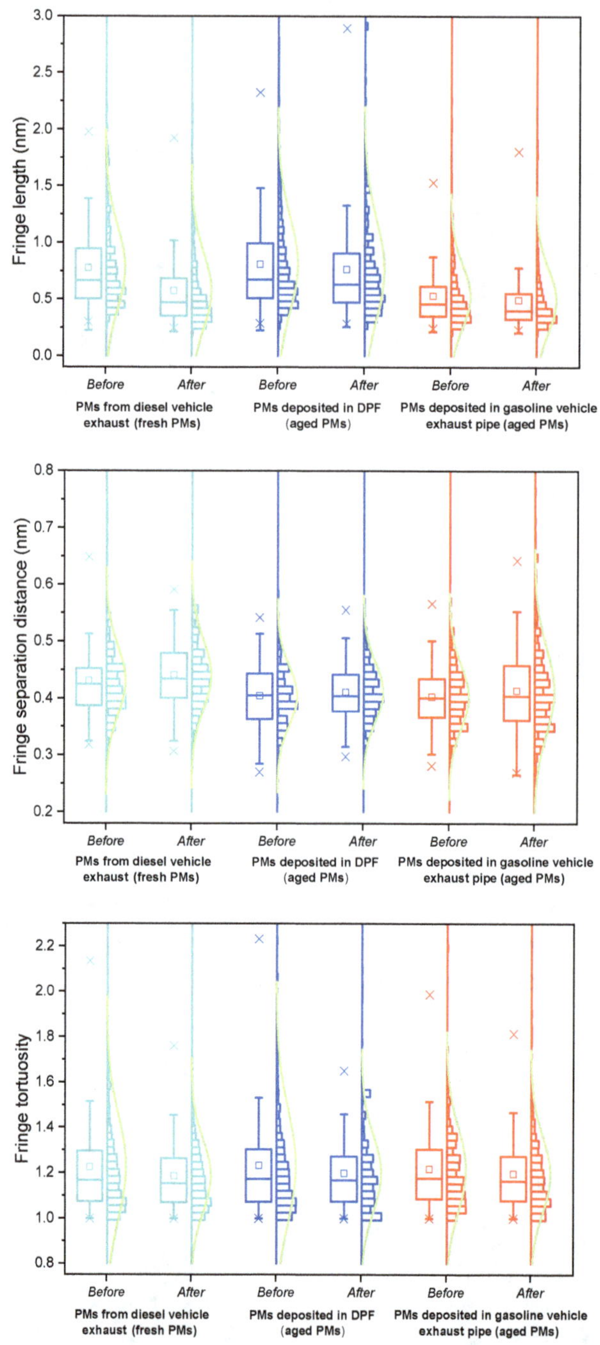

Figure 6: Distributions of fringe length, separation distance, and tortuosity along with box charts for different PM samples before and after exposure to vehicle light. Note: "error bar" stands for the outlier, "×" stands for 1 and 99% of the values, "line in the box" stands for the median value, "□" stands for the mean value, and "curve" stands for the normal distribution fitting curve.

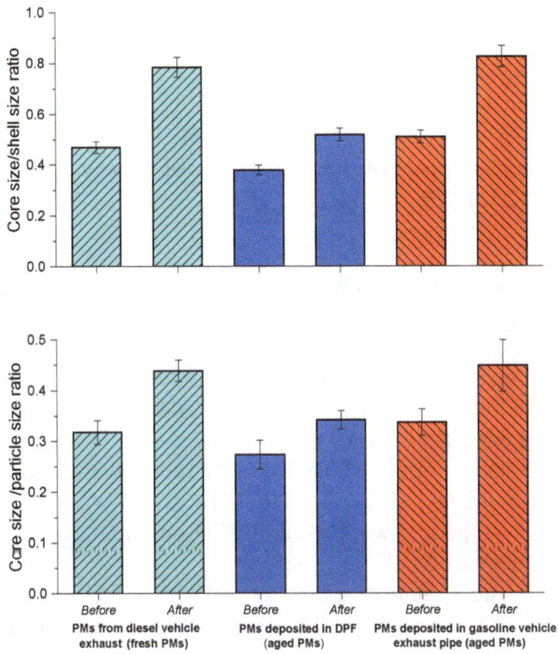

Figure 7: Core size/shell size and core size/particle size ratios for different PM samples before and after exposure to the vehicle light.

comparison with the aged PMs, receive the highest changes after their exposure to the vehicle light that could be mostly related to their highest volatile substances ratio and smallest particle size. Thus, according to analysing the diameters of more than 1000 particles for each sample, the average diameter of fresh PMs can be enlarged from 21.2 to 30.4 nm due to the impact of vehicle light, while the aged PMs of DPF and gasoline vehicle exhaust pipe receive less enlargement impact (32.3 to 33.5 nm and 27.1 to 29.8 nm, respectively).

4 CONCLUSIONS

We have seen that vehicle light has the potential to significantly change the nanostructure of primary particles by reductions in fringe length and fringe tortuosity (more straight carbon layers), and enlargements in fringe separation distance and core size/shell size and core size/particle size ratios for all three samples tested. Since the fresh PMs receive the highest changes in nanostructure as well as having the highest volatile components ratio and the smallest diameter, their average diameter can be enlarged from 21.2 to 30.4 nm due to the impact of vehicle light, while the aged PMs of diesel particulate filter and gasoline vehicle exhaust pipe receive less enlargement impact (32.3 to 33.5 nm and 27.1 to 29.8 nm, respectively). When there are changes in the nanostructure and size for the particles, the other properties (physicochemical and cell toxicity) of these particles could be also affected, resulting in having positive and negative impacts on human health and the environment. For instance, the latest research and review studies published in 2020–2021 [14], [24] observed the reduction in the diameters of the particles causes an increase in the negative impact of PM cell toxicity properties on human and animal health. Obtaining further information about the impact of different types of vehicle headlights under various conditions and PM sources on the physicochemical and cell toxicity properties of PMs to realize the best type of vehicle headlight regarding its lower negative influence on human health and the environment will be our future task.

ACKNOWLEDGEMENT

This work was supported in part by the University of Macau under UM Macao Talent Programme (UMMTP-2019-02).

REFERENCES

[1] World Health Organization (WHO), World health statistics overview 2019: Monitoring health for the SDGs, sustainable development goals. WHO/DAD/2019.1, 2019. https://www.who.int. Accessed on: 22 Jul. 2021.

[2] World Health Organization (WHO), Ambient (outdoor) air pollution, 2018. https://www.who.int/news-room/fact-sheets/detail/ambient-(outdoor)-air-quality-and-health. Accessed on: 22 Jul. 2021.

[3] World Health Organization (WHO), 9 out of 10 people worldwide breathe polluted air, but more countries are taking action, 2018. https://www.who.int/news-room/detail/02-05-2018-9-out-of-10-people-worldwide-breathe-polluted-air-but-more-countries-are-taking-action. Accessed on: 22 Jul. 2021.

[4] Health Effects Institute, State of global air 2019: A special report on global exposure to air pollution and its disease burden, 2019. www.stateofglobalair.org. Accessed on: 22 Jul. 2021.

[5] Ghadikolaei, M.A., Wong, P.K., Cheung, C.S., Ning, Z., Yung, K.F., Zhao, J., Gali, N.K. & Berenjestanaki, A.V., Impact of lower and higher alcohols on the physicochemical properties of particulate matter from diesel engines: A review. *Renewable and Sustainable Energy Reviews*, **143**, 110970, 2021.

[6] Gangwar, R.S., Bevan, G.H., Palanivel, R., Das, L. & Rajagopalan, S., Oxidative stress pathways of air pollution mediated toxicity: Recent insights. *Redox Biology*, **34**, 101545, 2020.

[7] Akhtar, U.S., Rastogi, N., McWhinney, R.D., Urch, B., Chow, C.W., Evans, G.J. & Scott, J.A., The combined effects of physicochemical properties of size-fractionated ambient particulate matter on in vitro toxicity in human A549 lung epithelial cells. *Toxicology Reports*, **1**, pp. 145–156, 2014.

[8] Shin, S.W., Song, I.H. & Um, S.H., Role of physicochemical properties in nanoparticle toxicity. *Nanomaterials*, **5**(3), pp. 1351–1365, 2015.

[9] Dockery, D.W., Pope, C.A., Xu, X., Spengler, J.D., Ware, J.H., Fay, M.E., Ferris, B.G. & Speizer, F.E., An association between air pollution and mortality in six U.S. cities. *The New England Journal of Medicine*, **329**, pp. 1753–1759, 1993.

[10] Gauderman, W.J., Avol, E., Gilliland, F., Vora, H., Thomas, D., Berhane, K., McConnell, R., Kuenzli, N., Lurmann, F., Rappaport, E., Margolis, H., Bates, D. & Peters, J., The effect of air pollution on lung development from 10 to 18 years of age. *The New England Journal of Medicine*, **351**, pp. 1057–1067, 2004.

[11] Katsouyanni, K., Touloumi, G., Samoli, E., Gryparis, A., Le Tertre, A., Monopolis, Y., Rossi, G., Zmirou, D., Ballester, F., Boumghar, A., Anderson, H.R., Wojtyniak, B., Paldy, A., Braunstein, R., Pekkanen, J., Schindler, C. & Schwartz, J., Confounding and effect modification in the short-term effects of ambient particles on total mortality: results from 29 European cities within the APHEA2 project. *Epidemiology*, **12**(5), pp. 521–531, 2001.

[12] Neer, A. & Koylu, U.O., Effect of operating conditions on the size, morphology, and concentration of submicrometer particulates emitted from a diesel engine. *Combustion and Flame*, **146**, pp. 142–154, 2006.

[13] Rohani, B. & Bae, C., Morphology and nano-structure of soot in diesel spray and in engine exhaust. *Fuel*, **203**, pp. 47–56, 2017.

[14] Chew, S., Kolosowska, N., Saveleva, L., Malm, T. & Kanninen, K.M., Impairment of mitochondrial function by particulate matter: Implications for the brain. *Neurochemistry International*, **135**, 104694, 2020.

[15] Finlayson-Pitts, B.J., Wingen, L.M., Perraud, V. & Ezell, M.J., Open questions on the chemical composition of airborne particles. *Communications Chemistry*, **3**, 108, 2020.

[16] Fuzzi, S., Baltensperger, U., Carslaw, K., Decesari, S., Denier van der Gon, H., Facchini, M.C., Fowler, D., Koren, I., Langford, B., Lohmann, U., Nemitz, E., Pandis, S., Riipinen, I., Rudich, Y., Schaap, M., Slowik, J.G., Spracklen, D.V., Vignati, E., Wild, M., Williams, M. & Gilardoni, S., Particulate matter, air quality and climate: lessons learned and future needs. *Atmospheric Chemistry and Physics*, **15**, pp. 8217–8299, 2015.

[17] Kumar, P., Kalaiarasan, G., Porter, A.E., Pinna, A., Kłosowski, M.M., Demokritou, P., Chung, K.F., Pain, C., Arvind, D.K., Arcucci, R., Adcock, I.M. & Dilliway, C., An overview of methods of fine and ultrafine particle collection for physicochemical characterisation and toxicity assessments. *Science of The Total Environment*, **756**, 143553, 2021.

[18] Weitekamp, C.A., Kerr, L.B., Dishaw, L., Nichols, J., Lein, M. & Stewart, M.J., A systematic review of the health effects associated with the inhalation of particle-filtered and whole diesel exhaust. *Inhalation Toxicology*, **32**(1), pp. 1–13, 2020.

[19] Wang, Y., Kamp, C.J., Wang, Y., Toops, T.J., Su, C., Wang, R., Gong, J. & Wong, V.W., The origin, transport, and evolution of ash in engine particulate filters. *Applied Energy*, **263**, 114631, 2020.

[20] Lapuerta, M., Rodríguez-Fernández, J. & Sánchez-Valdepeñas, J., Soot reactivity analysis and implications on diesel filter regeneration. *Progress in Energy and Combustion Science*, **78**, 100833, 2020.

[21] Mao, D., Ghadikolaei, M.A., Cheung, C.S., Shen, Z., Cui, W. & Wong, P.K., Influence of alternative fuels on the particulate matter micro and nano-structures, volatility and oxidation reactivity in a compression ignition engine. *Renewable and Sustainable Energy Reviews*, **132**, 110108, 2020.

[22] Ghadikolaei, M.A., Wei, L., Cheung, C.S., Yung, K.F. & Ning, Z., Particulate emission and physical properties of particulate matter emitted from a diesel engine fueled with ternary fuel (diesel-biodiesel-ethanol) in blended and fumigation modes. *Fuel*, **263**, 116665, 2020.

[23] Moffat, R.J., Describing the uncertainties in experimental results. *Experimental Thermal and Fluid Science*, **1**(1), pp. 3–17, 1988.

[24] Rossi, S., Buccarello, A., Malvezzi, C.C., Pinelli, S., Alinovi, R., Gerboles, A.G., Rozzi, G., Leonardi, F., Bollati, V., Palma, G.D., Lagonegro, P., Rossi, F., Lottici, P.P., Poli, D., Statello, R., Macchi, E. & Miragoli, M., Exposure to nanoparticles derived from diesel particulate filter equipped engine increases vulnerability to arrhythmia in rat hearts. *Environmental Pollution,* **284**, 117163, 2021.

SECTION 4
AIR POLLUTION MANAGEMENT, POLICY AND LEGISLATION

IMPACT OF COVID-19 PANDEMIC ON AIR QUALITY IN A TOURISTIC REGION

ANA CATARINA T. SILVA, PEDRO T. B. S. BRANCO & SOFIA I. V. SOUSA
LEPABE – Laboratory for Process Engineering, Environment, Biotechnology and Energy, Faculty of Engineering, University of Porto, Portugal

ABSTRACT

The rise of the COVID-19 pandemic led governments to impose mobility restrictions to their citizens and companies, having a greater impact on tourism. This can be used as a global-scale opportunity to study the impact of those restrictions on air quality, particularly in tourist sites. Hence, the aim of this work was to evaluate the impact of the COVID-19 pandemic on air quality in a touristic region. Air pollution was evaluated in a prominent touristic region in southern Europe (Algarve, Portugal). The concentrations of PM_{10}, NO_2, O_3 and CO were obtained from four air quality monitoring stations (traffic and background) in two lockdown and two lifting periods. The data collected was compared with historical data (2015–2019) and with reference limit values in the Portuguese legislation and the World Health Organization (WHO) guidelines. Overall, PM_{10} reduced in all studied periods and stations, with a higher decrease during the 1st lockdown and lifting (5.30%–73.6%) compared to the 2nd periods in the urban traffic and background stations, as well as NO_2 (35.8%–56.5%), in the urban background station evaluated. O_3 diminished in general (until 29.1%), while CO decreased during the 1st lockdown and lifting periods (64.5%–80.7%) and increased in the subsequent periods (9.66%–65.1%) in the urban traffic station. The percentage of exceedances to the reference limit values both in the legislation and in WHO guidelines mainly decreased or remained equal in all the studied periods when compared with the previous years. Thus, it was concluded that the mobility restrictions imposed during the COVID-19 pandemic contributed to a general reduction of air pollution in tourist sites. Since tourism is an important economic activity, some specific policies were proposed based on those restrictions, considering a sustainable equilibrium between the economy, the environment, and human health.

Keywords: air quality, COVID-19, lockdown, lifting, tourism, air pollution, SARS-CoV-2.

1 INTRODUCTION

On 11 March 2020 the World Health Organization (WHO) declared COVID-19 a global pandemic caused by the transmission of the SARS-CoV-2 virus, which forced the governmental entities to impose mobility restrictions on citizens and companies all over the world to prevent the spread of the virus [1]. Hence, investigating the influence of these restrictions on air quality can be a global-scale opportunity to understand their effects.

Among those mobility restrictions, the lockdowns (total or partial) consisted in one of the most restrictive policies adopted, enabling the reduction of traffic and socioeconomic activities [2]. As a consequence, air quality was claimed to improve in several studies conducted during the lockdown [3], [4]. Yet, some studies demonstrated that as soon as the lifting measures were adopted, the pollutants' concentrations increased nearly to the usual levels [5]. Regarding the socioeconomic sector, the pandemic had caused severe damages to it [6]. Attending the different socioeconomic activities, tourism was one of the most affected by the COVID-19 pandemic, especially during the 1st lockdown in 2020 which was the most restrictive of all the imposed lockdowns [7]. Currently, the touristic activity contributes largely to a country's socioeconomic dynamic. In Portugal, it is in the Algarve region where tourism contributes more to the economy since this is a worldwide recognised beach destination. Thus, it is important to study the impact of the pandemic on air quality in Algarve region in order to understand the extent of the air quality improvement in a region

WIT Transactions on Ecology and the Environment, Vol 252, © 2021 WIT Press
www.witpress.com, ISSN 1743-3541 (on-line)
doi:10.2495/AIR210131

predominantly touristic that was strongly affected by the pandemic, not only during the 1st lockdown but also in other subsequent restrictive periods.

As far as the authors' knowledge goes, only one study was published about the impact of the COVID-19 on air quality in Portugal, but it did not analyse data at a regional level, and it was only focused on PM_{10} and NO_2 during the 1st lockdown in 2020 [8]. Therefore, the present study aimed to: (i) evaluate the impact of the pandemic on air quality in the touristic region of Algarve (Portugal), specifically on the concentrations of suspended particles with an equivalent aerodynamic diameter smaller than 10 μm (PM_{10}), nitrogen dioxide (NO_2), tropospheric ozone (O_3) and carbon monoxide (CO), during the two total lockdowns and subsequent liftings imposed in 2020 and 2021; (ii) assess the impact in different types of air quality monitoring stations located in different environments, namely urban traffic, background and rural background stations; (iii) compare the data with the historical concentrations (2015–2019) and with the reference limit values in both the Portuguese legislation and the WHO guidelines; and (iv) define new policies for air quality improvement, considering a sustainable balance between the economy, the environment and the human health.

2 METHODOLOGY

2.1 Data collection and study periods

Hourly concentrations of PM_{10}, NO_2, O_3, and CO were obtained between 1 January 2020, and 4 April 2021, from the air quality monitoring network of the Algarve region (southern Portugal). That network is composed of four monitoring stations located at Portimão (David Neto, urban traffic), Albufeira (Malpique, urban background), Faro (Joaquim Magalhães, urban background), and Alcoutim (Cerro, rural background), as represented in Fig. 1 [9].

Figure 1: Location of the air quality monitoring stations from the Algarve region.

The hourly data from 1 January 2015 to 31 December 2019, were collected from the national database "QualAr" to be used as historical data for comparison with the 2020–2021 data. The five previous years were considered to minimise the influence of the meteorological conditions on air pollutants' concentrations [8].

The study periods were defined according to the restrictive measures adopted in Portugal during the COVID-19 pandemic. On 16 March 2020 the first restrictive measures to prevent the spread of the virus were adopted, and a few days after, on 19 March 2020, the first state of emergency begun, which lasted until 4 May 2020 (the first day of the 1st lifting) [10]–[12]. During this period, major restrictions were imposed on citizens' mobility, leading to a widespread closure of services, mandatory curfew with exits only allowed for essential needs and work whenever remote work was not feasible [13]. After this period, 2020 and 2021 were marked by different lockdown and lifting periods, with the measures' restrictiveness varying according to the epidemiologic situation. Table 1 summarises the study periods defined, and the main restrictive measures applied in each defined period.

Table 1: Study periods defined based on the Portuguese national government's restrictive measures adopted in each period.

Study period	Restrictive measures
I, 1st lockdown: 16 March–3 May 2020 [13]	• Closure of all services that provide non-essential needs and suspension of several activities, including schools and religious events; • All-day curfew for all citizens; exits allowed only for essential needs; • Circulation prohibited between municipalities (traffic related); • Suspension of flights and of the rail and fluvial means of transportation (only allowed for essential goods transport); • Mandatory remote work, whenever feasible.
II, 1st lifting: 4 May–15 June 2020 [12], [14]	• Permission for re-opening every type of commercial and cultural places, like restaurants, coffee shops, museums, etc.; • Allowed events with 20 people or less; • Re-opening schools for in person classes: nursery schools and 11th and 12th grades, and schools for disabled people; • Remote work still mandatory, but in-person work partially allowed (reduced number of workers simultaneously).
III, 2nd lockdown: 13 January–14 March 2021 [15], [16]	• Suspension of several activities and establishments that do not provide essential needs; • Circulation prohibited between municipalities (traffic related); • Only take-away regime or home delivery allowed for restaurants and similar activities; • Suspension of in-person educational activities, including nurseries and schools, and other activities/ establishments of social support.
IV, 2nd lifting: 15 March–4 April 2021 [17]	• Re-opening of in-person classes in nurseries, pre- and primary schools, and social support activities; • Permission of travelling to other countries.

2.2 Data analysis

Daily mean concentrations of PM_{10} and NO_2 were calculated based on a 50% data availability per day, while for O_3 and CO the daily maximum of the 8 hours running mean was calculated based on a minimum of 75% of data per day as defined by the Portuguese legislation (Decreto-Lei no. 102/2010). The daily historical mean concentration was calculated using at least three of the five years considered (> 50%).To be considered in the present study, a certain monitoring station was required to have at least 50% of the paired (historical – 2020/2021) daily mean concentrations for each pollutant and period, and only the monitoring stations with data for the four periods for each pollutant were considered.

Descriptive statistics were used to express the concentrations of each pollutant in each monitoring station. Normality was assessed using the Shapiro–Wilk normality test. The parametric Student's *t*-test and the non-parametric Wilcoxon Rank-Sum test were conducted to test the significance of the differences between the study periods and historical pollutants' concentrations. The level of statistical significance was set at 0.05. Furthermore, for each pollutant, the data was compared with the limit values defined in the Portuguese legislation (Decreto-Lei no. 102/2010) and in the WHO guidelines. Hence, PM_{10} concentrations were compared with the limit value/WHO guideline of 50 $\mu g/m^3$, while NO_2 were compared with the annual limit value/WHO guidelines of 40 $\mu g/m^3$ that was used as daily limit. O_3 was compared with the target value of 120 $\mu g/m^3$ which is based on the daily maximum of the 8 hours running mean, and CO was compared with the limit value of 10 mg/m^3.

Table 2 shows the identification code (ID), type, and the pollutants measured of each air quality monitoring station studied.

Table 2: Air quality monitoring stations evaluated from the Algarve region.

Monitoring station	ID	Type	Pollutants studied
Joaquim Magalhães	alg_01	Urban background	PM_{10}, NO_2, O_3
Malpique	alg_02	Urban background	PM_{10}, O_3
David Neto	alg_03	Urban traffic	PM_{10}, CO
Cerro	alg_04	Rural background	PM_{10}, O_3

3 RESULTS AND DISCUSSION

3.1 Pollutants' behaviour

The time-series plots of PM_{10}, NO_2, O_3, and CO are represented in Figs 2–5, respectively, for each monitoring station and pollutant evaluated, for the 1st lockdown and lifting in 2020 (green) and 2nd lockdown and lifting in 2021 (red), and respective historical periods (blue).

Overall, the time-series pattern along the four periods was similar to the historical time-series patterns for all the studied pollutants. Yet, some exceptions were denoted: (i) some peak concentrations of PM_{10} were observed during the I, II, III and IV periods, as a result of air mass intrusions from North Africa (Sahara and Sahel dust) that occurred on 19 March 2020, 29 May 2020, 15 February 2021, and 31 March to 4 April 2021, respectively [18]–[20]. Besides, at the beginning of January 2021 (winter season) an increase of PM_{10} levels occurred possibly due to the rise of residential wood combustion since this month was considered colder than usual over the last 4 years [21]; (ii) a relevant decrease of O_3 concentration occurred between 1 and 15 April 2020, during the Period I, and in the second

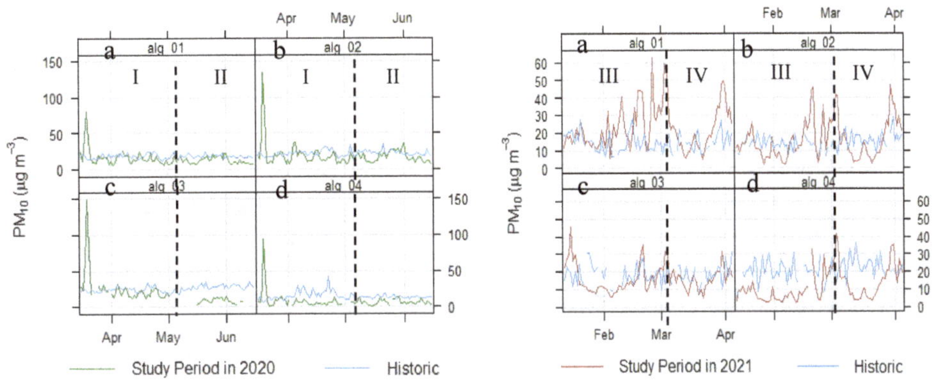

Figure 2: Time-series of PM_{10} for the historical and study periods in 2020 (1st lockdown – Period I, and 1st lifting – Period II) and 2021 (2nd lockdown – Period III, and 2nd lifting – Period IV) in the Algarve region air quality stations. (a) Joaquim Magalhães; (b) Malpique; (c) David Neto; and (d) Cerro.

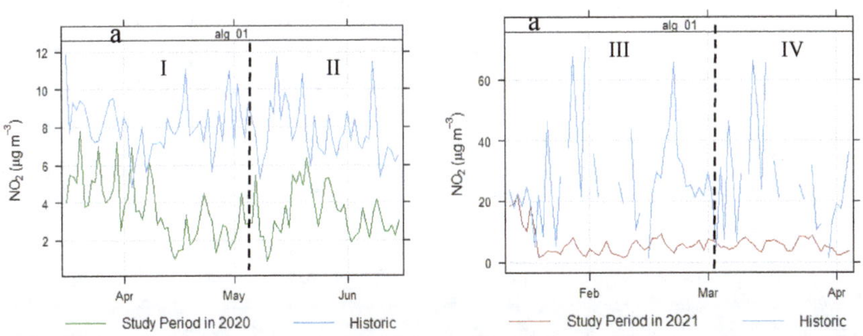

Figure 3: Time-series of NO_2 for the historical and study periods in 2020 (1st lockdown – Period I, and 1st lifting – Period II) and 2021 (2nd lockdown – Period III, and 2nd lifting – Period IV) in the Algarve region air quality stations. (a) Joaquim Magalhães.

half of January (Period III) due to anticyclones that occurred in Portugal with high levels of precipitation and low solar radiation, interfering with this pollutant formation [21], [22]. Besides, the reduction of O_3 observed might be connected to the reduction of its precursors, namely NO_2 at the urban background station (alg_01) studied for both.

3.1.1 Period I: 1st lockdown

During the 1st lockdown (Period I) the pollutants' concentrations generally reduced compared with the historical data. Hence, PM_{10}, NO_2, O_3, and CO had statistically significant reductions (respectively, 5.30%–46.3%, 54.2%, 0.75%–16.48%, and 80.7%, p-value < 0.05)

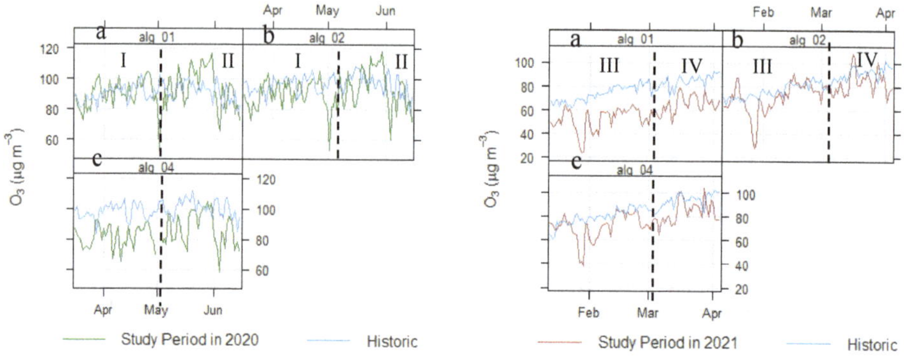

Figure 4: Time-series of O_3 for the historical and study periods in 2020 (1st lockdown – Period I, and 1st lifting – Period II) and 2021 (2nd lockdown – Period III, and 2nd lifting – Period IV) in the Algarve region air quality stations. (a) Joaquim Magalhães; (b) Malpique; and (c) Cerro.

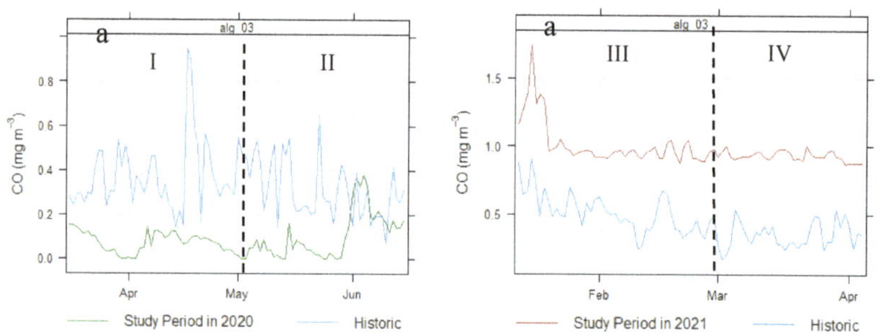

Figure 5: Time-series of CO for the historical and study periods in 2020 (1st lockdown – Period I, and 1st lifting – Period II) and 2021 (2nd lockdown – Period III, and 2nd lifting – Period IV) in the Algarve region air quality stations. (a) David Neto.

in all stations, with the highest decrease observed at the urban traffic station for CO. Although the touristic activity in Algarve region is usually higher during the spring/summer months, these reductions were expected especially at the urban traffic stations given the restrictive measures imposed during that period, specifically the mobility restrictions and all-day curfew. Other studies also verified a general reduction of those pollutants' concentrations during the 1st lockdown, namely Broomandi et al. [23] in Iran (compared with 2019 data) and Dantas et al. [24] in Brazil (compared with pre-lockdown and 2019 data), except for O_3 for which an increase was verified. However, Ordóñez et al. [25], who conducted a study in Europe, reported a decrease of O_3 only in the Iberian Peninsula due to the low solar radiation during this period (compared with 2015–2019 data).The exceedances to the reference limit values were generally reduced to zero or remained equal to the verified in the historical period, except in the case of PM_{10} in the 1st lockdown when exceedances almost doubled those in the historical period, yet it should be noted the air mass intrusion that occurred during this period, as previously stated. Table 3 shows the differences (%) between the concentrations in the study period and in the historical period and the percentage of exceedances in each monitoring station and pollutant.

Table 3: Differences (%) between concentrations in the study period and in the historical data, and percentage of exceedances for PM_{10}, NO_2, O_3, and CO for each monitoring station during the 1st lockdown (Period I).

		Differences between concentrations in the study and historical data (%)			
ID		PM_{10}	NO_2	O_3	CO
alg_01		−8.85*	−54.2*	−0.751*	NA
alg_02		−5.30*	NA	−6.38*	NA
alg_03		4.46*	NA	NA	−80.7*
alg_04		−46.3*	NA	−16.5*	NA
		Percentage of exceedances (%)			
ID		PM_{10}	NO_2	O_3	CO
alg_01	Study	2	0	0	NA
	Historical	0	0	5	NA
alg_02	Study	4	NA	0	NA
	Historical	2	NA	2	NA
alg_03	Study	6	NA	NA	0
	Historical	2	NA	NA	0
alg_04	Study	6	NA	0	NA
	Historical	3	NA	3	NA

NA = not applicable; * statistically significant (p-value < 0.05).

3.1.2 Period II: 1st lifting

In the 1st lifting (Period II) the reductions of the studied pollutants' concentrations were in general higher than in the Period I, with statistically significant reductions (p-value < 0.05) of 25.5%–73.6% (PM_{10}), 56.5% (NO_2), and 64.5% (CO). In this period, O_3 only evidenced a statistically significant reduction (13.5%, p-value < 0.05) at the rural background (alg_04, Cerro) station. The registered higher reductions in this period in comparison with Period I (the 1st lockdown) were expected. Although the restrictions regarding traffic mobility were lifted, the touristic activity usually increases largely during the spring/summer in the studied region. Thus, even though the 1st lifting had less restrictive measures than the 1st lockdown, the usual touristic activity (pre-pandemic situation) was not re-established as the flights were still suspended, therefore lowering the number of tourists. In some cases, it was observed a reduction of the exceedances to the reference limit values from the historical period to zero exceedances in this study period, and in other cases they remained zero. Table 4 represents the differences (%) between the concentrations in the study period and in the historical period and the percentage of exceedances in each monitoring station and pollutant.

3.1.3 Period III: 2nd lockdown

During the 2nd lockdown (Period III) the concentrations of the studied pollutants reduced less in general than in the 1st lockdown (Period I). Hence, PM_{10}, NO_2, and O_3 concentrations reduced by 5.1%–46.8%, 43.4%, and 8.54%–29.1%, respectively, statistically significant for all stations and pollutants (p-value < 0.05), with exception of two air quality monitoring

Table 4: Differences (%) between concentrations in the study period and in the historical data, and percentage of exceedances for PM_{10}, NO_2, O_3, and CO for each monitoring station during the 1st lifting (Period II).

		Differences between concentrations in the study and historical data (%)			
ID		PM_{10}	NO_2	O_3	CO
alg_01		–31.2*	–56.5*	2.94	NA
alg_02		–25.5*	NA	–3.84	NA
alg_03		–73.6*	NA	NA	–64.5*
alg_04		–54.6*	NA	–13.5*	NA
		Percentage of exceedances (%)			
ID		PM_{10}	NO_2	O_3	CO
alg_01	Study	0	0	0	NA
	Historical	0	0	0	NA
alg_02	Study	0	NA	0	NA
	Historical	1	NA	0	NA
alg_03	Study	0	NA	NA	0
	Historical	1	NA	NA	0
alg_04	Study	0	NA	0	NA
	Historical	0	NA	1	NA

NA = not applicable; * statistically significant (p-value < 0.05).

stations for PM_{10}, while CO concentrations significantly increased at the urban traffic station (9.66%, p-value < 0.05). Similarly to what happened in the 1st lockdown, it was expected a decrease of the pollutants' concentrations comparatively to the historical data, because the mobility of the citizens was restricted. Yet, when comparing the results between the two lockdowns, in the 1st lockdown the reductions were all statistically significant, while in the 2nd lockdown not all were significant. This may be globally explained by the more restrictive measures of the 1st lockdown in comparison with the 2nd. Moreover, in the 2nd lockdown (Period III), it was possible to travel between councils during the weekdays, which was forbidden in Period I. Hence, this may also explain the slight increase in CO concentrations observed. In most of the stations and pollutants the percentage of exceedances to the reference limit values reduced to zero in the study period. Table 5 represents the differences (%) between the concentrations in the study period and in the historical period and the percentage of exceedances in each monitoring station and pollutant.

3.1.4 Period IV: 2nd lifting

In the 2nd lifting (Period IV), the reduction of pollutants concentrations was lower than in the 1st. PM_{10}, NO_2, and O_3 reduced respectively by 14.9%–27.8%, 35.8%, and 7.89%–21.6%, whilst CO increased by 65.1%. Besides, the number of statistically significant (p-value < 0.05) reductions was lower in this period than in Period II (1st lifting). This was the period with the less restrictive measures considered in this study, with some restrictions being lifted like the allowance to fly and/or to travel between countries, which may have influenced the observed results, namely the increase of the CO concentrations at the urban traffic station. There were not found any exceedances to the reference limit values in this period, neither in the corresponding historical data. Table 6 represents the differences (%) between the

Table 5: Differences (%) between concentrations in the study period and in the historical data, and percentage of exceedances for PM_{10}, NO_2, O_3, and CO for each monitoring station during the 2nd lockdown (Period III).

<table>
<tr><td colspan="2"></td><td colspan="4">Differences between concentrations in the study and historical data (%)</td></tr>
<tr><td colspan="2">ID</td><td>PM_{10}</td><td>NO_2</td><td>O_3</td><td>CO</td></tr>
<tr><td colspan="2">alg_01</td><td>–5.10</td><td>–43.4*</td><td>–29.1*</td><td>NA</td></tr>
<tr><td colspan="2">alg_02</td><td>–35.4*</td><td>NA</td><td>–8.54*</td><td>NA</td></tr>
<tr><td colspan="2">alg_03</td><td>–5.58</td><td>NA</td><td>NA</td><td>9.66*</td></tr>
<tr><td colspan="2">alg_04</td><td>–46.8*</td><td>NA</td><td>–11.6*</td><td>NA</td></tr>
<tr><td colspan="2"></td><td colspan="4">Percentage of exceedances (%)</td></tr>
<tr><td colspan="2">ID</td><td>PM_{10}</td><td>NO_2</td><td>O_3</td><td>CO</td></tr>
<tr><td rowspan="2">alg_01</td><td>Study</td><td>5</td><td>0</td><td>0</td><td>NA</td></tr>
<tr><td>Historical</td><td>1</td><td>2</td><td>1</td><td>NA</td></tr>
<tr><td rowspan="2">alg_02</td><td>Study</td><td>0</td><td>NA</td><td>0</td><td>NA</td></tr>
<tr><td>Historical</td><td>5</td><td>NA</td><td>2</td><td>NA</td></tr>
<tr><td rowspan="2">alg_03</td><td>Study</td><td>0</td><td>NA</td><td>NA</td><td>0</td></tr>
<tr><td>Historical</td><td>6</td><td>NA</td><td>NA</td><td>0</td></tr>
<tr><td rowspan="2">alg_04</td><td>Study</td><td>0</td><td>NA</td><td>0</td><td>NA</td></tr>
<tr><td>Historical</td><td>2</td><td>NA</td><td>0</td><td>NA</td></tr>
</table>

NA = not applicable; * statistically significant (p-value < 0.05).

Table 6: Differences (%) between concentrations in the study period and in the historical data, and percentage of exceedances for PM_{10}, NO_2, O_3, and CO for each monitoring station during the 2nd lifting (Period IV).

<table>
<tr><td colspan="2"></td><td colspan="4">Differences between concentrations in the study and historical data (%)</td></tr>
<tr><td colspan="2">ID</td><td>PM_{10}</td><td>NO_2</td><td>O_3</td><td>CO</td></tr>
<tr><td colspan="2">alg_01</td><td>–14.9</td><td>–35.8*</td><td>–21.6*</td><td>NA</td></tr>
<tr><td colspan="2">alg_02</td><td>–20.1</td><td>NA</td><td>–7.89*</td><td>NA</td></tr>
<tr><td colspan="2">alg_03</td><td>27.6</td><td>NA</td><td>NA</td><td>65.1*</td></tr>
<tr><td colspan="2">alg_04</td><td>–27.8</td><td>NA</td><td>–8.52*</td><td>NA</td></tr>
<tr><td colspan="2"></td><td colspan="4">Percentage of exceedances (%)</td></tr>
<tr><td colspan="2">ID</td><td>PM_{10}</td><td>NO_2</td><td>O_3</td><td>CO</td></tr>
<tr><td rowspan="2">alg_01</td><td>Study</td><td>0</td><td>0</td><td>0</td><td>NA</td></tr>
<tr><td>Historical</td><td>0</td><td>0</td><td>0</td><td>NA</td></tr>
<tr><td rowspan="2">alg_02</td><td>Study</td><td>0</td><td>NA</td><td>0</td><td>NA</td></tr>
<tr><td>Historical</td><td>0</td><td>NA</td><td>0</td><td>NA</td></tr>
<tr><td rowspan="2">alg_03</td><td>Study</td><td>0</td><td>NA</td><td>NA</td><td>0</td></tr>
<tr><td>Historical</td><td>0</td><td>NA</td><td>NA</td><td>0</td></tr>
<tr><td rowspan="2">alg_04</td><td>Study</td><td>0</td><td>NA</td><td>0</td><td>NA</td></tr>
<tr><td>Historical</td><td>0</td><td>NA</td><td>0</td><td>NA</td></tr>
</table>

NA = not applicable; * statistically significant (p-value < 0.05).

concentrations in the study period and in the historical period and the percentage of exceedances in each monitoring station and pollutant.

3.2 Proposed measures

With the rise of the COVID-19 cases, several restrictions were imposed to prevent the spread of the infections by the SARS-CoV-2 virus, that led to a general reduction of the air pollutants' concentrations as evidenced in Section 3.1. Hence, some measures based on these results can be proposed to define new possible policies to improve air quality in the touristic region of Algarve. From all the restrictions stated in Section 2.1, the ones that mainly contributed to the pollutants' concentration reduction corresponded to the flights' suspension, all-day curfew, and the mobility restrictions, essentially adopted during the 1st lockdown and 1st lifting. Nevertheless, these restrictions also had a severe impact on the touristic activity and consequently on the economy of this region. Therefore, any proposed measures for improving air quality should consider a sustainable equilibrium between the environment, the economy, and human health. During the 2nd lockdown and 2nd lifting, the restrictions imposed were less restrictive, and even though the reduction of the pollutants' concentration was not so high, air pollution still decreased overall. Thus, the restrictions imposed during these periods can be adopted to reduce air pollution with lower constraints on the touristic activities. As a short-term measure, the application of mobility restrictions between municipalities (like in the Period III, 2nd lockdown) when a high air pollution event is predicted would highly contribute to the decrease of pollutants levels, vis. by reducing emissions from transports (cars, taxis, etc). Aside this, soft means of transportation should be favoured as a more long-term measure. Given the good weather conditions of this seasonal touristic site (tourism mostly occurs in warm months), places like this should encourage tourists to walk/cycle. Additionally, more bike lanes and bike-rentals should be created. These measures will have a positive direct impact not only on the environment, but also on human health and the economy, as tourists could still travel to and within the region but using more sustainable means. This measure was also proposed by Piccoli et al. [26], aside from the urban planning re-design, which is highly recommended, especially for safe cycling and walking. Another related measure that should be established as soon as possible is the transition from fossil fuels to renewable energies in aircraft engines. Some studies have been developed in this matter, and the use of hydrogen as fuel (that is under development) seems promising since its electrolysis only produces water and heat [27]. Hence, since Algarve is predominantly a touristic region with tourists arriving mainly by plane, the local air quality would benefit from cleaner technologies in the aviation sector. Although the flights' suspension is a measure that clearly benefits the environment, it causes a relevant negative socioeconomic impact in the region, drastically reducing the touristic activity, then other measures with lower socioeconomic impact should be favoured.

4 CONCLUSION

The present work assessed the impact of the COVID-19 pandemic on air quality in the touristic region of Algarve, Portugal, considering the two lockdowns and respective liftings imposed in 2020 and 2021. Overall, the air quality improved during the four periods in all air quality monitoring stations relative to the historical data (2015–2019). Thus, it was possible to conclude that the restrictions imposed, even the less restrictive ones, positively impacted the air quality. PM_{10}, NO_2, and O_3 concentrations consistently reduced in all the studied periods, while CO decreased in the 1st lockdown and lifting and increased in the 2nd lockdown and lifting. The reductions in the air pollutants' concentrations were higher in the

1st lockdown than in the 2nd and in the 1st lifting than in the 2nd, especially of CO at the urban traffic station, since the mobility was more restricted than in the other periods. The percentage of exceedances to the reference limit values also reduced in all study periods in comparison with the historical periods, mostly to zero exceedances. Based on the results, and considering a sustainable equilibrium between the environment, the economy and the human health, some measures were proposed to define new possible policies to improve air quality in a beach touristic area like Algarve. Restricting citizens' travelling to and from the region and between municipalities was proposed as a short-term measure to be applied when a high air pollution event is predicted while encouraging tourists to use soft means of transportation (walking, cycling, etc.) was suggested as a long-term measure. Instead of banning flights, providing the transition from fossil fuels to more sustainable energy sources would benefit the air quality without wrecking the economy.

ACKNOWLEDGEMENTS

This work was financially supported by: Base Funding – UID/EQU/00511/2020 of the Laboratory for Process Engineering, Environment, Biotechnology and Energy – LEPABE – funded by national funds through FCT/MCTES (PIDDAC). Sofia I.V. Sousa thanks the Portuguese Foundation for Science and Technology (FCT) for the financial support of her work contract through the Scientific Employment Stimulus – Individual Call – CEECIND/02477/2017.

REFERENCES

[1] Cucinotta, D. & Vanelli, M., WHO declares COVID-19 a pandemic. *Acta. Biomed.*, **91**(1), pp. 157–160, 2020. DOI: 10.23750/abm.v91i1.9397.

[2] Liu, F., Wang, M. & Zheng, M., Effects of COVID-19 lockdown on global air quality and health. *Sci. Total Environ.*, **755**(Pt 1), 142533, 2021. DOI: 10.1016/j.scitotenv.2020.142533.

[3] Filonchyk, M., Hurynovich, V. & Yan, H., Impact of COVID-19 lockdown on air quality in the Poland, Eastern Europe. *Environmental Research*, **198**, 110454, 2021. DOI: 10.1016/j.envres.2020.110454.

[4] Ganguly, R. A.-O., Sharma, D. & Kumar, P., Short-term impacts of air pollutants in three megacities of India during COVID-19 lockdown. 2021. DOI: 10.1007/s10668-021-01434-9.

[5] Kumari, P. & Toshniwal, D., Impact of lockdown on air quality over major cities across the globe during COVID-19 pandemic. *Urban Climate*, **34**, 100719, 2020. DOI: 10.1016/j.uclim.2020.100719.

[6] Kar, S. et al., Impact of coronavirus (COVID-19) outbreak on society, air quality, and economy in India: A study of three "p"s of sustainability in India. *Sustainability (Switzerland)*, **13**(5), pp. 1–27, 2021. DOI: 10.3390/su13052873.

[7] Porto Business School, O impacto da COVID-19 no setor do turismo. 2020.

[8] Gama, C., Relvas, H., Lopes, M. & Monteiro, A., The impact of COVID-19 on air quality levels in Portugal: A way to assess traffic contribution. *Environmental Research*, **193**, 110515, 2021. DOI: 10.1016/j.envres.2020.110515.

[9] Comissão de Coordenação e Desenvolvimento Regional do Algarve (CCDR Algarve), Qualidade do Ar na Região do Algarve – Relatório 2° Trimestre de 2015, 2015.

[10] Presidência da República, Decreto do Presidente da República no. 14-A/2020. *Diário da República* 1.ª série, **55**(March), pp. 13-(2)a-(4), 2020.

[11] República Portuguesa, Comunicado do Conselho de Ministros de 19 de março de 2020, 2020. https://www.portugal.gov.pt/pt/gc22/governo/comunicado-de-conselho-de-ministros?i=334. Accessed on: 22 Mar. 2021.
[12] República Portuguesa, Comunicado do Conselho de Ministros de 30 de abril de 2020, 2020. https://www.portugal.gov.pt/pt/gc22/governo/comunicado-de-conselho-de-ministros?i=344. Accessed on: 21 Mar. 2021.
[13] República Portuguesa, Comunicado do Conselho de Ministros eletrónico de 16 de março de 2020, 2020. https://www.portugal.gov.pt/pt/gc22/governo/comunicado-de-conselho-de-ministros?i=332. Accessed on: 22 Mar. 2021.
[14] República Portuguesa, Comunicado do Conselho de Ministros de 9 de junho de 2020, 2020.
[15] República Portuguesa, Comunicado do Conselho de Ministros de 13 de janeiro de 2021, 2021. https://www.portugal.gov.pt/pt/gc22/governo/comunicado-de-conselho-de-ministros?i=394. Accessed on: 4 Apr. 2021.
[16] República Portuguesa, Comunicado do Conselho de Ministros de 21 de janeiro de 2021, 2021. https://www.portugal.gov.pt/pt/gc22/governo/comunicado-de-conselho-de-ministros?i=397. Accessed on: 4 Apr. 2021.
[17] República Portuguesa, Comunicado do Conselho de Ministros de 11 de março de 2021, 2021. https://www.portugal.gov.pt/pt/gc22/governo/comunicado-de-conselho-de-ministros?i=407. Accessed on: 4 Apr. 2021.
[18] Macedo, C., APA alerta para ocorrência de fraca qualidade até esta sexta-feira. *Ambiente Magazine*, 19 Mar., 2020.
[19] Agência Portuguesa do Ambiente (APA), Previsão de transporte de partículas naturais com origem em regiões áridas, 2020.
[20] Agência Portuguesa do Ambiente (APA), Previsão de transporte de partículas naturais com origem em regiões áridas, 2021.
[21] Instituto Português do Mar e da Atmosfera (IPMA), Boletim Climatológico Portugal Continental Janeiro 2021, 2021.
[22] Instituto Português do Mar e da Atmosfera (IPMA), Abril 2020 Portugal Continental, 2020.
[23] Broomandi, P., Karaca, F., Nikfal, A., Jahanbakhshi, A., Tamjidi, M. & Kim, J.R., Impact of COVID-19 event on the air quality in Iran. *Aerosol and Air Quality Research*, **20**(8), pp. 1793–1804, 2020. DOI: 10.4209/aaqr.2020.05.0205.
[24] Dantas, G., Siciliano, B., França, B.B., da Silva, C.M. & Arbilla, G., The impact of COVID-19 partial lockdown on the air quality of the city of Rio de Janeiro, Brazil. *Science of The Total Environment*, **729**, 139085, 2020. DOI: 10.1016/j.scitotenv.2020.139085.
[25] Ordóñez, C., Garrido-Perez, J.M. & García-Herrera, R., Early spring near-surface ozone in Europe during the COVID-19 shutdown: Meteorological effects outweigh emission changes. *Science of The Total Environment*, **747**, 141322, 2020. DOI: 10.1016/j.scitotenv.2020.141322.
[26] Piccoli, A. et al., Modeling the effect of COVID-19 lockdown on mobility and NO_2 concentration in the Lombardy region. *Atmosphere*, **11**(12), pp. 1–18, 2020. DOI: 10.3390/atmos11121319.
[27] Airbus, Is this the next "clean" energy to power aviation? 2020. https://www.airbus.com/newsroom/stories/Is-this-the-next-clean-energy-to-power-aviation.html. Accessed on: 24 Aug. 2021.

HEALTH RISK ASSESSMENT AND BLACK CARBON: STATE OF ART AND NEW PROSPECTIVES

ORNELLA SALIMBENE[1], SALVATORE MORREALE[2] & FRANCESCO PILLA[3]
[1]Department of Environment, Land and Infrastructure Engineering (DIATI), Politecnico of Torino, Italy
[2]Information Technology for Humanitarian Assistance, Cooperation and Action (ITHACA), Italy
[3]School of Architecture Planning and Environment Policy, University College Dublin, Ireland

ABSTRACT

The characterization of health risks (HRA) in a population due to air pollution is crucial for the development of effective risk management policies and strategies. HRA is a formally required tool by nations as part of the decision-making process for new programs, regulations and policies that affect air quality, so it is important for decision makers to understand what resources are needed and what limitations may exist. In the hazard identification phase at local and urban level, concentration data provided by local monitors are used, but one of the main shortcomings is that population exposure estimates based on these data are often limited by the sparse geographic-temporal coverage of the measurements. Often the HRA evaluation does not take into account the complex chemical composition of the suspended particulate and therefore the different toxicity and oxidation potential associated with it. Current air quality standards for particulate matter (PM) use the mass concentration of PM (PM_{10}, $PM_{2.5}$) as a metric, but particles from combustion sources are more relevant to human health compared to particles from other sources and the impact of policies to reduce PM from combustion processes, is relatively small when estimating the effects for a reduction in total mass concentration. This research highlights the value of black carbon (BC) particles as a possible additional indicator in the management of air quality in Turin (north Italy), one of the most affected Italian cities by smog where a high number of asthmatics and allergic people are exposed every day to PM concentrations higher than the legal limits. According to the Council on Clean Transportation, Turin is among the top 100 cities in the world with a higher number of deaths due to respiratory diseases caused by transport. BC could be a valuable additional air quality indicator to assess human health risks caused by primary combustion particles.

Keywords: air pollution, black carbon, indicators.

1 INTRODUCTION

Air pollution was the fourth leading risk factor for early death worldwide in 2019, surpassed only by high blood pressure, tobacco use, and poor diet [1]. In 2019, air pollution moved up from the fifth to the fourth leading risk factor for death globally (Global Air 2020, www.who.airpollution.it), continuing to exceed the impacts of other widely recognized risk factors for chronic disease like obesity (high body-mass index), high cholesterol, and malnutrition (Fig. 1). Over the last several decades scientists have continued to build an extensive body of evidence on the risks that breathing poor-quality air poses to human health and our environment, perhaps the most extensive evidence that exists for any environmental risk factor. Exposure to air pollution has serious health consequences [2]. To better understand the burden of disease that air pollution places on society, there is scientific evidence of its effects on health. Short-term exposure to air pollution can damage health; for example, high-pollution days can trigger asthma symptoms, cause a local spike in hospitalizations or even deaths related to respiratory and cardiovascular diseases [3]. There is a broad scientific consensus that long-term exposures to air pollution contribute to an increased risk of illness and death from ischemic heart disease, lung cancer, chronic obstructive pulmonary disease (COPD), lower respiratory tract infections (e.g., pneumonia), stroke, diabetes and, more recently, adverse birth outcomes, and that the public health burden from these exposures is much greater than that from short-term exposures [4]. Ongoing

WIT Transactions on Ecology and the Environment, Vol 252, © 2021 WIT Press
www.witpress.com, ISSN 1743-3541 (on-line)
doi:10.2495/AIR210141

studies continue to explore the role of air pollution in the development of asthma, cognitive impairments and other effects (e.g., chronic kidney disorders). Air pollution is estimated to have contributed to 6.67 million deaths (95% IU: 5.90 to 7.49 million) worldwide in 2019, nearly 12% of the global total [5]. Air pollution is the main environmental risk factor for premature death, with its total impact exceeded only by hypertension (10.8 million, 95% IU: 9.51 to 12.1 million), tobacco use (8.71 million, 95% IU: 8.1 to 9.3 million) and food risks (7.94 million, 95% IU: 6.5 to 9.8 million). Every year, far more people around the world die from exposure to air pollution than from road traffic accidents, an estimated 1.28 million in 2019 [1]. This large disease burden reflects the substantial contribution that long-term exposures to air pollution make to chronic noncommunicable diseases, approximately 80% of the air pollution burden is attributed to noncommunicable diseases. $PM_{2.5}$ and ozone together contribute up to 40% of deaths from COPD, a debilitating lung disease [6]. As shown in Fig. 2, air pollution also contributes to 30% of lower respiratory tract infections and 20% of infant mortality in the first month of life [7]. The air pollution burden of each of these diseases is borne differently in different countries around the world. These variations reflect various social, economic and demographic factors (in addition to exposures) that affect the underlying health and vulnerability of populations to air pollution in specific regions. To reduce the disease burden attributable to each of these diseases and to each air pollutant, each country will need to explore and understand its own data. Some of this data is available to the international scientific community (e.g., exposures and mortality rates/days) on the State

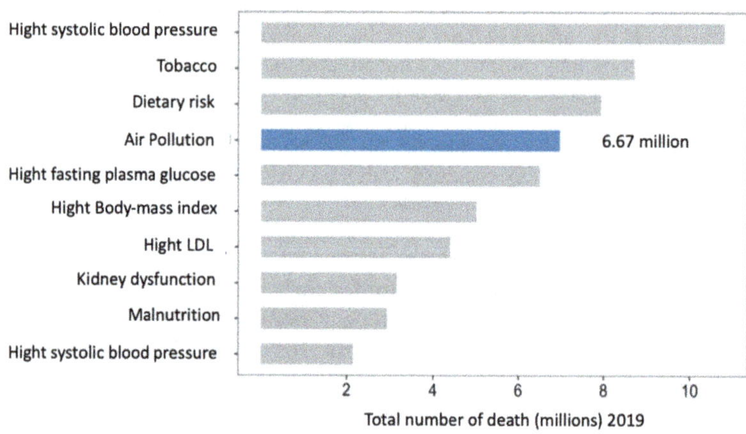

Figure 1: Global ranking of risk factors by total number of deaths from all causes in 2019 [8].

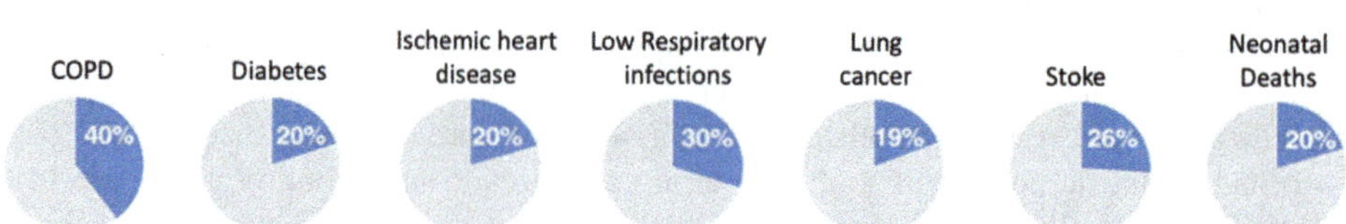

Figure 2: Percentage of global deaths from specific causes attributable to total air pollution [8].

of Global Air website, while other data can be found on the Global Burden of Disease website.

In 2006, the WHO published a report on the average levels of PM_{10} in Italian cities recorded from 2002 to 2004 [9]. The population covered in this report consists of residents of Italian cities with over 200,000 inhabitants for which the environmental data needed for the analysis were available, these cities are Turin, Genoa, Milan, Trieste, Padua, Venice-Mestre, Verona, Bologna, Florence, Rome, Naples, Catania and Palermo. The impact on health of air pollution was very large with 8,220 deaths per year, on average, attributable to concentrations of PM_{10} above 20 mg/m^3. This corresponds to 9% of all-cause mortality (excluding accidents) in the Italian population over 30 years of age. The impact on short-term mortality was calculated with 1,372 deaths, or 1.5% of total mortality in the entire population. Hospital admissions attributable to PM_{10} were of a similar magnitude. The presence of ozone at concentrations above 70 mg/m^3 represents 0.6% of all causes of mortality. The WHO study used an updated and extended dataset and considered 25 adverse health outcomes and various exposure scenarios. Compliance with European Union legislation can lead to substantial savings in terms of diseases avoided, therefore local authorities, through policies that mainly aim at reducing emissions from urban transport and energy production, can contribute to obtaining significant benefits in terms of health. In Italy, the VIIAS Project (Integrated Assessment of the Impact of Air Pollution on the Environment and Health) carried out in the framework of the initiatives of the Disease Control Center (CCM) of the Ministry of Health (www.viias.it), coordinated by the Department of Epidemiology of the SSR of Lazio, carried out the assessment of the health impact of atmospheric pollution by publishing interesting results: in 2005, an estimated 34,552 deaths from natural causes attributable to long-term exposure to $PM_{2.5}$ and 23,387 deaths from exposure to NO_2; in addition 1,707 deaths from diseases affecting the respiratory system are attributable to exposure to ozone. The report underlines the inequalities of health effects on the Italian territory: pollution affects the north and in general urban areas congested by traffic and proximity to industrial areas. Even domestic heating, especially due to the increase in the use of biomass (mainly wood and pellets) is responsible for the worsening of air quality and the consequent impact on health. It was possible to appreciate the health effects of pollution over time. In 2010 there was an important decrease in the effects of $PM_{2.5}$ (21,524 deaths) and NO_2 (11,993 deaths), due to the reduction of emissions and the reduction in consumption attributable to the economic recession. The results of the project indicate that the consumption of biomass for heating as an element of trade-off between climate policies and air quality, produce a health impact worse than that which has already occurred with the decrease in emissions due to the economic recession. It is currently not possible to have a systematic measurement of the effects of the city motor vehicle traffic reduction policies in terms of black carbon (BC) reduction as this pollutant is not measured in real time and in a timely manner by the entire monitoring network of Arpa Piemonte in Turin city. The importance of implementing specific multi-sectorial projects on urban scale would be of considerable scientific interest. Integrating the data of the Arpa control units with real-time data collected by low-cost instrumentation and with clinical information relating to lung function (Fig. 3) represents an interesting scientific approach to be implemented at local scale with specific urban projects dedicated to HRA.

2 TRADITIONAL PM INDICATORS FOR HEALTH RISK ASSESSMENT

Atmospheric particulate matter (PM) is the most widely used indicator for estimating the health impact of atmospheric pollution on the human population and is among the pollutants

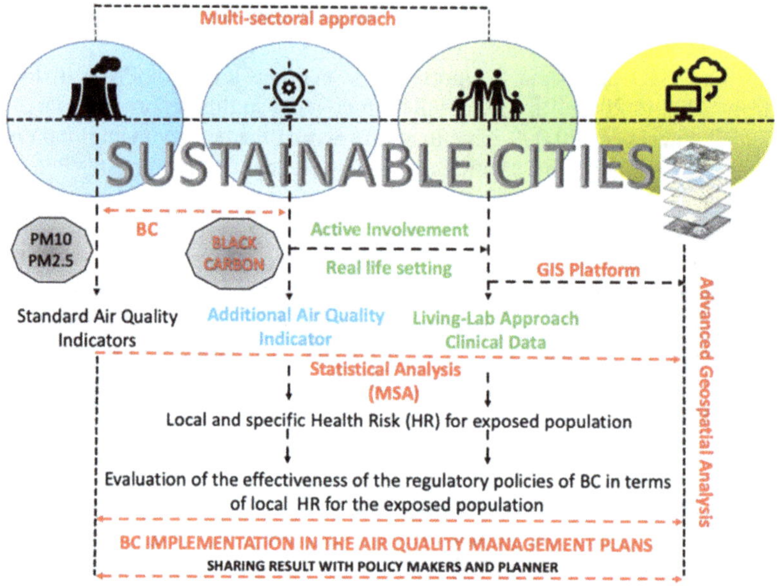

Figure 3: Multidisciplinary approach and inclusion of BC as proxy traffic indicator on an urban scale.

most frequently associated with various health outcomes [10]. According to some authors most risk assessment studies are still based on a two-decade-old system and neglect recent research on inhalation rate and PM size [10]. To overcome this problem, the level of risk must be explored by applying PM monitoring with high spatial and temporal resolution because it is of crucial importance to investigate methods of monitoring particulate inhalation in real time [10]. The commonly identified targets for exposure to atmospheric particulate matter are people who travel on foot, by bike, those who reside in highly trafficked areas or near specific sources. Children, the elderly and individuals already suffering from cardiovascular and respiratory diseases are usually classified as more vulnerable, in fact they present worsening and exacerbations of health conditions due to breathing and inhalation of particulate matter [11]. In particular, babies and children are more prone to respiratory problems such as cough, phlegm and bronchial asthma attacks [12]. Furthermore, children are the most vulnerable subjects to exposure to environmental dust because they are more likely to ingest large quantities of dust due to the habit of putting their hands in their mouths [13]. The PM, in addition to being inhaled directly by the person, can be deposited on the body and hands and it is therefore introduced directly into the body even following this path [14]. Furthermore, with regard to inhalation, it must be said that the respiratory rate in children (3–12 years) is double that of an adult, so this physiological parameter should be integrated in exposure studies [15]. Furthermore, people who tend to do more sports and be outdoors will also have higher rates of exposure to air pollution and the increases in active transport (bike/foot) leads to increased inhaled dose of air pollution [16]. Studies have shown a relationship between chronic exposure to particulate matter and increased risk of premature birth and low birth weight of newborns [8]. Older people, on the other hand, may already have undiagnosed heart/lung disease. The composition of PM is very complex because it may contain various toxic substances that have serious and different effects on human health.

Studies on the composition of atmospheric particulate matter are still lacking which should be encouraged especially in urban areas most affected by air pollution. In the specific case of Turin (north Italy), the fraction of carbonaceous pollutants of primary nature (IRIDE project, 2012, www.istituto-iride.com) constitutes (on average) a fraction of atmospheric particulate $PM_{2.5}$ between 9% and 16% (with a primary coefficient (OC/EC) of 0.67 and 1.8 respectively). Also including the inorganic primary components, the total primary fraction emitted directly as it is into the atmosphere and for which it is possible to act locally for remediation purposes, can be estimated on average between 33% and 40% of $PM_{2.5}$. This percentage refers to all the contributions of the different sources present in the specific study area of Turin (IRIDE Project). It is understandable the need for pervasive and cross-linked monitoring over the entire urban area for an assessment of the individual contributions of the most significant types of sources (traffic, heating and point emissions). The current health risk assessment from air pollution does not have the efficiency necessary to reduce the level of risk of exposure to particulate matter in European cities, precisely because the pervasive high resolution space–time monitoring is still missing and is more and more important to incentive pervasive and multidisciplinary projects. From the international studies [1] it can be seen that the research area relating to ambient air is more developed than those of other areas (occupational exposures) due to the sampling and the size of the PM studied. (health risk assessment of air pollution/general principles, www.euro.who.int).

3 BLACK CARBON AT LOCAL SCALE AS PROXY TRAFFIC INDICATOR

Current air quality standards for particulate matter (PM) use the mass concentration of PM (PM with aerodynamic diameters $\leq$ 10 μm (PM_{10}) or $\leq$ 2.5 μm ($PM_{2.5}$)) as the metric. It has been suggested that particles from combustion sources are more relevant to human health than particles from other sources, but the impact of policies to reduce PM from combustion processes is usually relatively small when estimating the effects. for a reduction in the total mass concentration [17]. Different particle sizes, composition, or characteristics can be related to specific emission sources better than other air pollutants and therefore may be considered a (more) suitable indicator. Thus, PM_{10} may be an appropriate indicator when considering the impact of resuspension of road dust, while black carbon (BC) is a more sensitive indicator for exhaust emissions from road traffic [17]. It is significant highlights the value of black carbon particles (BC) as an additional indicator in the management of air quality in a city heavily affected by smog such as Turin (northern Italy), in fact BC could be a valuable additional indicator of air quality to assess specific health risks of air quality dominated by primary combustion particles. Evidence on the health effects of black smoke (or British BS smoke) was identified decades ago and this pollution was used to recommend early guidelines for exposure limits in relation to public health protection. In the 1990s, BS was one of the most used air quality indicators in European time series studies linking mortality to pollution. The difficulty of standardizing measurements and appreciation of the health effects of non-black components of particulate matter (PM) have drawn researchers' attention to the mass concentration of inhalable or respirable suspended PM fractions such as PM_{10} and $PM_{2.5}$ (WHO Regional Office for Europe, 2000, www.whqlibdoc.who.it), therefore the BS (not regulated by air quality legislation) and the intensity of its monitoring have decreased over the years. Black carbon (BC) is a chemical component of fine particulate matter (PM $\leq$ 2.5 μm), is made up of pure carbon in several related forms and is formed through the incomplete combustion of fossil fuels, biofuels and biomass. In urban areas, road traffic is one of the main sources of PM [8] and it is likely that not all components of PM are equally important in causing health effects [1], in fact the attention of scientific community on the qualitative and not only quantitative aspect of PM, is always higher. Combustion particles also arise from sources from wood and coal combustion, shipping and industrial sources, and these sources can significantly contribute to the concentrations of combustion particles in the urban environment [1]. In a systematic review of the literature, Krzyzanowski et al. [17] concluded that transport-related air pollution contributes to an increased risk of death, particularly from cardiopulmonary causes, respiratory symptoms, and non-allergy-related diseases. In a more recent review of traffic-related air pollution, HEI [8] concluded that there was sufficient evidence to support a causal relationship between exposure to traffic-related air pollution and exacerbation of asthma and suggestive evidence a causal relationship with onset of childhood asthma, non-asthmatic respiratory symptoms, impaired lung function, total and cardiovascular mortality, and cardiovascular morbidity. Many authors agree that BC is a major component [9] in both anthropogenic and natural soot and causes human morbidity and premature mortality [18]. At the state of the art all the aspects concerning the BC are not yet well defined (definition, measurement methodologies and estimation of the effects) (Roberta Vecchi, www.air.unimi.it). There is growing concern in the scientific community that current mass-based PM standards are not suitable for characterizing the real health risks caused by air pollution (such as near areas with high motorized traffic on major roads or in communities dominated by wood smoke). Furthermore, emission reduction measures such as the use of particle traps or the introduction of environmental zones are believed to be effective in reducing exposure to traffic-related air pollution, but the estimated impact of such measures is relatively small when expressed in relation to a reduction in the mass concentration of PM [16]–[18]. Nitrogen dioxide (NO_2) is a regulated component of air pollution that is also used as an indicator of traffic-related air pollution in health impact assessment and air quality management. NO_2 is not a suitable indicator for assessing the effect of traffic abatement measures on exposure to combustion particles because some abatement measures, such as filters on diesel-powered vehicles, can increase NO_2 levels [7]. Furthermore, spatial gradients near roads are less pronounced for NO_2 than for black smoke (BS) and particle number due to the high background concentrations of NO_2 [17]. This is less of a concern for nitric oxide (NO) and nitrogen oxides (NO_x), which is the sum of NO and NO_2, but these components are currently unregulated and do not appear to be toxicologically important at current levels. Obviously not all the health effects associated with the mass of PM can be attributed to a marker of combustion particles for this reason it is possible to consider as a good indicator the measurements of the black carbon particles carried out at urban scale and with high spatio-temporal resolution (ion black carbon (BC), the absorption coefficient (Abs), elementary carbon (EC) and organic carbon (OC), particle number concentration, the surface area of the particles (which is of great interest on alveolar-capillary diffusion)) and specific PM components. The extent of the data available to support the health relevance of these measures varies widely. Probably the cost and complexity of monitoring to estimate personal exposure to ultrafine particles, which is characterized by particle number concentrations rather than particle mass, limits the feasibility of particle number as an additional metric. In 2003, a WHO working group recommended re-evaluating BS as an indicator of traffic-related air pollution, *but there is still no systematic comparison using PM and BC indicators to estimate health effects.* Grahame et al. examined the evidence for the effects of BC on cardiovascular health endpoints and concluded that it may be desirable to enact a BC-$PM_{2.5}$ standard [19]. Smith et al. [20] noted that although the results of their time series meta-analysis suggest greater effects per unit mass of sulfate than BC personal exposure, this distinction was less clear in the few studies that directly compared the estimated effects of both indicators. This observation points to the need to critically compare studies that measured PM mass as well as BC. BC could be a useful indicator in addition to particle mass if the health risks associated with BC are quantitatively or qualitatively different from those associated with PM mass on a unit mass basis. Many epidemiological studies base estimates of health effects on BC measurements from all sources of combustion, not exclusively from traffic, instead it would be interesting to focus on specific sources. In this basic case study, it was observed the BC concentrations from July 2018 to August 2021 in a residential area to investigate the trends of BC concentrations in relation to PM_{10} and $PM_{2.5}$ monitored in the urban station of Lingotto (Torino city).

4 CONCLUSION

Black carbon represents an important indicator in residential sites exposed to vehicular traffic but is not sufficiently monitored. In Torino city, three years of data were collected in an urban station (Fig. 4) located in Lingotto area. From the basic statistical analysis the trend of BC was observed as a more sensitive indicator to the increase in vehicular traffic, compared to PM_{10} and $PM_{2.5}$.

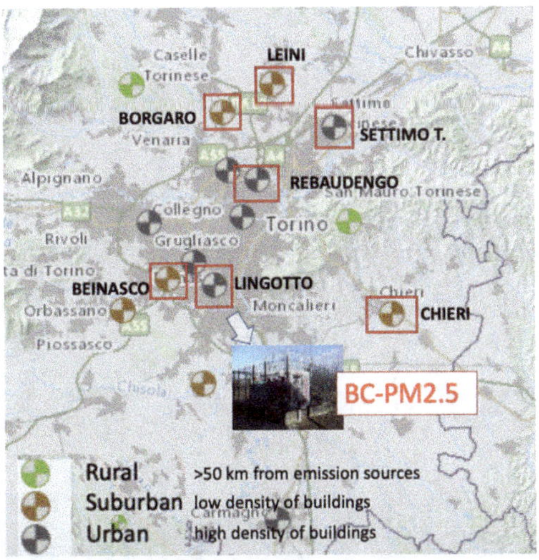

Figure 4: Location of monitoring stations for the measurement of PM and BC, managed by Arpa Piemonte (Torino, Italy).

The concentrations of PM_{10}, $PM_{2.5}$ and BC were detected by an aethalometer installed in a fixed control unit managed by the Piedmont ARPA and located in a densely populated area. Concentration trends, maximum, minimum and average, were observed for the duration of three years at a site representative of the average residential/pedestrian exposure for the city of Turin. BC measurements started in July 2018, therefore there are not complete databases on BC concentrations in Turin. The only station present in the city is Lingotto control unit and it provides hourly data that can be directly correlated to vehicle traffic data. Fig. 5 shows the average concentrations of $PM_{2.5}$ and PM_{10} detected from 2018 to 2020. The graphs show the highest concentration of polluting particulate matter during the winter period, corresponding to the focus of domestic heating. As regards PM_{10}, winter concentrations from December to February 2018 reached maximum values between 80 and 100 $\mu g/m^3$, between December and February 2019 maximum values are between 80 and 90 $\mu g/m^3$. In 2020, on the other hand, during the lockdown caused by the COVID-19 epidemic, maximum concentrations result between 60 and 100 µg/m^3. The maximum concentration of $PM_{2.5}$ recorded in Lingotto was higher during winter 2020, compared to winter 2019–2018; on the contrary, during the spring–summer season 2020 the maximum concentrations of $PM_{2.5}$ were lower than those recorded in 2018 and 2019. In fact, from April to August 2020 the maximum concentrations of $PM_{2.5}$ remained below 20 µg/m^3; perhaps the explanation is to be attributed to the lockdown and reduced vehicular mobility in the city.

Figure 5: Monthly concentrations of PM_{10}, $PM_{2.5}$, maximum, minimum and average values for the years 2018, 2019, 2020.

Fig. 6 shows the BC concentrations from July 2018 to 2020, the different reference scale can be observed for the different pollutants: BC constitutes a small fraction of $PM_{2.5}$ while representing a significant tracer of toxicity. The highest concentrations of BC result recorded at the winter months. During the 2020 lockdown, higher concentrations of BC were recorded during the winter season, probably due to the increased use of home heating, as citizens spent more time indoors. Fig. 7 shows a comparison with the dispersive meteorological parameters and the concentrations of PM_{10}, $PM_{2.5}$ and BC in 2020. The trend of the concentrations of fine dust is inversely proportional to the parameters of meteorological dispersion, specifically in Fig. 7 it’s possible to observe the relationship between dust and rainfall for the year 2020.

The most dangerous component of $PM_{2.5}$, black carbon, follows the trends of the other two particulate matrices and maximum values are observed in November (low rainfall) and January (average monthly rainfall: 100 mm).

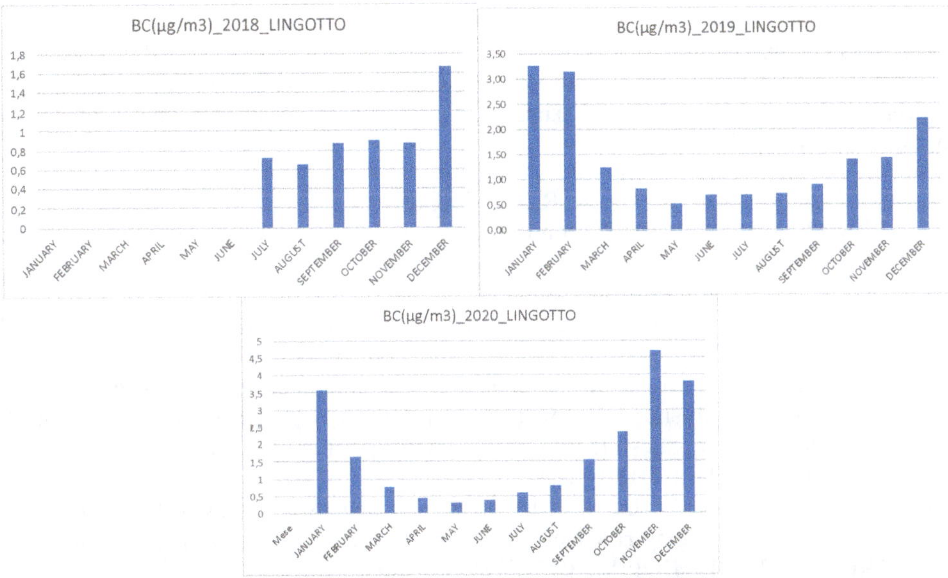

Figure 6: Average concentrations of BC from 2018 to 2020 monitored in Torino Lingotto.

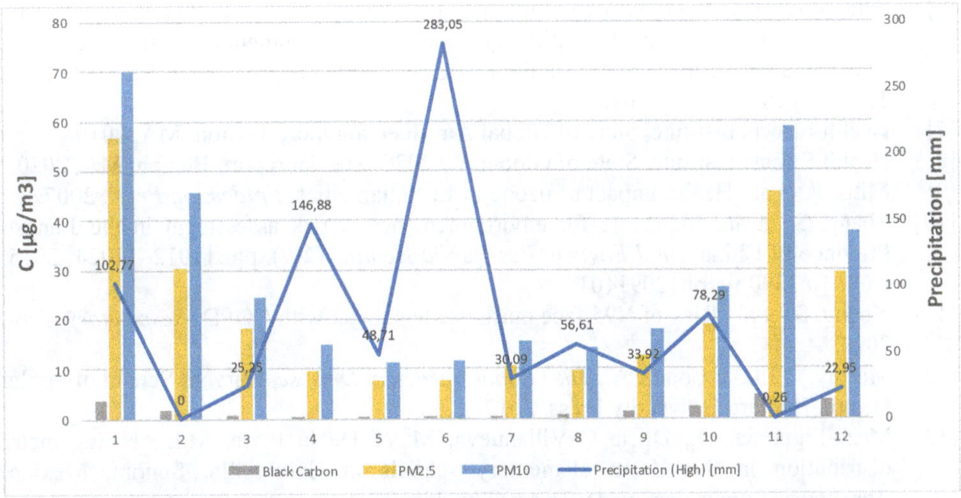

Figure 7: Relations between PM_{10}, $PM_{2.5}$, BC and the rainfall recorded in the year 2020 in Torino Lingotto.

It would be interesting to deepen the study of BC exposure precisely in the winter semester, in relation to the meteorological variables in the Turin area, but since no measurements are available, it is impossible to study the continuous trend of BC in the various points of the city. Hence the need to promote targeted projects on a local scale that can encourage low cost monitoring of the urban fabric starting with local experiments in specific urban sites (highly trafficked and green areas) in order to better understand traffic regulation policies and real exposure to BC possibly not based only on models and projections but on real-time data.

Promoting the use of BC as an additional indicator of the state of air quality would allow to deepen the knowledge on the real exposure of citizens to this marker still not regulated by air quality control policies.

REFERENCES

[1] WHO Report 2016, Air pollution is a mayor threat to health. *The Lancet Respiratory Medicine*, **4**(5), p. 351, 2016.

[2] Cohen, A.J. et al., Estimates and 25-year trends of the global burden of disease attributable to ambient air pollution: An analysis of data from the Global Burden of Diseases Study 2015, **389**(10082), 13–19 May, pp. 1907–1918, 2017.

[3] Schraufnagel, D.E., et al., Air pollution and noncommunicable diseases: A review by the Forum of International Respiratory Societies' Environmental Committee, Part 2: Air pollution and organ systems. *Chest*, **155**(2), pp. 417–426, 2019. DOI: 10.1016/j.chest.2018.10.041.

[4] Rao, X., Montresor-Lopez, J., Puett, R., Rajagopalan, S. & Brook, R.D., Ambient air pollution: An emerging risk factor for diabetes mellitus. *Current Diabetes Reports*, **15**(6), p. 603, 2015. DOI: 10.1007/s11892-015-0603-8.

[5] Boogaard, H., Walker, K. & Cohen, A.J., Air pollution: The emergence of a major global health risk factor. *Int Health*, **11**(6), pp. 417–421, 2019. DOI: 10.1093/inthealth/ihz078. PMID: 31613318.

[6] Schraufnagel, D.E. et al., Air pollution and noncommunicable diseases: A review by the Forum of International Respiratory Societies' Environmental Committee, Part 1: The damaging effects of air pollution. *Chest*, **155**(2), pp. 409–416, 2019. DOI: 10.1016/j.chest.2018.10.042.

[7] Health Effects Institute, State of Global Air. Special report, Boston, MA, 2019.

[8] Health Effects Institute, State of Global Air 2020. Special report. Boston, MA, 2020.

[9] Mitis, F. et al., Health impact of ozone in 13 Italian cities. *Epidemiol. Prev.*, 2007.

[10] Zhang, S. et al., Indicators for environment health risk assessment in the Jiangsu Province of China. *Int J Environ Res Public Health*, **12**(9), pp. 11012–11024, 2015. DOI: 10.3390/ijerph120911012.

[11] Young, S. et al., Risk of N95 face mask use in subject with COPD. *Respiratory Care*, 2020.

[12] Murray, C.J.L. & Lopez, A., *The Global Burden of Disease,* Harvard School of Public Health, Harvard University Press, 2013.

[13] Meza-Figueroa, D., De la O-Villanueva, M. & De la Parra, M.L., Heavy metal distribution in dust from elementary schools in Hermosillo, Sonora, México. *Atmospheric Environment*, **41**(2), pp. 276–288, 2015.

[14] Ferreira-Baptista, L. & De Miguel, E., Geochemistry and risk assessment of street dust in Luanda, Angola: A tropical urban environment. *Atmospheric Environment*, **39**(25), pp. 4501–4512, 2005.

[15] Zhou, H. et al., Ambient air pollution and daily hospital admissions for respiratory disease in children in Guiyang, China. *Front Pediatr.*, **7**, p. 400, 2019. DOI: 10.3389/fped.2019.00400.

[16] Tainio, M. et al., Air pollution, physical activity and health: A mapping review of the evidence. *Environment International*, **147**, 105954, 2021.
[17] Krzyzanowski, M. et al., Air pollution in the mega-cities. *Curr. Envir. Health Rpt.*, **1**, pp. 185–191, 2014. DOI: 10.1007/s40572-014-0019-7.
[18] Wang, X., Chen, R., Meng, X., Geng, F., Wang, C. & Kan, H., Associations between fine particle, coarse particle, black carbon and hospital visits in a Chinese city. *Science of The Total Environment*, **458–460**, pp. 1–6, 2013.
[19] Grahame, T.J., Klemm, R. & Schlesinger, R.B., Public health and components of particulate matter: The changing assessment of black carbon. *Journal of the Air and Waste Management Association*, **64**(6), pp. 620–660, 2014. DOI: 10.1080/10962247.2014.912692.
[20] Smith, K.R. et al., Public health benefits of strategies to reduce greenhouse-gas emissions: Health implications of short-lived greenhouse pollutants. *The Lancet*, **374**(9707), pp. 2091–2103, 2009.

SECTION 5
MONITORING, MEASURING AND DATA

INDOOR RADON ASSESSMENT IN SCHOOL BUILDINGS: WHERE, WHEN AND HOW TO MEASURE?

PAULO BARROS[1*], ANTÓNIO CURADO[1,2†] & SÉRGIO I. LOPES[3,4‡]
[1]proMetheus, Instituto Politécnico de Viana do Castelo, Portugal
[2]CONSTRUCT – LFC, Faculty of Engineering (FEUP), University of Porto, Portugal
[3]ADiT – Instituto Politécnico de Viana do Castelo, Portugal
[4]IT – Instituto de Telecomunicações, Campus Universitário de Santiago, Aveiro, Portugal

ABSTRACT

A pre-selection procedure for sample choice in an extensive experimental campaign for indoor radon assessment is mandatory to ensure the homogeneity of the case study to be monitored. The implementation of a simple methodology for room selection guarantees more homogenized measurement results for later comparison and analysis. Hence, the case study for in-situ instrumentation must include a representative and homogeneous sample of the whole, with a particular focus on rooms occupied daily during the working day, namely offices and cabinets for staff. All buildings under monitoring hold a substantial number of rooms that should be assessed for radon potential evaluation, however, due to the reduced time frame for in-situ instrumentation and the shortage of radon detectors, only a representative sample can be picked for monitoring. Given the constraints, a straightforward methodology for sample selection based on the evaluation of the most relevant criteria that influence radon risk exposure – occupancy, site geology, room type, room size and floor location, air renovation, building materials, etc. – is a key process to ensure the quality of the experimental campaign. An experimental campaign for indoor radon assessment will be carried out during Spring of 2021 for evaluating the impact on indoor air quality (IAQ) during school reopening, after a long lockdown period due to COVID-19, which lasted from 15 January 2021 to 19 April 2021. The higher education institution to be evaluated includes six distinct schools spread in several geographically dispersed locations in the north region of Portugal and will be monitored to assess radon concentration, atmospheric pressure, and other hygrothermal parameters. In the end, the result will rely on a matrix that prioritizes the most relevant parameters on radon potential analysis by applying a three-level scale for scoring: low, moderate, and high. Each of these levels will be associated with a score scale allowing the design of a multicriteria analysis for the selected criteria, thus building a methodology for sample selection in complex scenarios for in-situ monitoring.

Keywords: IAQ, radon, assessment, sample selection, multicriteria analysis matrix.

1 INTRODUCTION

School buildings across Europe often present insufficient ventilation that is associated with high levels of chemical pollutants, mould problems and dampness, and Portugal is no exception [1]. The lack of air renovation in school buildings is linked with respiratory problems, infectious diseases, and reduced learning outcomes [2]. To tackle the problem, the Portuguese regulation for IAQ in school buildings demands a simplified annual assessment [3].

Concerning indoor radon, the regulation established a limit concentration of 300 Bq/m^3, without indicating the occupation period to be considered for measurement (full time or work time). Radon gas is a radioactive element of natural origin, colourless, odourless, and tasteless [4]. It results from the decay of uranium naturally present in the mineral grains of

* *ORCID: https://orcid.org/0000-0003-4287-5078*
† *ORCID: https://orcid.org/0000-0002-5828-6086*
‡ *ORCID: https://orcid.org/0000-0001-6944-7757*

WIT Transactions on Ecology and the Environment, Vol 252, © 2021 WIT Press
www.witpress.com, ISSN 1743-3541 (on-line)
doi:10.2495/AIR210151

soils and granitic rocks and some construction materials and tends to accumulate inside buildings since it is heavier than air [4]. Very often, indoor radon levels tend to increase with buildings' energy efficiency reinforcement. The gas arising from subsoil enters the building through the ground and is captured inside of airtight constructions, most of them built or refurbished with well-insulated walls, roofs, and floors, and equipped with double-glazed windows, contributing to low airflow rates and aggravating IAQ [5]. Furthermore, indoor radon concentration varies widely throughout the day and the year, tending to reach higher values during the night period, for a daily cycle, and during the Winter season, for an annual cycle [2]. Indoor radon detection is separated into short-term and long-term assessments by using, more recently, advanced technology supported in digital solutions that have made radon measurements more convenient and accurate.

Due to seasonal fluctuations, long-term measurements for periods longer than 3 months are more suited for detailed characterization. Long-term radon detectors are based on alpha-track detection (AT) method by using containers that enclose small sheets of special plastic material to detect radiation [6]. In short, the alpha particles emitted by the radon gas will leave tracks on these plastic sheets for later evaluation, which may take anything from 3 months up to a year. Digital solutions, known as continuous radon monitors (CRM), are more convenient for measuring radon concentration by digitizing radon detection. This solution combines the advantages of old methods and brings flexibility, consistency, long lifetime, high accuracy, quick response time, continuous readings, and long-term and short-term capabilities. Short-term radon detection is suitable when a straightforward diagnostic is needed, and measurements will take between 2 days to 3 months. These types use charcoal-based methods to absorb the radon gas which is later resealed and returned to a lab for evaluation. Short-term measurements can roughly estimate the prevailing indoor radon concentration in a specific room under monitoring and are usually adopted to provide a crude indication of radon level for further analysis at a later stage [6].

In-situ measurements are essential to assess indoor radon concentration and are based on standardized protocols to ensure accurate and consistent results [6]. Although there are many published works focused on how, when, and where to assess [5], [7], a simple and straightforward methodology for sample selection based on a set of criteria with distinct weights that depend on their relevance – that can be integrated into clear and objective decision-making process, is, to the best of our knowledge, yet to be implemented. In short, the main goal of this study is to outline a methodology based on short-term measurements that aims to identify the most critical rooms to be instrumented in an extensive assessment campaign that needs further validation based on the instrumentation results [8], [9].

The structure of this paper is as follows. In Section 2, the materials and methods are described, presenting the proposed methodology for sample selection. Section 3 presents the results, discusses the advantages and disadvantages of sample characterization, and presents a multicriteria analysis matrix, based on the selected parameters. Finally, the conclusions and future work of this article are addressed in Section 4.

2 MATERIALS AND METHODS

Indoor radon concentration is prominent in the north region of Portugal, due to the soil lithology and to the use of granite blocks as a raw material in the construction industry [10]. Therefore, issues related to bad IAQ are particularly acute in this region, which is evidenced by a study carried out by Veloso et al. in 2011 that estimates radon to be associated with between 18% to 28% of lung cancer cases in the north region of Portugal, causing an average of around 215 deaths per year [11].

In scenarios of considerable diversity in the building stock for in-situ monitoring, a pre-selection procedure for sample choice is mandatory to ensure a representative and homogeneous sample selection process. Besides the geographic spread and the type of construction, other differences that influence in-situ monitoring are particularly relevant:

- Lithological characterization and soil type.
- Room occupancy (high occupancy rate versus vacant rooms).
- Type of ventilation (natural ventilation versus mechanical ventilation).

The case study for in-situ measurements must include a representative and homogeneous sample of the whole, so that experimental results can be used for data analysis and comparison, with a particular focus on rooms occupied daily during working time, namely administrative offices, and cabinets for staff [12]. For reasons related to increased exposure to radon gas, priority will be given to ground floor compartments. However, to check how indoor radon concentration is reduced with the ground clearance of the instrumented room, the analysis over a building vertical alignment will be implemented. This specific analysis allows checking how radon level is dissipating as it moves from one building floor to another building floor.

All buildings under monitoring hold many rooms that need to be assessed, however, due to the reduced time frame for instrumentation and the shortage of available radon detectors, only a few rooms will make part of the case study to be monitored. Given the constraints, an adequate procedure for sample selection can be a very effective operational tool, to allow the triage of the most determinant rooms to be instrumented based on a previous criteria analysis. A stage related to sample selection is a key process part of a straightforward methodology, whose first step is the evaluation of the most relevant criteria that influence indoor radon potential (Fig. 1), so that the quality of the experimental campaign can be assured. Fig. 1 depicts the methodology adopted for in-situ measurements implementation based on a set of five sequential steps that comprehend a range of tasks to be undertaken for each stage.

Sample selection is the first step to be carried out by adopting a multicriteria analysis for radon potential assessment. For this, several influencing criteria concerning indoor radon potential are plotted on a matrix linking them to a three-level scale: low, moderate, and high. Each risk level will be associated with a score allowing to design a grade scale used to rank the most relevant criteria to be selected. The model will be used to assess radon potential and applied to all buildings that make part of the case study. This stage implies a previous analysis to choose the most relevant criteria for sample selection. Criteria related to occupancy schedules, site geology and lithological characterization based on geological maps, room type, room size or floor location, air renovation systems and ventilation, prevalent building materials, etc. are of utmost importance to evaluate indoor radon potential for a further stage experimental campaign.

After sample selection, the second stage to be implemented will be the so-called preliminary works. In this stage, preparatory activities for continuous in-situ measurements will be carried out by using a digital CRM instrument with an incorporated data logger, to register indoor air temperature, relative humidity, atmospheric pressure, and indoor radon concentration. To implement the experimental campaign, a certified digital CRM instrument with data logging capability must be selected for accurate and reliable average readings. Although long-term measurements are the most accurate procedure to assess potential radon health hazards, in this work short-term assessment will be used by applying a 1 week measurement to detect evidence of harmful radon levels. These 7 days represent a

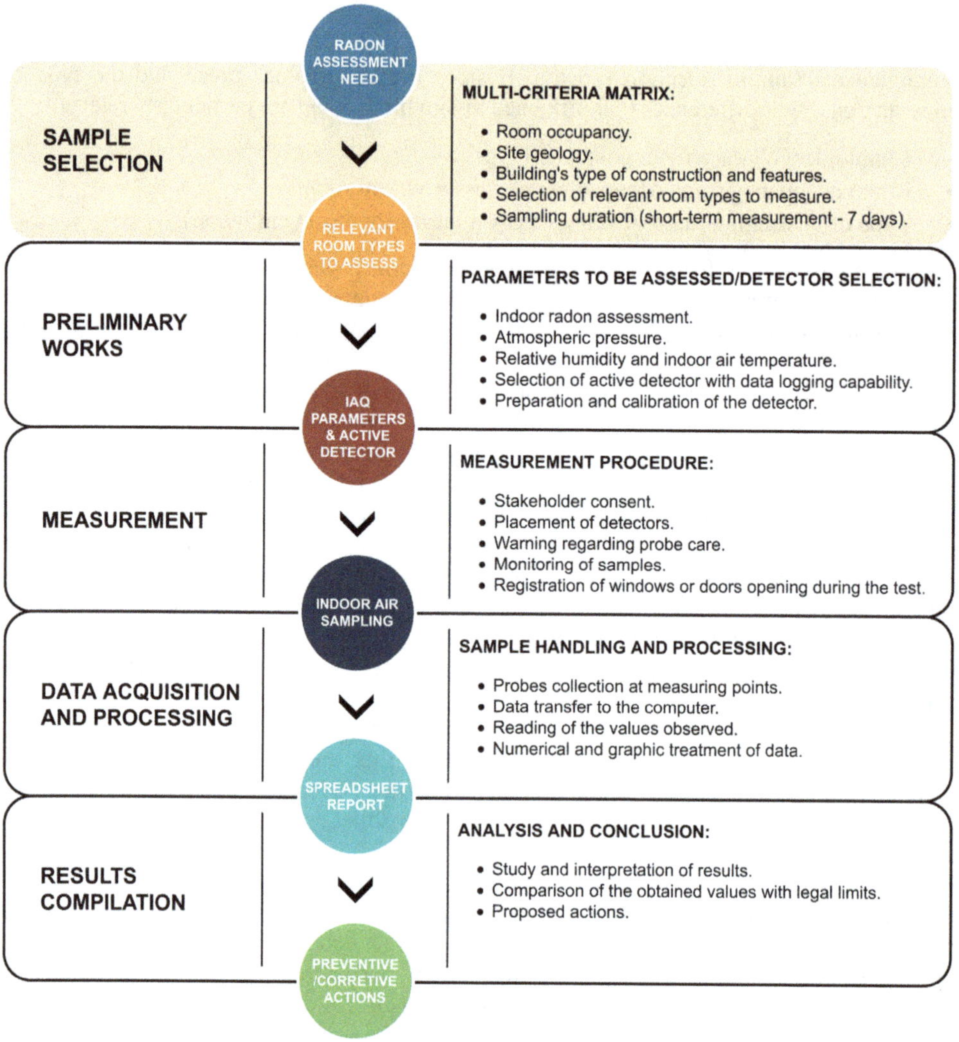

Figure 1: Proposed methodology for in-situ measurements.

cycle that allows us to see the radon concentration pattern in the period of occupation and out of occupation (labour days plus weekend). If by any means the detected levels are higher than initially expected, it will be necessary to reconsider the approach to long-term measurements.

The third stage of the proposed methodology relates to the measurement stage. As described in Fig. 1, this study must be carried out with the informed consent of all staff involved. All of them were previously informed of the precautions to be taken during the sampling, namely, to not tamper or disconnect the CRM instrument to not compromise the results. During this stage, the correct CRM instrument placement is very important to get reliable measurements, since placement mistakes could lead to false-positive readings, like for instance placing a CRM instrument too close to a known radon source, or in a very small room of an area in an unoccupied basement. As buildings have multiple areas and floors, CRM instruments should be placed on the lowest occupied level of the building, and then go from there to other occupied floors [13]. The most important recommendations to follow in the CRM instrument placement are as follows [14]: it should not be exposed to direct sunlight or electromagnetic radiation; the probe should be on a shelf, or table positioned lying flat, against an interior wall that is free from ventilation sources like vents or air returns; it must be kept at least 150 cm away from all windows, doorways, ventilation devices, and exterior walls, to avoid fresh air: it should be at least 50 cm above the floor and below the ceiling and 25 cm away from other objects. Radon probes cannot be moved or touched during the measurement procedure.

The fourth stage concerns data acquisition and processing. In-situ assessment will take place under normal conditions of room "operation", i.e., people will use the room like they normally do during daily work time, within the regular conditions of use of the spaces, so opening windows or going in and out of the compartments is interpreted as normal usage behaviour and will be not valued. The statistical treatment of the measured parameters will be inferential statistics, and the parameters will be interpreted with caution and correlated with other relevant criteria. This will allow us to draw conclusions and make statements that will serve as a reference in similar future procedures of radon risk potential assessment.

The fifth stage is related to the results' presentation. After the testing period is complete, results are examined, and conclusions are extracted. Two possible actions can be then implemented: a first action regarding the decision on the need for mitigation must be undertaken by analysing short-term testing results. It will be possible to decide if long-term testing is advisable. The results will allow implementing a set of preventive or corrective actions, if necessary, and to decide how quickly to mitigate; a second action concerning retesting after implementing corrective actions. It is planned to retest all rooms yearly after renovation. In the future it will be important to define strategies for reviewing and maintaining the measures implemented based on the retesting results.

To implement the proposed methodology, a set of six schools spread through the north region of Portugal will be assessed to monitor rooms' IAQ, mainly related to radon concentration. Education institutions play a decisive responsibility in awakening the awareness regarding issues towards IAQ and a healthy indoor environment to create a safe and sustainable campus [15], [16]. The rooms to be monitored are occupied by staff members, administrative employees, grantees, researchers, and teaching staff, i.e., all enclosed space under in-situ monitoring share a permanent occupancy during working hours. The school buildings that make part of the case study to be instrumented are geographically dispersed throughout a large region, and differences in schools' characteristics are substantial, both in terms of building period, as in terms of building process, and in types of materials used for construction and finishing: for instance, the oldest school to be monitored is a public interest monument recently retrofitted, built with thick granite walls, and converted into a school building [17]. On the other hand, the most recent school building is a recent high-tech building equipped with several automation mechanisms and air conditioning centralized systems.

3 RESULTS AND DISCUSSION

Nowadays concerns with IAQ are directly linked to the time spent inside buildings, since today's society spends 80% to 90% of time indoors in homes, at workplaces or schools, in shopping and leisure, etc., where IAQ is two to five times worse than outdoor air [18]. Most of the contributions to a poor IAQ arise from construction materials, cladding and coating, furniture, cleaning products, and poor ventilation and air renewal, just to name a few [19].

Concerning radon, as is already well known, most of the contribution comes from the soil, and when radon enters an enclosed room, its accumulation can lead to high concentrations, representing a possible health harm that must be assessed by using a well-defined methodology [20].

Given the need to optimize resource usage, save time and to reduce costs, the in-situ assessment must include a representative sample of enclosed spaces in each of the 6 monitored schools, with a particular focus on rooms with long-term daily occupancy, namely administrative offices, professors' rooms, and staff cabinets. This strategy intentionally leaves classrooms out of assessment, given the fact that presential classes never last more than 2 hours for a particular group or class, then switching to another distinct group or class. Therefore, occupants' radon exposure risk is considerably lower than the risk in an administrative office, or a staff cabinet, where the occupation period is always considerably larger for the same people who work there.

After a thorough analysis of all schools that make part of the case study including buildings analysis for rooms' selection, based on the experience acquired in previous experimental campaigns conducted in the same geographic region, three main criteria were attended to build up the methodology for radon potential assessment:

- Room occupancy (occupancy schedule, number of occupants and room location).
- Site geology and lithological characterization based on official geological maps (soil type and layers structure).
- Building's type of construction and features (construction age, ventilation regime, heating and air conditioning systems installed).

Complementary to this criteria's choice, some main assumptions were taken previously to carry out the sample selection:

- Rooms for instrumentation are selected among enclosures occupied by school professors and staff over a period that can reach 35 hours per week, the total number of working hours for civil servants.
- Selected rooms must be representative of different working environments – administrative rooms or offices, professor, and staff cabinets, etc.
- Ground floor rooms are prioritized for indoor radon assessment.
- Measurements across a vertical alignment will be undertaken to assess radon variation in height throughout the building.

Table 1 shows among all occupied rooms, the selected rooms that will be used for in-situ monitoring. Tested rooms will be about 20% of all total rooms, involving the same percentage of employees, on average.

Table 1: Number of relevant rooms and staff by school (#1 to #6).

Rooms and staff	Schools					
	#1	#2	#3	#4	#5	#6
All rooms\Tested rooms	41/9	34/18	76/16	34/9	33/9	34/9
All staff\Tested staff	70/9	62/18	200/16	82/9	45/9	31/9

The implemented procedure for sample selection involves the analysis of the most influencing criteria to assess radon potential in a scenario of shortage of active radon detectors for in-situ assessment, bringing out the following achievable for future applications:

- Allows selecting a representative sample for in-situ monitoring.
- Privilege rooms with an occupancy schedule for in-situ monitoring.
- Allows a broad evaluation of IAQ by assessing indoor radon concentration, atmospheric pressure, air temperature and relative humidity.
- Serves as a guide to help schools to assess radon potential.

Tables 2 and 3 present results concerning the use of a multicriteria analysis for sample selection. Both matrices relate the selected criteria and sub-criteria (in rows), with the relevant room types for the instrumented schools (in columns). The results derived from the cross-analysis of information will allow the selection of the critical rooms that should be measured in each school, by understanding whether a room located in a particular floor shows a higher radon potential when compared to others with a distinct location.

The quantification of the parameter "radon assessment need" for each criterion is divided into four levels within a score scale from 0 to 3, as follows: non-applicable (0 points), low (1 point), moderate (2 points), and high (3 points). Each score is attributed based on the sub-criteria analysis and related to the increasing level of radon potential exposure. All sub-criteria that do not have direct application to quantify the assessed parameter should be scored with 0 points or marked with "–" for a straightforward simplification of the results.

After scoring, the checksum for all relevant rooms is added. If the cross-sectional analysis of these criteria returns a checksum equal to or below 22 that indicates that the radon potential is moderate or low, therefore these room types should be ignored. On the other hand, if the value is above 22 it will be necessary to select those compartments for radon assessment according to these steps:

1. Checksum scores above 22 should be placed in descending order.
2. At least four room types with the highest scores should be chosen.
3. The number of available radon detectors should be evenly distributed among the selected room types.

By the analysis of the results presented in Tables 2 and 3, it stands out, for all schools to be assessed, that the adoption of a score value of 22, referred to as the "border score", tends to favour monitored areas like receptions, offices, cabinets, and laboratories in comparison to libraries, canteens, bars, or workshops. This happens because the 6 schools have similar organizational profiles and layouts, except for the oldest school that differs from the other in many architectural characteristics, building materials, devices and equipment installed. Hence, the remaining 5 schools share similar scores on various sub-criteria, in terms of building features, layout, and occupancy patterns. This aspect is also verified regarding site geology, which is the most important main criteria, as schools are all located in the same region.

In overall terms, these results will trigger the following actions:

- Implement in-situ measurements by using a 7-day short-term procedure, starting with all frequently occupied rooms placed on the ground floor, during the same period.
- If the measurement results are higher than 300 Bq/m^3, a new follow-up procedure must be implemented for all monitored rooms. The new measurements should follow the same procedures as the previous ones, mainly regarding the locations, the conditions of the essays. If the new experimental results match with the initial, short-term measurements strategy seems to be enough.

Table 2: Multicriteria analysis matrix.

Main criteria	Sub-criteria	Assessment need (score)	School 1 – Relevant room types								School 2 – Relevant room types								School 3 – Relevant room types							
			Reception	Management board offices	Administrative offices	Professors and researchers cabinets	Laboratories	Library	Canteen/Bar	Workshops	Reception	Management board offices	Administrative offices	Professors and researchers cabinets	Laboratories	Library	Canteen/Bar	Workshops	Reception	Management board offices	Administrative offices	Professors and researchers cabinets	Laboratories	Library	Canteen/Bar	Workshops
Room occupancy	**Occupation profile:**																									
	Work time	Moderate (2)	–	2	2	2	–	2	2	–	–	2	2	2	–	2	2	–	–	2	2	2	–	2	2	–
	Full time (24 hours)	High (3)	3	–	–	–	–	–	–	–	3	–	–	–	–	–	–	–	3	–	–	–	–	–	–	–
	Partial time (morning/afternoon)	Low (1)	1	1	1	1	1	–	–	1	1	1	1	1	1	–	–	–	1	1	1	1	1	–	–	1
	Occupancy profile:																									
	Staff members	High (3)	3	3	3	–	–	3	3	3	3	3	3	–	–	3	3	–	3	3	3	–	–	3	3	3
	Grantee	Low (1)	–	–	–	–	1	–	–	–	–	–	–	–	1	–	–	–	–	–	–	–	1	–	–	–
	Professor/Researcher	High (3)	–	–	–	3	3	–	–	–	–	–	–	3	3	–	–	–	–	–	–	3	3	–	–	–
	Room occupancy:																									
	Low (1)	Low (1)	1	–	–	–	–	–	–	–	1	–	–	–	–	–	–	–	1	–	–	–	–	–	–	–
	Medium (1–5)	Moderate (2)	–	2	2	2	2	2	–	2	–	2	2	2	2	2	2	–	–	2	2	2	2	2	–	2
	High (5+)	High (3)	–	–	–	–	–	–	3	–	–	–	–	–	–	–	–	–	–	–	–	–	–	–	3	–
	Room size (area):																									
	Below 100 m^2	High (3)	3	3	3	3	3	–	–	–	3	3	3	3	3	–	–		–	3	3	3	3	–	–	–
	Between 100–250 m^2	Moderate (2)	–	–	–	–	–	–	–	2	–	–	–	–	–	2	–	–	–	–	–	–	–	–	–	2
	Above 250 m^2	Low (1)	–	–	–	–	–	1	1	–	–	–	–	–	–	–	1	–	1	–	–	–	–	1	1	–
Site geology	**Lithological characterization:**																									
	Igneous	High (3)	–	–	–	–	–	–	–	–	3	3	3	3	3	3	3	–	3	3	3	3	3	3	3	3
	Metamorphic or sedimentary	Low (1)	1	1	1	1	1	1	1	1	–	–	–	–	–	–	–	–	–	–	–	–	–	–	–	–
	Soil type and layer's structure:																									
	Granitic/schist soils	High (3)	3	3	3	3	3	3	3	3	3	3	3	3	3	3	3	–	3	3	3	3	3	3	3	3
	Non granitic/schist soils	Low (1)	–	–	–	–	–	–	–	–	–	–	–	–	–	–	–	–	–	–	–	–	–	–	–	–

Table 2: Continued.

Main criteria	Sub-criteria	Assessment need (score)	School 1 – Relevant room types								School 2 – Relevant room types								School 3 – Relevant room types							
			Reception	Management board offices	Administrative offices	Professors and researchers cabinets	Laboratories	Library	Canteen/Bar	Workshops	Reception	Management board offices	Administrative offices	Professors and researchers cabinets	Laboratories	Library	Canteen/Bar	Workshops	Reception	Management board offices	Administrative offices	Professors and researchers cabinets	Laboratories	Library	Canteen/Bar	Workshops
Building's type of construction and features	**Buildings constructive typology:**																									
	Stone masonry construction	High (3)	–	–	–	–	–	–	–	–	–	–	–	–	–	–	–	–	–	–	–	–	–	–	–	–
	Concrete construction	Moderate (2)	–	–	–	–	–	–	–	–	2	2	2	2	2	2	2	–	2	2	2	2	2	2	2	2
	Prefabrication	Low (1)	1	1	1	1	1	1	1	1	–	–	–	–	–	–	–	–	–	–	–	–	–	–	–	–
	Compartment location:																									
	Basement/Cellar	High (3)	–	–	–	–	–	–	–	3	–	–	–	–	–	–	–	–	–	–	–	–	–	–	–	–
	Ground/Main floor	Moderate (2)	2	–	2	2	2	2	2	–	–	–	–	–	–	–	–	–	2	2	2	2	2	2	2	2
	Any Upper floor	Low (1)	–	1	–	1	1	–	–	–	1	1	1	1	1	1	1	–	–	–	–	1	1	–	–	–
	Prevalent buildings materials:																									
	Granite/schist, solid rock, concrete	High (3)	–	–	–	–	–	–	–	–	–	–	–	–	–	–	–	–	–	–	–	–	–	–	–	–
	Stone coverings	Moderate (2)	–	–	–	–	–	–	–	–	–	–	–	–	–	–	–	–	–	–	–	–	–	–	–	–
	Others	Low (1)	1	1	1	1	1	1	1	1	1	1	1	1	1	1	1	–	1	1	1	1	1	1	1	1
	Type of ventilation:																									
	Natural	High (3)	3	3	3	3	3	–	–	3	3	3	3	3	3	–	–	–	3	3	3	3	3	–	–	–
	Mechanical/Air conditioning	Low (1)	–	–	–	–	–	1	1	–	–	–	–	–	–	1	1	–	–	–	–	–	–	1	1	1
	Heating system:																									
	Yes	Low (1)	1	1	1	1	1	1	1	–	1	1	1	1	1	1	1	–	1	1	1	1	1	1	1	–
	No	Moderate (2)	–	–	–	–	–	–	–	2	–	–	–	–	–	–	–	–	–	–	–	–	–	–	–	2
Checksum (Border score = 22)			23	22	23	24	23	18	19	22	25	25	25	25	24	21	20	–	24	26	26	27	26	21	22	22

Table 3: Multicriteria analysis matrix.

Main criteria	Sub-criteria	Assessment need (score)	School 4 – Relevant room types								School 5 – Relevant room types								School 6 – Relevant room types							
			Reception	Management board offices	Administrative offices	Professors and researchers cabinets	Laboratories	Library	Canteen/Bar	Workshops	Reception	Management board offices	Administrative offices	Professors and researchers cabinets	Laboratories	Library	Canteen/Bar	Workshops	Reception	Management board offices	Administrative offices	Professors and researchers cabinets	Laboratories	Library	Canteen/Bar	Workshops
Room occupancy	**Occupation profile:**																									
	Work time	Moderate (2)	–	2	2	2	–	2	2	–	–	2	2	2	–	2	2	–	–	2	2	2	–	2	2	–
	Full time (24 hours)	High (3)	3	–	–	–	–	–	–	–	3	–	–	–	–	–	–	–	3	–	–	–	–	–	–	–
	Partial time (morning/afternoon)	Low (1)	1	1	1	1	1	–	–	–	1	1	1	1	–	–	–	–	1	1	1	1	1	–	–	1
	Occupancy profile:																									
	Staff members	High (3)	3	3	3	–	–	3	3	3	3	3	3	–	–	3	3	–	3	3	3	–	–	3	3	3
	Grantee	Low (1)	–	–	–	–	1	–	–	–	–	–	–	–	–	–	–	–	–	–	–	–	1	–	–	–
	Professor/Researcher	High (3)	–	–	–	3	3	–	–	–	–	–	–	3	–	–	–	–	–	–	–	3	3	–	–	–
	Room occupancy:																									
	Low (1)	Low (1)	1	–	–	–	–	–	–	–	1	–	–	–	–	–	–	–	1	–	–	–	–	–	–	–
	Medium (1–5)	Moderate (2)	–	2	2	2	2	2	–	2	–	2	2	2	–	2	–	–	–	2	2	2	2	2	–	2
	High (5+)	High (3)	–	–	–	–	–	–	3	–	–	–	–	–	–	–	3	–	–	–	–	–	–	–	3	–
	Room size (area):																									
	Below 100 m^2	High (3)	3	3	3	3	3	–	–	3	3	3	3	3	–	–	–	–	3	3	3	3	3	–	–	3
	Between 100–250 m^2	Moderate (2)	–	–	–	–	–	2	2	–	–	–	–	–	–	–	2	–	–	–	–	–	–	2	2	–
	Above 250 m^2	Low (1)	–	–	–	–	–	–	–	–	–	–	–	–	–	1	–	–	–	–	–	–	–	–	–	–
Site geology	**Lithological characterization:**																									
	Igneous	High (3)	3	3	3	3	3	3	3	3	–	–	–	–	–	–	–	–	3	3	3	3	3	3	3	3
	Metamorphic or sedimentary	Low (1)	–	–	–	–	–	–	–	–	1	1	1	1	–	1	1	–	–	–	–	–	–	–	–	–
	Soil type and layer's structure:																									
	Granitic/schist soils	High (3)	3	3	3	3	3	3	3	3	3	3	3	3	–	3	3	–	3	3	3	3	3	3	3	3
	Non granitic/schist soils	Low (1)	–	–	–	–	–	–	–	–	–	–	–	–	–	–	–	–	–	–	–	–	–	–	–	–

Table 3: Continued.

Main criteria	Sub-criteria	Assessment need (score)	School 4 – Relevant room types								School 5 – Relevant room types								School 6 – Relevant room types							
			Reception	Management board offices	Administrative offices	Professors and researchers cabinets	Laboratories	Library	Canteen/Bar	Workshops	Reception	Management board offices	Administrative offices	Professors and researchers cabinets	Laboratories	Library	Canteen/Bar	Workshops	Reception	Management board offices	Administrative offices	Professors and researchers cabinets	Laboratories	Library	Canteen/Bar	Workshops
Building's type of construction and features	**Buildings constructive typology:**																									
	Stone masonry construction	High (3)	3	3	3	3	3	3	–	3	–	3	3	–	–	–	–	–	–	–	–	–	–	–	–	–
	Concrete construction	Moderate (2)	–	–	–	–	–	–	2	–	2	–	–	2	–	2	2	–	2	2	2	2	2	2	2	2
	Prefabrication	Low (1)	–	–	–	–	–	–	–	–	–	–	–	–	–	–	–	–	–	–	–	–	–	–	–	–
	Compartment location:																									
	Basement/Cellar	High (3)	–	–	–	–	–	–	–	–	–	–	–	–	–	–	–	–	–	–	–	–	3	–	–	3
	Ground/Main floor	Moderate (2)	2	2	2	–	–	–	2	2	2	2	2	–	–	2	2	–	–	–	–	–	–	–	–	–
	Any Upper floor	Low (1)	–	–	–	1	1	1	–	–	–	–	–	1	–	–	–	–	1	1	1	1	–	1	1	–
	Prevalent buildings materials:																									
	Granite/schist, solid rock, concrete	High (3)	3	3	3	3	3	3	3	3	–	3	3	–	–	–	–	–	–	–	–	–	–	–	–	–
	Stone coverings	Moderate (2)	–	–	–	–	–	–	–	–	–	–	–	–	–	–	–	–	–	–	–	–	–	–	–	–
	Others	Low (1)	–	–	–	–	–	–	–	–	1	–	–	1	–	1	1	–	1	1	1	1	1	1	1	1
	Type of ventilation:																									
	Natural	High (3)	3	3	3	3	3	–	–	3	3	3	3	3	–	–	–	–	3	3	3	3	3	–	–	3
	Mechanical/Air conditioning	Low (1)	–	–	–	–	–	1	1	–	–	–	–	–	–	1	1	–	–	–	–	–	–	1	1	–
	Heating system:																									
	Yes	Low (1)	1	1	1	1	1	1	1	–	1	1	1	1	–	1	1	–	1	1	1	1	1	1	1	–
	No	Moderate (2)	–	–	–	–	–	–	–	2	–	–	–	–	–	–	–	–	–	–	–	–	–	–	–	2
Checksum (Border score = 22)			29	29	29	28	27	24	25	27	24	27	27	23	–	19	21	–	25	25	25	25	26	21	22	26

When these actions are done, it is time to analyse the collected data and decide if a follow-up long-term assessment is better to understand the average radon concentrations for the entire school year. If after a long-term assessment the results are still critical then mitigation actions should be considered. When major renovations are implemented in buildings it is essential to retest all previous rooms to check the effectiveness of mitigation measures.

4 CONCLUSIONS AND FUTURE WORK

The implemented methodology for sample selection is a preliminary tool concerning radon potential evaluation, designed to be replicated in similar experimental scenarios towards indoor radon testing, as well as for other IAQ parameters.

Particularly designed for school buildings, the methodology relies on a multicriteria analysis matrix designed for selecting the most relevant spaces to be tested in an indoor radon campaign. The focus is set on rooms with high occupancy, typically occupied by staff in regular working periods where the likelihood of being exposed to indoor radon gas increases. The scenario with extended occupancy is mostly relevant for school staff in a 5-day workweek, mainly for people working in administrative offices, or researchers and professors in laboratory rooms or cabinets.

To develop the multicriteria analysis matrix, the most relevant criteria, sub-criteria, and room types were selected, case by case, for each monitored room for the six instrumented schools. In-situ monitoring will consider short-term assessment based on a 1-week monitoring cycle, including weekend days.

The experimental campaign selected during the presented case study is ongoing during the spring of 2021 and its experimental results will be applied in future radon mitigation studies, where applicable.

Lastly, the proposed methodology, relying on a multicriteria matrix, is meant to be a decision-making tool for complex monitoring scenarios, holding many criteria to deal with and where a multicriteria analysis can help to solve problems related to criteria selection.

ACKNOWLEDGEMENTS

This work is a result of the project TECH – Technology, Environment, Creativity and Health, Norte-01-0145-FEDER-000043, supported by Norte Portugal Regional Operational Program (NORTE 2020), under the PORTUGAL 2020 Partnership Agreement, through the European Regional Development Fund (ERDF). This work was developed within the scope of the project proMetheus – Research Unit on Materials, Energy and Environment for Sustainability, FCT Ref. UID/05975/2020, financed by national funds through the FCT/MCTES.

REFERENCES

[1] Fraga, S. et al., Indoor air quality and respiratory symptoms in Porto schools. *Revista Portuguesa de Pneumologia*, **14**(4), pp. 487–507, 2008. DOI: 10.1016/S2173-5115(08)70277-X.

[2] World Health Organization, Regional Office for Europe, Methods for monitoring indoor air quality in schools: Report of a meeting, Bonn, Germany, pp. 4–5, Copenhagen, 2011. https://apps.who.int/iris/handle/10665/108596.

[3] Decreto-Lei n.º 101-D/2020, de 7 de dezembro, Diário da República n.º 237/2020, 1º Suplemento, Série I de 2020-12-07, 7–2(21–75), 2020. https://dre.pt/application/conteudo/150570704.

[4] Radon and Health Minister of Public Works and Government Services Canada, 2011. https://nuclearsafety.gc.ca/pubs_catalogue/uploads/Jan.-2012-Radon-and-Health-eng.pdf. Accessed on: 4 May 2021.

[5] Fojtíková, I. & Navrátilová Rovenská, K., Methodology for measurement in schools and kindergartens: Experiences. *Radiation Protection Dosimetry,* **164**(4), pp. 612–617, 2015. DOI: 10.1093/rpd/ncv340.

[6] Zeeb, H. & Shannoun F., *WHO Handbook on Indoor Radon: A Public Health Perspective*, World Health Organization, 2009. https://apps.who.int/iris/bitstream/handle/10665/44149/9789241547673_eng.pdf.

[7] Ielsch, G., Ferry, C., Tymen, G. & Robé, M.C., Study of a predictive methodology for quantification and mapping of the radon-222 exhalation rate. *Journal of Environmental Radioactivity*, **63**(1), pp. 15–33, 2002. DOI: 10.1016/S0265-931X(01)00133-3.

[8] Ielsch, G., Cushing, M.E., Combes, P. & Cuney, M., Mapping of the geogenic radon potential in France to improve radon risk management: Methodology and first application to region Bourgogne. *Journal of Environmental Radioactivity*, **101**(10), pp. 813–820, 2010. DOI: 10.1016/j.jenvrad.2010.04.006.

[9] Azevedo, R., Silva, J.P., Lopes, N., Curado, A. & Lopes, S.I., Short-term indoor radon gas assessment in granitic public buildings: A multi-parameter approach, pp. 415–418, 2020. DOI: 10.1007/978-3-030-35533-3_50.

[10] Curado, A. & Lopes, S.I., Impact of human occupation on indoor radon concentration: A study based on in-situ measurements for a set of households in alto-Minho, Portugal. *Behave 2016 – 4th European Conference on Behaviour and Energy Efficiency*, Coimbra, Portugal, 8–9 Sep., 2016. https://www.researchgate.net/publication/313059490_Impact_of_the_human_occupation_on_indoor_radon_concentration_a_study_based_on_in-situ_measurements_for_a_set_of_households_in_Alto-Minho.

[11] Veloso, B., Nogueira, J.R. & Cardoso, M.F., Lung cancer and indoor radon exposure in the north of Portugal: An ecological study. *Cancer Epidemiology*, **36**(1), pp. e26–e32, 2012. DOI: 10.1016/j.canep.2011.10.005.

[12] Azevedo, R., Silva, J.P., Lopes, N., Curado, A. & Lopes, S.I., Short-term indoor radon gas study in a granitic school building: A comparative analysis of occupation periods. *Science and Technologies for Smart Cities. SmartCity 360 2019. Lecture Notes of the Institute for Computer Sciences, Social Informatics and Telecommunications Engineering*, vol. 323, eds H. Santos, G. Pereira, M. Budde, S. Lopes & P. Nikolic, Springer: Cham, pp. 80–89, 2020. DOI: 10.1007/978-3-030-51005-3_9.

[13] Guide for Radon Measurements in Public Buildings (Workplaces, Schools, Day Cares, Hospitals, Care Facilities, Correctional Centres), 2015. https://www.canada.ca/content/dam/hc-sc/migration/hc-sc/ewh-semt/alt_formats/hecs-sesc/pdf/pubs/radiation/radon_building-edifices/27-15-1468-Radon Measurements_PublicBuildings-EN13.pdf. Accessed on: 4 May 2021.

[14] Corentium Plus, by Airthings – Professional Digital Radon Monitor, User Manual. https://cdn2.hubspot.net/hubfs/4406702/Website/Product%20Sheets/Plus/Plus%20User%20Manual.pdf. Accessed on: 4 May 2021.

[15] Martins, P., Lopes, S.I., R. da Cruz, A.M. & Curado, A., Towards a smart and sustainable campus: An application-oriented architecture to streamline digitization and strengthen sustainability in academia. *Sustainability MDPI AG*, **13**(6), 3189, pp. 1–25, 2021. DOI: 10.3390/su13063189.

[16] Mendes, J., Curralo, A., Curado A. & Lopes, S.I., The sustainable smartbottle: A proposed design methodology to minimize plastic pollution. *Advances in Design and Digital Communication. Digicom 2020. Springer Series in Design and Innovation*, vol. 12, eds N. Martins & D. Brandão, Springer: Cham, pp. 1–13, 2021. DOI: 10.1007/978-3-030-61671-7_27.

[17] Lopes, S.I., Silva, J., Antão, A. & Curado, A., Short-term characterization of the indoor air radon concentration in a XII century monastery converted into a school building. *Energy Procedia*, **153**(2018). *5th International Conference on Energy and Environment Research, ICEER2018 – 23–27 Jul. 2018*, Prague, Czech Republic, pp. 303–308, 2018. DOI: 10.1016/j.egypro.2018.10.036.

[18] Indoor air pollution: New EU research reveals higher risks than previously thought. European Commission Press release: 22 Sep. 2003. https://ec.europa.eu/commission/presscorner/detail/en/IP_03_1278. Accessed on: 4 May 2021.

[19] United States Environmental Protection Agency, Controlling pollutants and sources: Indoor air quality design tools for schools. https://www.epa.gov/iaq-schools/controlling-pollutants-and-sources-indoor-air-quality-design-tools-schools. Accessed on: 4 May 2021.

[20] United States EPA – National Radon Training Centers, *Radon Measurement in Schools: Self-Paced Training Workbook*, 1994. https://nepis.epa.gov/Exe/ZyPDF.cgi/000001WZ.PDF?Dockey=000001WZ.pdf. Accessed on: 4 May 2021.

AIR QUALITY OBSERVATIONS IN THE EAST OF QUITO, ECUADOR IN 2018–2020: COMPARISONS BETWEEN PRE- AND POST-COVID-19 CONDITIONS

MARIA CAZORLA*
Universidad San Francisco de Quito USFQ, Colegio de Ciencias e Ingenierías,
Instituto de Investigaciones Atmosféricas, Ecuador

ABSTRACT

Ozone, NO, NO_2 as well as meteorological observations are continuously monitored at the Atmospheric Measurement Station (EMA, Spanish acronym) at Universidad San Francisco de Quito (USFQ) in Ecuador. The monitoring station is sited in Cumbayá, an Andean urban area in the valley east of Quito. In 2020, namely after the onset of the COVID-19 pandemic, primary emissions decreased due to various mobility restrictions and strategies applied in the city to try to curb the spread of the disease. These changes had an impact in ambient levels of NO and NO_2 (collectively called NO_x), which were recorded at EMA USFQ. On the other hand, an increase in ozone was detected when compared with measurements from previous years. In this work, the differences in NO_x and ozone in 2020 with respect to levels observed in 2018 and 2019 are explored. To this end, an analysis that includes 10-minute measurements of the three species is presented along with meteorological data for 2018–2020. In addition, the typical diurnal patterns of the three species are analysed under regular traffic conditions as well as under different mobility restrictions adopted in 2020. Moreover, the seasonality of the three species is characterized within the context of weather patterns in the study region. Finally, an explanation of the observed levels of ozone, based on insight gained recently in regard to ozone production mechanisms in Quito, is presented.
Keywords: ozone, NO_x, air quality, Quito, Ecuador, COVID-19.

1 INTRODUCTION

In the wake of the COVID-19 sanitary emergency, many countries applied restrictions to citizens' free mobility in an effort to curb the spread of the disease, while waiting for more permanent solutions such as massive vaccination. In Ecuador, different mobility restrictions were applied as early as March 2020. Restrictions were imposed for the entire country by a national emergency committee (COE, Spanish acronym), although local governments also applied measures according to each city's needs. In this work, different mobility restrictions imposed in the capital city of Quito in year 2020 and their impact on levels of NO, NO_2, and ozone are examined.

Quito is a city located within the complex topography of the Ecuadorian Andes. It spreads North to South, along the base of Pichincha volcano, with an altitude that varies from about 2,400 masl (meters above sea level) in the eastern valleys to about 2,800 masl at higher regions. Busy residential neighbourhoods as well as the business centre and the historical centre occupy this active urban area. Detailed topographic maps and cross sections can be found in previous publications [1], [2].

In Quito, a mid-size city of about 2.78 million inhabitants [3], the public transportation is mainly composed by a fleet of diesel buses. Public transportation is used by most of the population (over 60%) for their daily mobility needs [4]. Nevertheless, the number of private vehicles, which mostly use gasoline, have increased substantially in the past decade.

* *ORCID: https://orcid.org/0000-0001-5295-2968*

WIT Transactions on Ecology and the Environment, Vol 252, © 2021 WIT Press
www.witpress.com, ISSN 1743-3541 (on-line)
doi:10.2495/AIR210161

Crowded by busy traffic, Quito is a city where regulations on mobility of private vehicles are not foreign to citizens. In 2010, the city issued a rule to restrict mobility during the morning and evening rush hours. Ever since, owners of private vehicles became familiar with not driving one day a week from 07:00 to 09:30 and from 16:00 to 19:30. In July 2019, the local government extended the restriction to 15 hours to ease work of repaving streets [5]. As the year ended and 2020 began with disturbing news about the COVID-19 pandemic, the new mobility regulations in Quito merged with national decrees and strict measures to restrict citizen's free mobility, which we discuss in this work.

The footprint of traffic in Quito is observable in the diurnal patterns of air quality observations, for example of NO and NO_2 [1], [6]. During the morning rush hour, emissions become trapped in a shallow boundary layer, whose evolution varies seasonally depending on surface conditions [2]. Under regular traffic conditions, the makeup of primary pollutants in the ambient air is such that high ambient NO_x induce ozone levels that are generally below 50–55 ppbv (10-minute data). This effect is due to atmospheric chemistry in a NO_x-saturated regime as has been proposed in previous work [1]. However, in 2020, there was a change. As the urban fleet of buses and private vehicles abandoned the streets due to a variety of COVID-19-related restrictions, the proportion of urban emissions (VOCs to NO_X) shifted. A recent study [6] demonstrates that mobility restrictions caused the chemistry of ozone production in Quito to shift towards a more NO_x-limited regime. This latter study was done upon the onset of the COVID-19 pandemic in Ecuador, with a subset of data taken in March and April 2020, which is usually a rainy period.

The present work contains not only the rainy season, but the entire set of conditions observed in 2020 and documents the signature of NO_x observations in relation to the main mobility restrictions. Moreover, the meteorological context within which restrictions took place is discussed. Finally, ozone levels in 2020 are examined in connection to ambient NO_x and observations are compared to pre-pandemic years 2018 and 2019.

2 METHODS

Air quality and meteorological observations are continuously measured on the main campus of Universidad San Francisco de Quito (USFQ) at EMA station (Spanish acronym for Atmospheric Measurement Station). The coordinates of the measuring site are 0.196°S, 78.4°W, and 2,414 masl.

Meteorological data used in this work correspond to solar radiation, temperature, and precipitation. Solar radiation is measured with a certified second class Kipp & Zonen pyranometer model CMP3, while temperature is measured with a Vaisala HUMICAP probe that also detects humidity, although those data are not presented. The rate of acquisition for the pyranometer is 1 s^{-1}, while for the temperature sensor it is 0.033 s^{-1}. Precipitation is measured with a Texas Electronic rainfall sensor model TR-525 M. Data quality is ensured through periodic calibrations of the entire meteorological station. More details on instrumentation can be found in Cazorla and Tamayo [7].

NO_x measurements at EMA are taken with a Teledyne chemiluminescence sensor model T200. Quality assurance of NO_x data is done through periodic in-lab calibration using a certified NO standard (4.794 ppmv) and preparing calibration mixtures with zero air for a range observed in the ambient air between 0 to 200 ppbv. Linearity of the calibration curve is carefully checked. Data is collected at a rate of 1 s^{-1}. Data are presented for the time period 2018 to 2020. In 2019, NO_x data was lost from June to October due to a damage of the sensor that needed part replacement.

Ambient ozone is continuously measured at EMA with a Thermo 49i UV photometer. Data quality assurance is done through continuous intercomparisons against independent methods. First, ozone measurements are intercompared at EMA against a 2B Technologies ozone monitor. Second, ozone readings are checked against electrochemical concentration cells that are used periodically for vertical profiling [8]. Finally, measurements are compared against those publicly available from a neighbouring station (Tumbaco) run by Quito's Air Quality Network [9]. Data from the 49i sensor are collected at a 1 s^{-1} rate. Ozone data are presented for years 2018 to 2020. There are missing data in June as well as in September and part of October 2019 due to a halt in EMA operations related to expansion work.

To process data, the original resolution of the different measurements was reduced to 10-minute. Typical patterns of variables during a season or a time period were obtained as median diurnal variations (MDV). To this end, 10-minute data for the entire time period (for example a month) were overlapped in a 24-hour plot and the median was computed for every hour. Thus, the MDV of solar radiation for every month of years 2018 to 2020 is presented. Ambient temperature is depicted in a yearly plot with the average of daily maxima in every month of years 2018 to 2020. Precipitation was calculated as cumulative rainfall every month of the study time period.

In regard to air quality data, the MDV of NO_x was obtained from 10-minute data for every month of 2018 to 2020 as the addition of NO and NO_2. NO_x levels were discussed in light of a table in which the main mobility restrictions in 2020 were summarized. Ozone data are presented in monthly plots in which 10-minute observations for years 2018 to 2020 are overlapped in order to visualize main differences in fine resolution data. Finally, years with the most complete data sets (2018 and 2020) were used to compare main differences before and after COVID-19 conditions. Hence, MDV curves for NO, NO_2, and ozone from March to December in years 2018 and 2020 were overlapped to compare pre- and post-pandemic features as diurnal patterns of ozone precursors and ozone.

3 RESULTS AND DISCUSSION

3.1 Meteorological context

Meteorological conditions are critical to understand air quality data due to their influence in the mixing volume where air pollutants become mixed, the effect that solar radiation has in photochemistry, and the cleansing action of precipitation.

At high altitude on the equator, solar radiation is intense year-round. Seasonality of solar radiation and sky conditions at the measuring site are determined by the time of year and the sudden formation of clouds at this highly complex mountainous site. Fig. 1 depicts the solar radiation MDV for every month during years 2018–2020, which illustrates the range of variability in sky conditions. Usually, two cloudy and rainy seasons take place around March and November, while summer starts in July and extends to mid-September. Occasionally, a "little summer" happens at the end of December into January.

In 2020, the pandemic year, January was a sunny month along with August. July and November were atypical in the sense that in July was mostly cloudy, while November had mostly sunny skies in the morning until noon, while storms developed in the early afternoon. These latter conditions are usually observed in mid-September and October, while November is usually a cloudier month. Sunny mornings and middays in November had an influence in air quality under pandemic mobility restrictions. A discussion is presented in the following section.

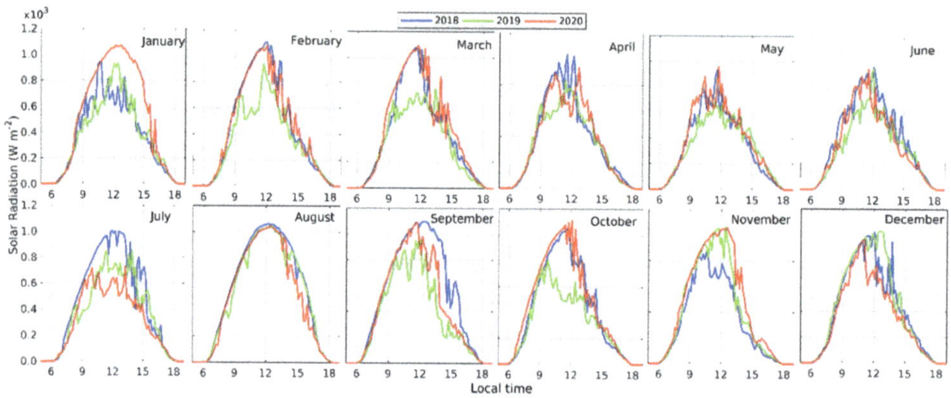

Figure 1: Median diurnal variation (MDV) of solar radiation for every month of years 2018 (blue), 2019 (green), and 2020 (red) observed at EMA USFQ in Quito, Ecuador.

Solar radiation impacts surface temperature, but this variable is influenced by high altitude at this Andean site, which keeps ambient temperatures at pleasant levels in spite of the equatorial latitude. The left panel of Fig. 2 depicts the average of daily temperature maxima per month for years 2018 to 2020. As indicated, month-to-month variability is restricted to a range between 22°C to just over 25°C (1-hour data) with cooler temperatures in April–May and higher temperatures in August–September. Cloud cover conditions, presented in Fig. 1, are reflected in mean ambient temperature maxima in Fig. 2. In regard to precipitation, year 2020 had the most cumulative rainfall when compared to 2018 and 2019. Also, the main temperature in the summer was cooler than, for example, in 2018.

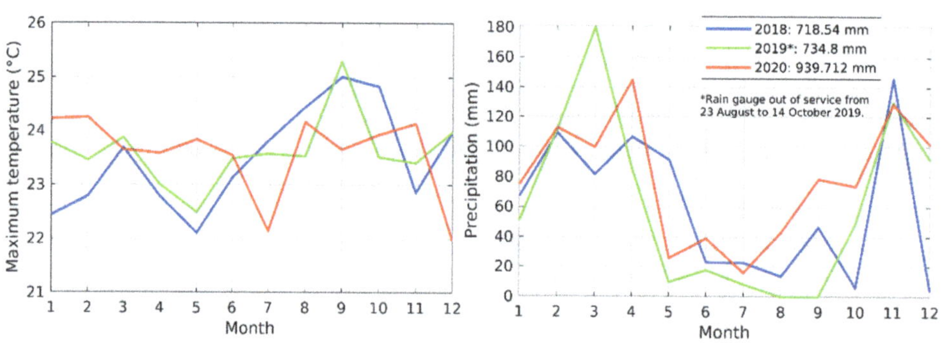

Figure 2: Left: average of maximum daily temperature per month for years 2018 to 2020. Right: total monthly precipitation.

3.2 NO_x observations in connection with mobility restrictions

The signature of traffic emissions at the study site in years 2018 to 2020 is observable in ambient measurements of NO_x depicted in monthly panels in Fig. 3. As indicated before, monthly NO_x MDV curves were obtained from 10-minute NO and NO_2 data. Thus, under pre-pandemic conditions (2018, 2019 and January to February 2020), the morning rush hour NO_x signal is very distinctive and usually ranges between 15 to 45 ppbv, on average, although in October 2018 it reached 80 ppbv. It is important to remark that curves in Fig. 3 are monthly MDVs, while individual 10-minute data could be higher than 100 ppbv.

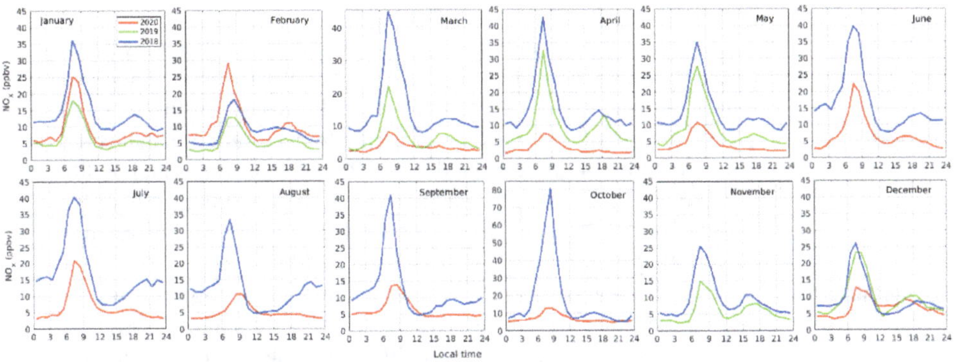

Figure 3: NO_x median diurnal variation (MDV) for every month of years 2018 (blue), 2019 (green), and 2020 (red) obtained from 10-minute data taken at EMA USFQ.

Starting March 2020, ambient NO_x substantially decreased as mobility restrictions and school closures took place. There was a variety of measures applied in the city to try to curb the COVID-19 contagion. Table 1 summarizes the main regulations adopted by authorities, which had an effect observable in NO_x measurements. Thus, on 17 March mobility was generally restricted and public transportation suspended with the exception of vehicles for essential activities. This confinement in Quito lasted 78 days. As a result, morning NO_x substantially decreased to below 10 ppbv as can be observed in Fig. 1 in panels that depict data for March, April, and May. On 3 June, the city opened up some activities and public transportation resumed with controls in the number of passengers per bus. This latter measure is evident in air quality data as the morning rush hour peak increased to over 20 ppbv, on average. There was a new outbreak of COVID-19 infections in July and August in Quito. Thus, in August new restrictions were adopted that shortened the hours of service of municipal transportation. Measures in this month had the effect of shifting the morning rush hour peak of NO_x from 07:00 to 09:00 as can be observed in the August panel in Fig. 3. In September, the decree to limit the constitutional right to free mobility expired. Nevertheless, regulations to reduce traffic continued in the city by applying different strategies that used the last digit of the plate number.

3.3 Ambient ozone observations

As stated in the introduction section, 10-minute ozone observations in Quito are usually below 55 ppbv, which leads to hourly averages that most of the time meet the Ecuadorian air quality standard for ozone (50.9 ppbv, 8-hour mean) [1], [6]. Such ozone levels at an urban site with equatorial insolation has been attributed to a NO_x-saturated environment [1], [6]. However, as discussed before, NO_x emissions substantially decreased after March 2020

Table 1: Summary of some of the main mobility restrictions imposed in Quito in 2020.

17 March–May	School closures in effect. Public transportation suspended in Quito. The national government issued a decree to restrict citizens' constitutional right to free mobility [10]. Curfew: 14:00–05:00. The general lockdown lasted 78 days.
3 June	Public transportation resumes. Mobility of private vehicles is restricted by the last digit of the plate number in a way that cars are only allowed to drive three times a week Curfew: 21:00–05:00 [11].
August	New restrictions in schedule of public transportation managed by the municipality of Quito from 05:00–19:30 on weekdays and 06:00–17:30 on Friday and Saturday, no circulation on Sunday. Curfew: 19:00–05:00. Private vehicles could drive only on authorized days according to the last digit of the plate number [12].
September–December	End of decree to generally restrict citizens' free mobility. Bus schedule extended. Curfew: 23:00–05:00. Circulation of private vehicles still restricted through different strategies that used the last digit of the plate number [13]. Other restrictions were applied during holidays.

when compared to pre-pandemic conditions. Being an ozone precursor, this significant decrease in NO_x had an impact in ozone production. Recently published work that used a chemical box model and ozone data until April 2020 [6] showed that with less abundant NO_x, the chemistry of ozone production shifted to a more NO_x-limited regime. It was proposed that such shift in precursors had the potential to increase the chemical production of ozone, which could accumulate in the ambient air given favourable meteorological conditions. Looking into 10-minute ozone data in year 2020 after April (Fig. 4) and comparing with data in 2018 and 2019, the prediction proposed in Cazorla et al. [6] was met in June, October, and mainly in November 2020.

The month of November 2020 presents a true departure from the usual annual pattern of ozone levels in Quito. As depicted in Fig. 4 (2018, 2019 data) and as documented previously [1], under regular traffic and weather conditions, the month of November

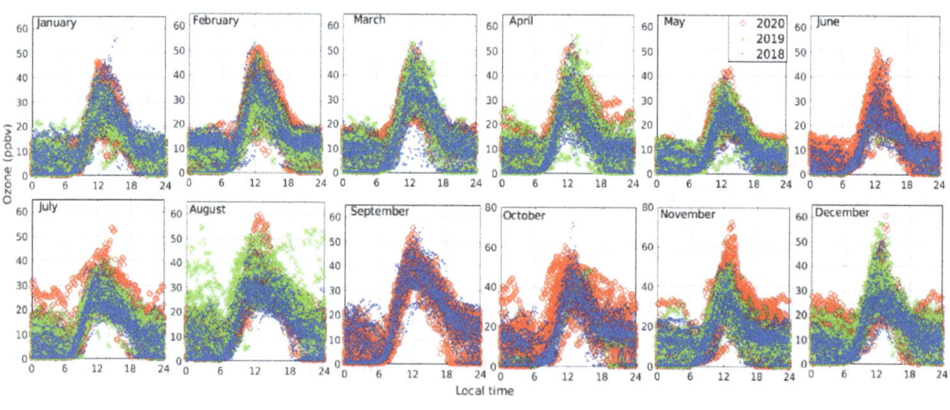

Figure 4: Ambient ozone (10-minute data) measured at EMA USFQ every month of years 2018 (blue), 2019 (green), and 2020 (red).

(usually rainy) is not a time period when events of high ozone are expected. Usually, if a high ozone episode is to be observed, it takes place in September and occasionally the first days of October, as happened in 2018 (a few 10-minuted ozone readings on the 1st of October 2018 surpassed 60 ppbv). As discussed in Section 3.1, the month of November 2020 was peculiar in the sense that it was a seasonally rainy month (over 120 mm of precipitation), but had atypical mornings and middays that were very sunny. This meteorological factor combined with a shift in the proportion of ozone precursors led to increased photochemical rates of ozone production according to Cazorla et al. [6]. As a result, in the 10-minute time series, there is a new record of 15 days in November 2020, when ozone maxima ranged between 55 and 72 ppbv (the latter reading took effect on 14 November).

As an additional comparison, Fig. 5 illustrates overlapped mean levels of NO, NO_2 and ozone from March to December 2018 (left) and 2020 (right). Thus, abundant NO during regular traffic (2018) rapidly titrates ozone in the morning rush hour. With daylight, NO_2 undergoes photolysis. However, ozone production is non-linear with respect to levels of precursors [14]. As investigated in previous work [6], [14], a high NO_x environment leads to lower ozone production rates because precursors (NO_x and free radicals hydroxyl, OH, and hydroperoxy, HO_2) are lost to the formation of other by-products such as nitric acid [14]. This mechanism explains usual (rather low) ozone levels in a busy equatorial urban area such as Quito. In contrast, when COVID-19 mobility restrictions were applied, fresh NO levels in the morning rush hour decreased five times and stayed low in the day along with NO_2, as can be observed in the right panel in Fig. 5. The ozone titration effect with NO lessened not only in the morning, but also in the evening. With less abundant NO_x in the environment, there is more availability of radicals to produce secondary NO_2 followed by immediate photolysis and more rapid ozone production. This latter statement was proposed in Cazorla et al. [6] with the clarification that meteorological conditions are key to induce ozone accumulation in the ambient air. In this work, it is proposed that higher ozone in 2020, in particular in November, is the result of an enhanced ozone production mechanism whose chemistry is shifted to a more NO_x-limited regime. At the moment, observational evidence of higher levels of ozone in connection with predictions stated in previous work [6] in regard to potential scenarios for increased ozone is presented. However, additional

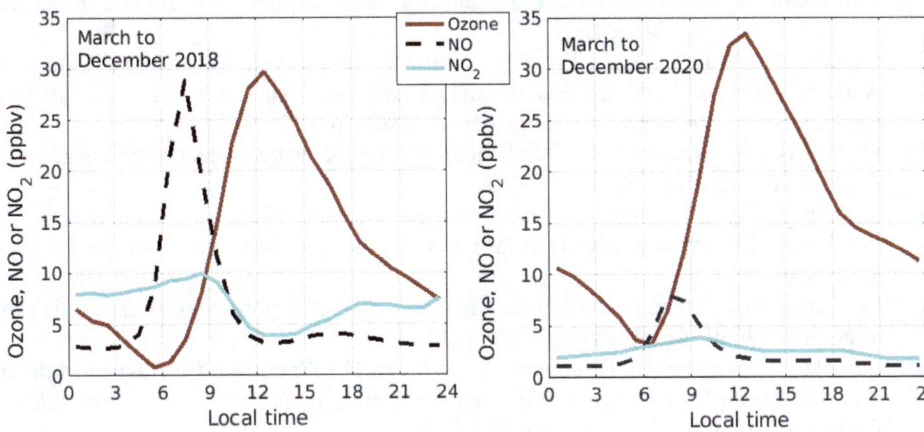

Figure 5: Levels of NO, NO_2, and ozone from March to December 2018 (left) and 2020 (right) as median diurnal variations from data taken at EMA USFQ.

research that incorporates chemical and transport modelling is needed in the future to scrutinize the underlying chemistry and the influence of meteorological factors particularly in November 2020 in Quito.

4 CONCLUSIONS

In 2020, the COVID-19 pandemic forced nations to confine their citizens to somewhat constrain massive contagion. Ecuador was not an exception. In the capital city of Quito, different mobility restrictions were imposed since mid-March and lasted, to different extents, the entire year. From data collected at Universidad San Francisco de Quito, the overall decrease in the abundance of NO_x in the ambient air from March to December 2020, when compared to 2018, was by a factor of 5. Limitations in schedules for public transportation and authorized days to drive private vehicles caused a shift in the rush hour peak of NO_x from 07:00 to 09:00 in August 2020. Work that analysed the effect on ozone photochemistry in Quito, done with initial lockdown data (March–April 2020) [6], predicted the potential for increased ozone with reduced NO_x due to a shift in the chemical regime of ozone production. From an observational perspective, this prediction was met mainly in November 2020, when there was a record of 15 days when ozone maxima ranged from 55 to 72 ppbv in the 10-minute time series. This consistent departure from the peak baseline for this month (below 50 ppv) points to atmospheric chemistry as the underlying reason. At the moment this unseasonal change is reported and it is proposed that, based on previous work [6], a shift to a more NO_x-limited regime of ozone production explains observations in November 2020, but additional research needs to be performed in the near future to prove this hypothesis.

ACKNOWLEDGEMENTS

Since 2020 ozone observations at EMA USFQ were significantly augmented thanks to equipment and funds under the support of the Vienna Convention Trust Fund and through an agreement (No. 224581) with the World Meteorological Organization. Thanks to Edgar Herrera for continuous operations and data processing at EMA USFQ.

REFERENCES

[1] Cazorla, M., Air quality over a populated Andean region: Insights from measurements of ozone, NO, and boundary layer depths. *Atmospheric Pollution Research*, **7**, pp. 66–74, 2016.

[2] Cazorla, M. & Juncosa, J., Planetary boundary layer evolution over an equatorial Andean valley: A simplified model based on balloon-borne and surface measurements. *Atmospheric Science Letters*, **19(8)**, p. e775, 2018.

[3] INEC Instituto Nacional de Estadísticas y Censos, www.ecuadorencifras.gob.ec/. Accessed on: 15 May 2021.

[4] Gestión Digital, https://www.revistagestion.ec/sociedad-analisis/quito-por-que-privilegiar-al-transporte-privado-si-la-mayoria-usa-el-publico. Accessed on: 15 May 2021.

[5] El Comercio, https://www.elcomercio.com/actualidad/quito/pico-placa-municipio-jorge-yunda.html. Accessed on: 10 May 2021.

[6] Cazorla, M., Herrera, E., Palomeque, E. & Saud, N., What the COVID-19 lockdown revealed about photochemistry and ozone production in Quito, Ecuador. *Atmospheric Pollution Research*, **12**(1), pp. 124–133, 2020.

[7] Cazorla, M. & Tamayo, E., Atmospheric measurement station at universidad San Francisco de Quito (EMA): Ground-based physical meteorology instrumentation and assessment of initial measurements. *ACI Avances en Ciencias e Ingenierías*, **6**(2), pp. C21–C30, 2014.
[8] Cazorla, M., Ozone structure over the equatorial Andes from balloon-borne observations and zonal connection with two tropical sea level sites. *Journal of Atmospheric Chemistry*, **74**, pp. 377–398, 2017.
[9] Secretaría de Ambiente de Quito, http://www.quitoambiente.gob.ec/ambiente/index.php/descarga-datos-historicos. Accessed on: 1 May 2021.
[10] El Comercio, https://www.elcomercio.com/tendencias/sociedad/restriccion-movilidad-excepcion-ecuador-coronavirus.html. Accessed on: 20 May 2021.
[11] El Comercio, https://www.elcomercio.com/actualidad/quito/coronavirus-movilidad-quito-amarillo.html. Accessed on: 20 May 2021.
[12] El Comercio, https://www.elcomercio.com/actualidad/quito/transporte-municipal-quito-cambio-horario.html. Accessed on: 20 May 2021.
[13] El Comercio, https://www.elcomercio.com/actualidad/quito/vigencia-hoy-circula-vehiculos-quito.html. Accessed on: 20 May 2021.
[14] Kleinman, L.I. et al., Sensitivity of ozone production rate to ozone precursors. *Geophysical Research Letters*, **28**(15), pp. 2903–2906, 2001.

IMPACT OF COVID-19 PANDEMIC ON N AND S ATMOSPHERIC DEPOSITION FLUXES IN A MEGACITY IN THE NORTH EAST OF MEXICO

JOSÉ A. SOLÍS CANUL[1], ROSA M. CERÓN BRETÓN[1], JULIA G. CERÓN BRETÓN[1], SIMÓN CARRANCO LOZADA[2], EVANGELINA RAMÍREZ LARA[3], MARIA DE LA LUZ ESPINOSA FUENTES[4], ROCÍO GARCÍA MARTÍNEZ[4], JAIR R. CARRILLO AVILA[5], ARMANDINA VALDEZ CAVAZOS[5], REYNA DEL CARMEN LARA SEVERINO[1], MARCELA RANGEL MARRÓN[1], ANA P. CORTES PÉREZ[1] & MIRANDA MARTÍNEZ ORTEGA[1]

[1]Faculty of Chemistry, Autonomous University of Carmen, México

[2]León Campus, National Politechnique Institute, México

[3]Faculty of Chemistry, Autonomous University of Nuevo León, México

[4]Atmospheric Sciences Center, National University of México, México

[5]Atmospheric Monitoring System (SIMA) of Monterrey, México

ABSTRACT

During 2020–2021, the COVID 19 pandemic resulted in a lockdown in many countries around the world. This work assessed the impact of the restrictions to economical and industrial activities, and population mobility on N and S atmospheric deposition fluxes in the Metropolitan Zone of Monterrey (MZM). Trends of atmospheric deposition fluxes were analyzed and compared during two periods. The first period before pandemic from 27 February 2017 to 15 October 2018; the second one, from 20 October 2020 to 5 May 2021. Both periods including two climatic seasons (dry and rainy) and considering the same sampling points along MZM. Passive collectors of throughfall deposition (hydrologic flux of ions to floor contained within a solution) were used to estimate concentrations for SO_4^{2-} and NO_3–. From a meteorological analysis, by using back-air masses trajectories and wind roses, the influence on transport and deposition of pollutants was assessed. The results showed that the change rates of N and S before and during the lockdown ranged from 63%–67% and 90% for N and S, respectively. The response of S was more sensitive to the lockdown than N. Since, during pandemic, industrial and vehicular emissions, and forest fires were reduced; it had a positive impact on atmospheric deposition. However, besides the lockdown due to COVID-19, the observed changes in spatial and temporal deposition patterns also could be attributed to meteorological conditions associated to the sampling season, particularly to precipitation patterns. Mean atmospheric deposition fluxes for N and S for the two studied periods were compared with critical load values established for sensitive ecosystems in Europe. Exceedances to critical load values before pandemic were significant higher than those found for the period during pandemic, suggesting that the lockdown improved the air quality in MZM.

Keywords: COVID-19, nitrogen, sulphur, atmospheric deposition, Monterrey.

1 INTRODUCTION

Atmospheric deposition is the main process to transfer materials (particles and gases) from the atmosphere to terrestrial surface; this process can occur by two ways: wet deposition and dry deposition. Wet deposition is a process in which atmospheric pollutants are incorporated to rain drops and transferred to the terrestrial surface by precipitation, mainly in the form of rain, snow or fog. Dry deposition depends on the type of component (gas or particle), whose deposition occurs on the terrestrial surface in the absence of precipitation [1]. Inputs of acidity by deposition to aquatic and terrestrial ecosystems have gained considerable public and scientific interests [2]. Dry and wet deposition contribute to the biogeochemistry cycle, and can explain nutrient inputs to the ecosystems [3].

Total emissions of N and S in Mexico have increased since 1980. In spite of S emissions have been reduced as a result of the implementation of new regulations and changes in the

WIT Transactions on Ecology and the Environment, Vol 252, © 2021 WIT Press
www.witpress.com, ISSN 1743-3541 (on-line)
doi:10.2495/AIR210171

composition of fuels, deposition rate of N in comparison with those reported for S have increased significantly [4]. Saturation phenomena of N constitutes a threat for natural resources and ecosystems. Besides, the effects of N deposition on water quality constitute a concern due to some diseases of human health are associated to high levels of nitrates. In the other hand, the high deposition of S cause severe damages to forests, sulfur dioxide cause damages when its derivative sulfite is absorbed faster than receptors are detoxified. In spite of their importance, N and S atmospheric deposition is not considered as a criteria pollutant, and therefore, standardized sampling devices and reference values or regulations are not available. Regarding this, critical loads can be used as a valuable tool by decision makers to establish public policies focused to protect aquatic and terrestrial ecosystems in polluted regions. However, the establishment and application of critical loads requires accurate measurements of N and S deposition [5]. In Mexico, there are some individual stations of specific research groups measuring the major components in wet deposition, but they are limited and are not enough to calibrate empirical models, to monitor status, trends and exceedances to critical loads. Therefore, cost-effective sampling devices which integrate wet and dry deposition are required for the understanding of the distribution and effects of N and S deposition on ecosystems. Regarding this, throughfall deposition is widely used to estimate atmospheric deposition inputs to ecosystems, since it includes both, wet and dry deposition, and it is a non-expensive collector [6].

The comprehension of deposition mechanisms of N and S and their interactive effects is primordial not only to aspects related to deposition patterns at regional scale, but also to national and global scale. However, these patterns can be affected by changes in human activities, regarding this, it is expected that several atmospheric pollutants showed levels significantly lower during the COVID-19 pandemic as a result of restrictions in human activities in comparison with the observed patterns before the pandemic. Therefore, this work had as objective to study the spatial and temporal patterns of N and S atmospheric deposition in Metropolitan Zone of Monterrey (MZM) during two periods, the first occurring before COVID-19 pandemic (2017–2018), and the second occurring during the COVID-19 pandemic (2020–2021). Both results were compared to assess the effect of the restrictions on population mobility and industrial activities on atmospheric deposition patterns.

2 STUDY AREA

MZM is located at the north east of Mexico, grouping a total of 5,341,171 inhabitants, and it is considered as the second Metropolitan Area in Mexico, after Metropolitan Zone of Mexico Valley. In Fig. 1, 10 sampling sites selected for the installation of passive sampler are presented, their location is the same of atmospheric monitoring stations of Atmospheric Monitoring System of Nuevo León (SIMA). Sampling sites were grouped considering their land use as follows: industrial sites 1,3 and 5; urban sites 2,4,7,8,9 and 10; rural site 6.

3 METHODOLOGY

3.1 Sampling

Sampling was carried out in MZM considering two periods. The first one, occurring before COVID-19 pandemic from 27 February 2017 to 15 October 2018; the second one, occurring during the COVID-19 pandemic from 20 October 2020 to 5 May 2021. Both periods including two climatic seasons (dry and rainy) and considering the same sampling points along MZM. To measure wet and dry deposition fluxes is very complex and expensive, for this reason, throughfall collectors constitute a good choice to obtain reliable estimations of

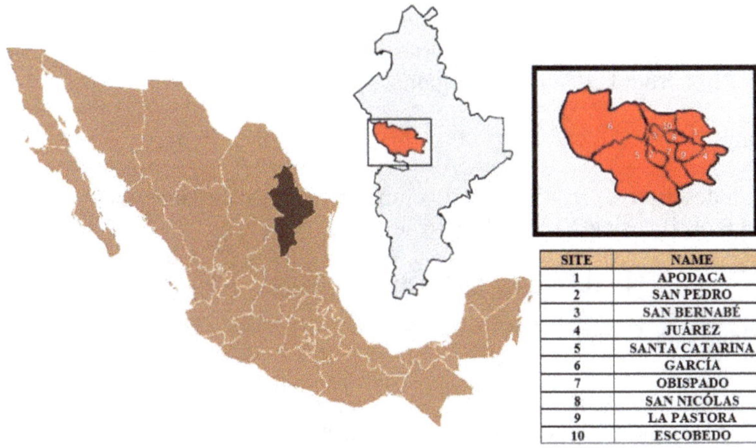

SITE	NAME	SOIL TYPE
1	APODACA	INDUSTRIAL
2	SAN PEDRO	URBAN
3	SAN BERNABÉ	INDUSTRIAL
4	JUÁREZ	URBAN
5	SANTA CATARINA	INDUSTRIAL
6	GARCÍA	RURAL
7	OBISPADO	URBAN
8	SAN NICÓLAS	URBAN
9	LA PASTORA	URBAN
10	ESCOBEDO	URBAN

Figure 1: Location of sampling sites in MZM.

N and S atmospheric inputs in a given place [6]. Throughfall deposition is defined as the hydrologic flux of ions to floor contained within a solution [7]. Throughfall sampling devices were installed in each atmospheric monitoring station of SIMA by using passive collectors assembled and operated according to Simkin et al. [8]. Passive collectors based on ionic exchange resin (IER) were used, consisting of a funnel connected to a column which contains a IER bed. The main advantage of this kind of devices is that can be used during long periods and at a low cost. NO_3^- and SO_4^{2-} ions can be exchanged by IER and trapped by functional groups with electric charges opposed [9]. IER columns consist of a funnel covered with a mesh (to avoid the fall of solid material as leaves and insects) connected to a PVC tube; inside the tube, 30 g of IER are poured. The resin used by the collectors is a mixed resin of polystyrene for anions and cations. The PVC tube is sealed with glass fiber in the bottom (as a platform of support for the resin bed) and in the top (as a filter). This tube is collocated inside of an exterior PVC tube which serves as a shadow tube to protect the resin from solar radiation and to avoid changes in its physical and chemical properties. The bottom edge of the column opens or closes by using a standard valve of PVC to let or avoid the drainage.

3.2 Chemical analysis and IER extraction procedure

An extraction system specially designed for this propose was constructed with a PVC tube of 5 mm ID and 15 cm length which is connected to the resin tubes. The extraction is carried out adding 100 ml of KCI 2N, and letting rest during 20 minutes, after that, the valve is open. Once, the samples extracts are obtained, are refrigerated and stored until their analysis. Nitrate ions were determined by a colorimetric technique [5] and sulfate ions were determined by a turbidimetric technique [6]. Finally, from nitrate and sulfate concentrations, and considering the sampling area and time period of sampling, atmospheric deposition fluxes for N and S (kg Ha^{-1} yr^{-1}) were estimated.

3.3 Meteorological parameters and wind analysis

During the two sampling periods, meteorological parameters were measured in the atmospheric monitoring stations of SIMA. In addition, an analysis of the influence of winds on ionic concentrations was carried out calculating wind roses by using WRPLOT software at local scale, and calculating the back trajectories of air masses by using the Lagrangian hybrid model HYSPLIT from US NOAA at regional scale.

3.4 Statistical analysis

It was carried out a descriptive, relational and comparative statistical analysis by applying different statistical tools by using XLSTAT 2016 software.

4 RESULTS AND DISCUSSION

4.1 Nitrate deposition fluxes

4.1.1 Before COVID-19 pandemic

The mean deposition flux for nitrate during the rainy season 2018 was 6.54 kg Ha^{-1} yr^{-1} with a maximum value of 7.39 kg Ha^{-1} yr^{-1} corresponding to site 8, San Pedro, located at SW from MZM; and a minimum flux of 5.62 kg Ha^{-1} yr^{-1}, for site 10, Juárez, located at SE from MZM. In the other hand, the mean deposition flux for NO_3^- for the dry season 2017 was 3.30 kg Ha^{-1} yr^{-1}, with a maximum value of 4.38 kg Ha^{-1} yr^{-1} in site 8, San Pedro, located at SW from MZM, and a minimum flux of 0.40 kg Ha^{-1} yr^{-1}, corresponding to site 9, La Pastora, located at SE from MZM. According to Fig. 2, where analysis by season is presented, it can be observed that the mean deposition flux of nitrate was higher for the rainy season 2018. In addition, the maximum value was found during this season. Nitrate levels during the period before COVID-19 pandemic were higher during the rainy season, suggesting that chemical reactions in aqueous phase played an important role in the removal process of pollutants during this season. By applying a Friedman test, it was found that there were significant differences between nitrate deposition fluxes between seasons. In the analysis by sampling point, a mean nitrate deposition flux of 4.37 kg Ha^{-1} yr^{-1} was found, with a maximum value of 7.39 kg Ha^{-1} yr^{-1} during the rainy season 2018 in site 8, San Pedro, located at SE from ZMZ; and a minimum flux of 0.41 kg Ha^{-1} yr^{-1}, observed during dry season 2017 in site 9, La Pastora, located at SE from MZM.

In Fig. 3, the analysis by sampling point can be observed, the higher mean flux and the maximum flux of nitrate were found in site 8, followed by site 1, that corresponds to San Pedro located at SW, and Escobedo, located at NE from MZM. In addition, the minimum fluxes were observed in sites 9 and 10, corresponding to La Pastora and Juárez, both sites located at SE from MZM. Therefore, it suggests that the main emission sources of nitrogen oxides were vehicular sources, since the higher fluxes were found in sites grouped as urban sites. From the Friedman test, it was concluded that there were significant differences between sampling sites. From Fig. 4, it can be observed that nitrate deposition fluxes were higher in sites with a land use type urban (sites 1, 4, 6, 8, 9 and 10) that corresponds to Escobedo, San Nicolás, Obispado, San Pedro, La Pastora and Juárez, located at N, NE, in the center, SW, and SE from MZM, respectively. However, sites grouped as rural and industrial showed also high levels of nitrate. Since, a greater variability was observed, a possible explanation can be the variability in the strength of the sources, since the main source of nitrate precursors is the high vehicular traffic. It was supported by a Friedman test applied to grouped sites, concluding that there were not significant differences in nitrate deposition fluxes by land use, suggesting that the main sources were local sources, in this case vehicular sources in urban areas.

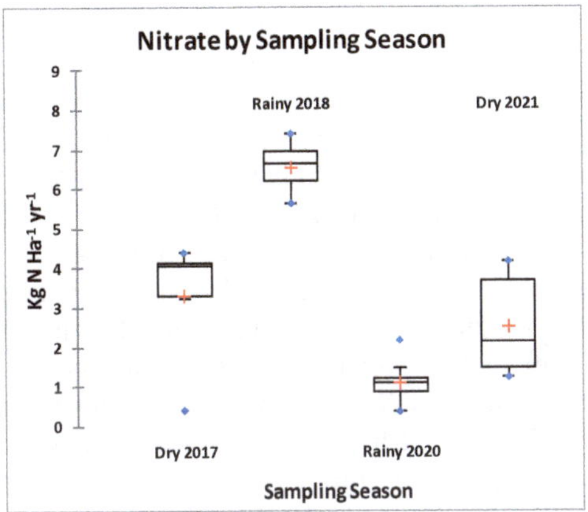

Figure 2: N atmospheric deposition fluxes by season before COVID-19 (dry 2017 and rainy 2018) and during COVID-19 (rainy 2020 and dry 2021).

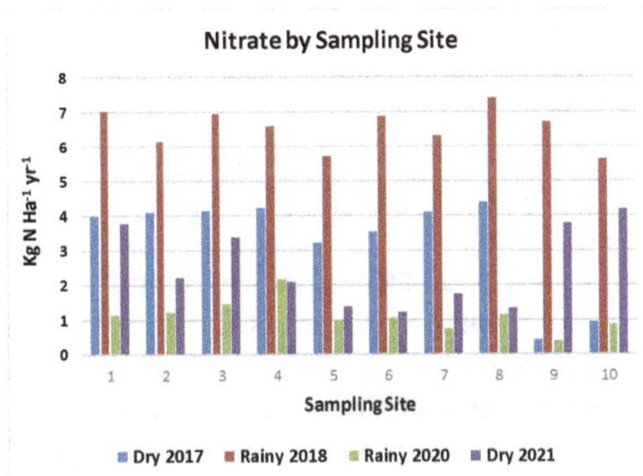

Figure 3: N atmospheric deposition fluxes by sampling site before COVID-19 (dry 2017 and rainy 2018) and during COVID-19 (rainy 2020 and dry 2021).

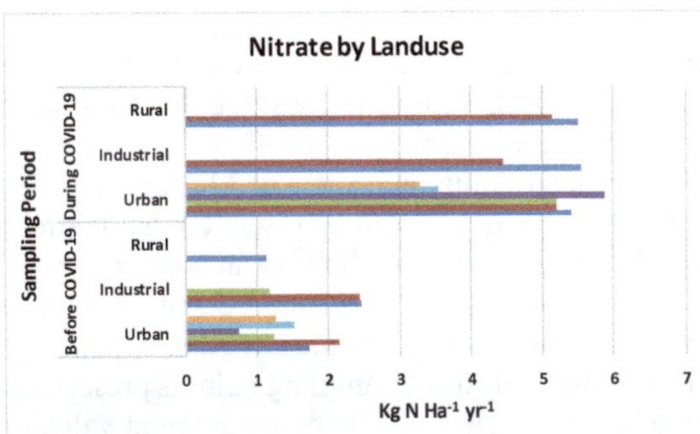

Figure 4: N atmospheric deposition fluxes by land use before COVID-19 and during COVID-19.

4.1.2 During COVID-19 pandemic

The mean value obtained during rainy season 2020 was 1.10 kg Ha^{-1} yr^{-1}, with a maximum values of 2.17 kg Ha^{-1} yr^{-1} corresponding to site 4 (Juárez), located at SE from MZM; and a minimum of 0.36 kg Ha^{-1} yr^{-1}, corresponding to site 9 (La Pastora), both with a land use type urban (Fig. 2). The mean value obtained during the dry season 2021 was 2.51 kg Ha^{-1} yr^{-1} with a maximum value of 4.17 kg Ha^{-1} yr^{-1} corresponding to site 10 (Escobedo), located at N from MZM; and a minimum value of 1.23 kg Ha^{-1} yr^{-1} in the site 6 (García) (Fig. 2). From Fig. 2, it can be observed that the mean N deposition flux was higher during the dry season in comparison with that observed during the rainy season. Even the minimum deposition flux during dry season was above of maximum deposition flux observed during the rainy season, suggesting that, a dilution phenomenon (as a result of the washing of the atmospheric column during the rainy season) played an important role in the nitrate deposition patterns in MZM.

From Friedman tests, it was found that, there were significant differences in nitrate levels between seasons, therefore, the occurrence of precipitation played an important role in the nitrate deposition patterns in MZM. Analyzing the differences between sites, the higher mean value was registered for site 10 (Escobedo) with a value of 2.50 kg Ha^{-1} yr^{-1} located at N from MZM; whereas the minimal value was found for site 6 (García) with a value of 1.13 kg Ha^{-1} yr^{-1} (Fig. 3). From Fig. 3, it can be observed that the minimal concentration was obtained for site 9 (La Pastora), whereas the maximum concentration was found in site 10 (Escobedo). From the Friedman test, it was found that there were not significant differences in nitrate deposition fluxes between sites; it suggests that nitrate levels were uniformly distributed along ZMZ, with origin in local sources. It can explain the differences in the seasonal pattern in comparison with the sampling period occurring before COVID-19 pandemic. Finally, from Fig. 4, it can be observed that nitrate deposition fluxes were higher in sites with a land use type industrial, followed by sites with a land use type urban. From Friedman tests, it was found that there were not significant differences in nitrate levels by land use, suggesting that nitrate levels were uniformly distributed along MZM with sources in common, in this case, vehicular sources.

4.2 Sulfate deposition fluxes

4.2.1 Before COVID-19 pandemic

The mean deposition flux of sulfate obtained during the rainy season 2018 was 23.65 kg Ha^{-1} yr^{-1}, with a maximum value of 28.63 kg Ha^{-1} yr–^{1} in site 1, Escobedo, located at N from MZM; and a minimum value of 17.31 kg Ha^{-1} yr^{-1}, corresponding to site 8, San Pedro, located at SW from MZM. In the other hand, the mean deposition flux for sulfate for dry season 2017 was 27.30 kg Ha^{-1} yr^{-1} with a maximum value of 47.69 kg Ha^{-1} yr^{-1} in site 6, Obispado, located at the center of MZM, and a minimum value of 7.86 kg Ha^{-1} yr^{-1}, in site 9, La Pastora, located at SE from MZM. From Fig. 5, it can be observed that the mean sulfate deposition flux was higher during the dry season, and showing in some cases values too high. From Friedman test it was found that, there were significant differences between seasons suggesting that, besides local sources, sulfate could have its origin in regional sources located at SE from MZM. In Fig. 6, the analysis by sampling point is presented, it was found a mean sulfate deposition flux of 25.03 kg Ha^{-1} yr^{-1}, with a maximum value of 47.69 kg Ha^{-1} yr^{-1}, during the dry season in site 6, Obispado; and a minimum value of 7.86 kg Ha^{-1} yr^{-1}, during the dry season in site 9, La Pastora, located at SE from MZM. From Fig. 6, it can be observed that the mean deposition flux of sulfate was higher in site 6, followed by sites 4 and 5, Obispado, San Nicolás and Apodaca, respectively. Since the higher sulfate deposition fluxes were found in the center and at N of MZM, it can be established that vehicular sources contributed in a significant way on the emission process of primary pollutants precursors of sulfate. From a Friedman test, it was found that there were not significant differences in sulfate deposition fluxes by sampling site, suggesting that sulfate levels were uniformly distributed along MZM.

In the other hand, in Fig. 7, it can be observed that the mean deposition fluxes were higher in sites with a land use type industrial (sites 3 and 5), that corresponds to San Bernabé and Apodaca, located at NW and NE from MZM; followed by sites with and urban land use (sites 1, 4, 6, 8, 9 and 10) that corresponds to Escobedo, San Nicolás, Obispado, San Pedro, La Pastora, and Juárez, located at N, NE, in the center, SW and SE from MZM. Applying a Friedman test, it was found that there were not significant differences in sulfate deposition fluxes by land use.

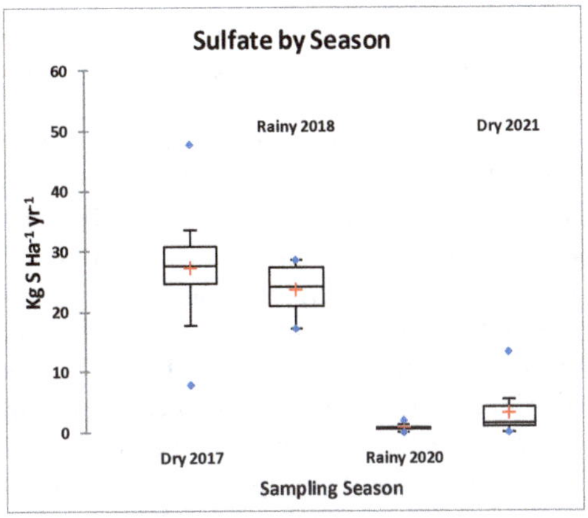

Figure 5: S atmospheric deposition fluxes by season before COVID-19 (dry 2017 and rainy 2018) and during COVID-19 (rainy 2020 and dry 2021).

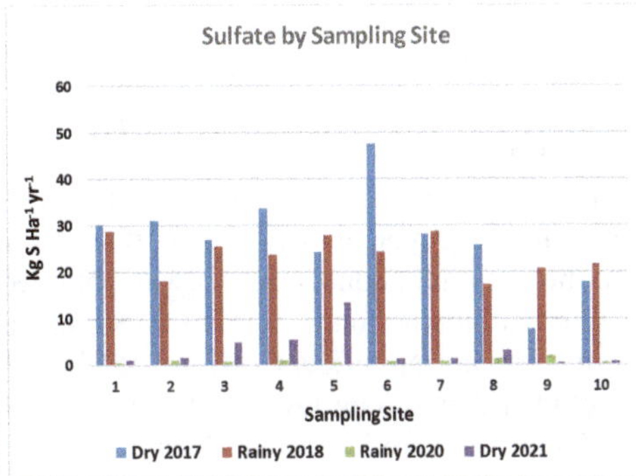

Figure 6: S atmospheric deposition fluxes by sampling site before COVID-19 (dry 2017 and rainy 2018) and during COVID-19 (rainy 2020 and dry 2021).

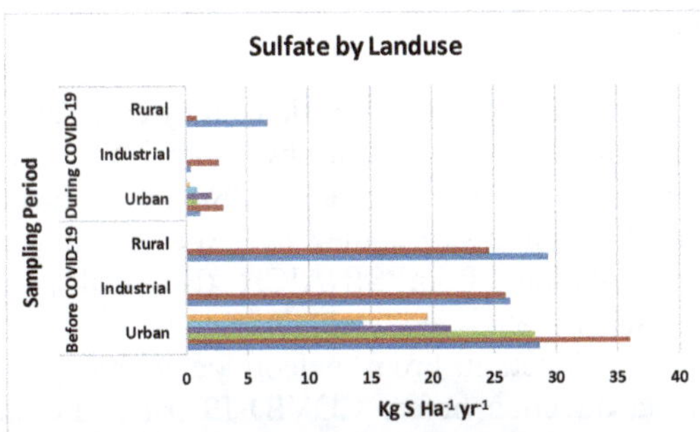

Figure 7: S atmospheric deposition fluxes by land use before COVID-19 and during COVID-19.

4.2.2 During COVID-19 pandemic

The mean value obtained for the rainy season 2020 was of 0.71 kg Ha^{-1} yr^{-1}, with a maximum value of 1.83 kg Ha^{-1} yr^{-1} corresponding to site 9 (La Pastora) located at E from MZM; and a minimum value of 0.045 kg Ha^{-1} yr^{-1}, corresponding to site 1 (Apodaca), with a land use type urban and industrial, respectively (Fig. 5). The mean value obtained for the dry season 2021 was of 3.18 kg Ha^{-1} yr^{-1} with a maximum value of 13.18 kg Ha^{-1} yr^{-1} corresponding to site 5 (Santa Catarina), located at SW from MZM; and a minimum value of 0.0021 kg Ha^{-1} yr^{-1}, corresponding to site 9 (La Pastora), with a land use type industrial and urban, respectively. From Fig. 5, analyzing the differences by season, it can be observed that the mean deposition flux for sulfate was considerably higher during the dry season in comparison with the rainy season, suggesting that during the rainy season, a dilution phenomenon as a result of the washing of the atmospheric column could contribute to obtain concentrations significantly lower in comparison with those obtained during the dry season. By applying a Friedman tests, it was concluded that, there were significant differences between seasons in S atmospheric deposition fluxes.

Analyzing the differences between sites, the highest mean value was registered for site 5 (Santa Catarina) with a value of 6.8336 kg Ha^{-1} yr^{-1} located at SW from MZM; whereas the minimum mean value was registered in site 10 (Escobedo) with a value of 0.39 kg Ha^{-1} yr^{-1}, with a land use type industrial and urban, respectively (Fig. 6). In addition, it can be observed that the minimum concentration was obtained for site 9 (La Pastora), whereas the maximum concentration was found for site 5 (Santa Catarina). From Friedman test, it was found that there were not significant differences in sulfate levels between sites, therefore, it can be concluded that sulfate levels were uniformly distributed along MZM; it suggests that only local sources were important during this period of pandemic, and regional sources did not contribute in a significant way on S deposition patterns considering that restrictions on mobility and industrial activities were applied not only in Mexico but also USA.

According to Fig. 7, it can be observed that mean sulfate deposition fluxes were higher with an industrial land use, however, the values obtained with an urban land use were also high, suggesting that anthropogenic local sources played an important role in S deposition. From Friedman test, it was found that there were not significant differences in sulfate levels by land use, suggesting that S deposition is uniformly distributed along MZM; this is in accordance with those found when differences between sampling sites were analyzed.

4.3 Meteorological analysis

Meteorological analysis at regional scale was carried out by estimating 24 h back air-masses trajectories at 500, 1000 and 1500 m altitude, by using HYSPLIT model from NOAA (National Oceanic and Atmospheric Administration, USA), by choosing one of the sampling sites located at NE from MZM, in this case, the site 1 Apodaca. In addition, meteorological parameters measured by stations from SINAICA (National System of Air Quality Information) were processed and analyzed to obtain wind roses for each sampling period to identify the prevailing wind direction. From meteorological analysis at regional and local scale, the prevailing wind direction before COVID-19 pandemic and during COVID-19 pandemic was E–SE, suggesting that the seasonal patterns found are related not only to the occurrence or not of precipitation (dilution phenomena) but also to the influence of regional sources, specially before COVID-19 pandemic, when regional sources contributed in a significant way on S deposition patterns in MZM. In the other hand, the spatial patterns found indicate that N and S levels were uniformly distributed along MZM; it means that local sources were important, however, since prevailing wind direction was the same during both sampling periods, differences found between sampling sites can be attributed to the strength of emission from the local sources in each specific case, since there were sites with a higher vehicular traffic in comparison with other sites; and the strength of emission was higher before the COVID-19 pandemic since regional contribution was higher than during the pandemic.

4.4 Reference values

4.4.1 Reference values

It has been proposed a critical load value of 5 kg N Ha^{-1} yr^{-1} for alpine ecosystems, which are more sensitive that ecosystems in low lands [7]. In the case of S, it has been proposed a critical load value of 3 kg S Ha^{-1} yr^{-1} for sensitive areas in Europe; whereas for natural forests, it has been proposed a range between 2 and 5 kg S Ha^{-1} yr^{-1} [8]. Regarding to Mexico, reference values are not available to compare the current N and S deposition fluxes in MZM.

4.4.1.1 Before COVID-19 pandemic

Considering the sampling period corresponding to 2 years before pandemic (2017–2018), the mean deposition fluxes for N and S were 4.88 and 25.03 kg $Ha^{-1}yr^{-1}$, respectively. N deposition fluxes did not exceed the reference value reported for alpine ecosystems, however, these values are already in the upper limit of this range. In addition, N deposition fluxes found in MZM were almost two times than those reported by Escoffie et al. [4] in Carmen Island (2.15 kg N $Ha^{-1}yr^{-1}$) in Campeche; by Sánchez [14] in Orizaba Valley (1.44 kg N $Ha^{-1}yr^{-1}$) in Veracruz; and almost 4 times than those reported by García [15] in Atasta-Xicalango (1.15 kg N $Ha^{-1}yr^{-1}$) in Campeche.

In the other hand, S deposition fluxes found 2 years before COVID-19 pandemic exceeded almost eight times the reference value proposed as critical load for sensitive areas, and almost five times the upper limit of the exceedance range established for natural forests in Europe. S deposition fluxes were almost six times than those found by Escoffie et al. [4] in Carmen Island (4.7 kg S Ha^{-1} yr^{-1}); and three times higher than those reported by García [15] in Atasta-Xicalango (8.57 kg S Ha^{-1} yr^{-1}); both sites in Campeche.

4.4.1.2 During the COVID-19 pandemic

Considering the sampling period corresponding to 1.5 years during pandemic (2020–2021), the mean deposition fluxes for N and S for MZM were 1.81 kg N Ha^{-1} yr^{-1} and 1.95 kg S Ha^{-1} yr^{-1}, respectively. In the case of N, this value is below of the reference value reported for alpine ecosystems; whereas in the case of S, the mean value is already close to the threshold value for sensitive areas and forests in Europe. However, besides the mean value, it is necessary to consider that in the case of S, maximum values occurring in site 5 (Santa Catarina) with 13.18 kg S Ha^{-1} yr^{-1} and in site 4 (Juárez) with 5.37 kg S Ha^{-1} yr^{-1} during the dry season 2021 exceeded almost 4 and 2 times the reference values proposed.

N levels found in MZM were in some cases lower and in other cases comparable to those reported by Escoffie et al. [4] in Carmen Island, in Campeche (2.15 N kg Ha^{-1} yr^{-1}); by Sánchez [14] in Orizaba Valley, in Veracruz (1.44 N kg Ha^{-1} yr^{-1}); and by García [15] in Atasta-Xicalango, in Campeche (1.15 kg Ha^{-1} yr^{-1}). In the other hand, in the case of S, mean S deposition fluxes in MZM during the COVID-19 pandemic were lower than those reported by Escoffie et al. [4] in Carmen Island (4.7 S kg Ha^{-1} yr^{-1}) and by García [15] in Atasta-Xicalango (8.57 S kg Ha^{-1} yr^{-1}); both sites in Campeche. However, it is necessary to consider that extreme values of S deposition occurred in sites 5 and 4 with 13.18 and 5.37 kg Ha^{-1} yr^{-1}, respectively; and these values were higher than those found in Carmen island, and Atasta. In addition, values reported by Sánchez [14] in Orizaba Valley (55.16 S kg Ha^{-1} yr^{-1}) in Veracruz were significantly higher than values reported by this study.

5 CONCLUSIONS

Comparing results found in both sampling periods (before and during COVID-19 pandemic) in MZM, before COVID-19 pandemic, N levels (4.88 kg N Ha^{-1} yr^{-1}) were 2.7 times higher than those found during COVID-19 pandemic (1.81 kg N Ha^{-1} yr^{-1}), it means that N deposition fluxes were 63%–67% lower during the pandemic, suggesting that due to a decrease in economical and industrial activities, and the restrictions of population mobility, nitrate levels were reduced significantly. Comparing S deposition fluxes obtained two years before COVID-19 pandemic (25.03 kg S Ha^{-1} yr^{-1}) with those values obtained during the pandemic (1.95 kg S Ha^{-1} yr^{-1}) in MZM, S deposition were 12.83 times higher than those found during the pandemic, it means that S deposition fluxes were 90% lower during the COVID-19 pandemic, suggesting that the restrictions in economical, industrial and vehicular activities had a positive effect not only on air quality but also S deposition patterns. The response of S was more sensitive to the lockdown than N, it can be explained from the regional origin of sulfate since the restrictions during the global pandemic not only were applied in Mexico but also in USA and Central America resulting in a marked decrease in the regional contribution. However, besides the lockdown due to COVID-19 pandemic, the observed changes in temporal deposition patterns for N and S also can be attributed to meteorological conditions associated to each sampling season, particularly to precipitation patterns well marked that occur in semi-arid zones of Mexico as Monterrey. Mean atmospheric deposition fluxes for N and S for the two studied periods were compared with critical load values established for sensitive ecosystems in Europe; exceedances to critical load values before pandemic were significant higher than those found during pandemic period, suggesting that the lockdown improved the air quality in MZM.

ACKNOWLEDGEMENT

The authors thank the Atmospheric Monitoring System (SIMA) of Monterrey in Nuevo León State for the facilities and the support for this project.

REFERENCES

[1] Avila, A. & Aguillaume, L., Monitorización y tendencias de la deposición de N en España, incluyendo polvo sahariano. *Revista Científica de Ecología y Medio Ambiente,* **26**(1), pp. 16–24, 2017.

[2] Zhang, G., Zeng, G., Du, Ch., Jiang, Y., Su, X., Xiang, R., Huang, L., Xu, M. & Zhang, Ch., Deposition patterns in bulk precipitation and throughfall in a subtropical mixed forest in Central-South China. *Forestry*, **80**(2), pp. 211–221, 2007.

[3] Cosme de Oliveira, R., Meier, M., Da Fonseca, J.F., Beldini, T., Crill, P.M., Batista de Camargo, P. & Van Huren, J., Chemical analysis of rainfall and throughfall in the Tapajós National Forest, Belterra, Pará, Brazil. *Rev. Ambient. Água*, **10**(2), pp. 1–20, 2015.

[4] Escoffie, R., Cerón, R.M. & Cerón, J.G., Throughfall deposition of N and S to mangrove ecosystems in the Southeast of Mexico. *Latest Trends in Energy, Environment and Development,* **7**, pp. 59–65, 2014.

[5] Root, H.T., Geiser, L.H., Fenn, M.E., Jovan, S., Hutten, M.A., Ahuja, S., Dillman, K., Schirokauer, D., Berryman, S. & McMurray, J.A., A simple tool for estimating throughfall nitrogen deposition in forests of Western North America using lichens. *Forest Ecology and Management*, **306**, pp. 1–8, 2013.

[6] Butler, T.J. & Likens, G.E., A direct comparison of throughfall plus stemflow to estimates of dry and total deposition for sulphur and nitrogen. *Atmos. Env.*, **129**, pp. 1253–1265, 1995.

[7] Fenn, M.E. & Poth, M.A., Monitoring nitrogen deposition in throughfall using ion exchange resin columns: A field test in the San Bernardino Mountains. *J. Environ. Qual*, **30**, pp. 2007–2014, 2004.

[8] Simkin, S.S., Lewis, D.N., Weathers, K.C., Lovett, G.M. & Schwarz, K., Determination of sulfate, nitrate and chloride in throughfall using ino-exchange resins. *Water, Air and Soil Pollution,* **153**, pp. 343–357, 2004.

[9] Arenas, G.G., Distribución espacial y temporal de los flujos de depósito atmosférico de N y S en el Área Metropolitana de Monterrey. *Tesis de Maestría,* Universidad Autónoma del Carmen, 2018.

[10] NMX-AA-079-SCFI-2001, Determinación del ion sulfato en Aguas naturales, residuales y residuales tratadas. Análisis de aguas, 2001.
[11] NMX-AA-074-SCFI-1981, Determinación de nitratos en aguas naturales, residuales y residuales tratadas. Análisis de aguas, 1981.
[12] Fenn, M.E., Poth, M.A. & Arbough, M.J., A throughfall collection method using mixed bed ion exchange columns. *Sci. World J.*, **2**, pp. 122–130, 2002.
[13] Grennfelt, O. & Nilsson, J., Critical loads for sulfur and nitrogen. Report from a workshop held at Skokloster, Sweden, 19–24 March, The Nordic Council of Ministers Report, 1988.
[14] Sánchez, V., Evaluación del depósito atmosférico de nitrógeno y azufre en la región de Córdoba-Orizaba. *Tesis de licenciatura*, Universidad Autónoma del Carmen, 2016.
[15] García, A., Estimación del depósito atmosférico de nitrógeno y azufre en la región de Xicalango: Atasta usando muestreadores pasivos. *Tesis de Maestría*, Universidad Autónoma del Carmen, 2016.

SOURCE APPORTIONMENT AND TOXIC EVALUATION OF PARTICLE-BOUND POLYCYCLIC AROMATIC HYDROCARBONS (PAHS) IN $PM_{1.1}$ OF BAOSHAN INDUSTRIAL AREA, SHANGHAI

WEIQIAN WANG & QINGYUE WANG
Graduate School of Science and Engineering, Saitama University, Japan

ABSTRACT

Ambient particles with five size-stages ($PM_{1.1}$, $PM_{1.1–2.0}$, $PM_{2.0–3.3}$, $PM_{3.3–7.0}$, coarse particles (>7.0μm) in aerodynamic diameter) were collected in the Baoshan industrial area of Shanghai (China) during spring and early autumn, 2017. Thirteen species of polycyclic aromatic hydrocarbon (PAH) compounds were measured by gas chromatography mass spectrometry (GC-MS) to measure size distribution characteristics, possible source apportionment and health risk assessment. Average total PAHs (TPAHs) in total suspended particles (TSP) were 4.65 (1.71–12.1) ng/m^3 in spring and 3.85 (2.07–5.75) ng/m^3 in early autumn and they mainly massed in $PM_{1.1}$ at each sampling period, especially in early autumn periods (about 57.0%–79.0%). Combustion-derived PAHs (COMPAHs), carcinogenic PAHs (CANPAHs) were also main associated in fine particles, such as $PM_{1.1}$. Average lifetime lung cancer risk (LLCR) of PAH exposure by inhalation in $PM_{2.0}$ were under 1×10^{-6} to indicate no carcinogenic risk but LLCR from $PM_{1.1}$ in early autumn were higher than that in spring. Carcinogenic equivalent (BaP-TEQ), mutagenic equivalent (BaP-MEQ) and potency equivalent (BaP-PEQ) analysis showed that the toxic PAHs belonged to high molecular weight (HMW) PAHs, especially Benzo[a]pyrene(BaP). Diagnostic ratios of (BaA)/(BaA+Chy) under 0.50 and IndP/(IndP+BghiP) value (about 0.20–0.50) in $PM_{1.1}$ indicated that diesel missions and petroleum combustion might be key PAHs sources. Principal component analysis (PCA) pronounced that stationary sources and diesel emission are the main sources, and higher and BaP showed the great effects from developed steel industry group at local industrial area. Air masses backward trajectories also showed the main local industry efforts and some long-distance transport effects. In short, compared with $PM_{1.1–2.0}$, COMPAHs and CANPAHs could be considered that they were easier to gather in $PM_{1.1}$, especially in early summer periods.

Keywords: particle-bound PAHs, $PM_{1.1}$, health risk, possible sources, steel industry, Shanghai.

1 INTRODUCTION

Polycyclic aromatic hydrocarbons (PAHs) as extensive organic contaminants which comprised of two or more benzene rings were primarily generated from the combustion of biomass and fossil fuels and the pyrosynthesis of species organic materials [1]–[3]. Due to ambient temperature, nature of aerosol, and species of PAHs, PAHs in ambient exist either or in the vapor phase or absorb into ambient particles and it is reported that the species of water-soluble inorganic ions in ambient particles could increase the solubility of toxic organic compounds, such as PAHs by acting as surface-active agents [4]–[6]. Most of the heavier PAHs presence in the particle-phase in the atmosphere because of their low vapor pressure and absorbed in on the fine inhalable particles and the 5–6 rings PAHs were usually associated with $PM_{2.0}$ [4], [5], [7], [8]. PAHs were reported as one of the first atmospheric pollutants which designated as suspected carcinogen and PAHs are highly lipid-soluble and rapidly distributed in a wide variety of tissues with a marked tendency for location in body fat, the most significant health effect to be expected from inhalation exposure to PAHs is an excess risk of lung cancer to all living organisms which has reported that about 1.6% of lung

WIT Transactions on Ecology and the Environment, Vol 252, © 2021 WIT Press
www.witpress.com, ISSN 1743-3541 (on-line)
doi:10.2495/AIR210181

cancer morbidity might be got the effects by inhalation exposure to PAHs in atmosphere of the urban area in China [3], [9], [10].

Shanghai as a typical metropolis located on the eastern coast in China is the commercial and financial center with a population of more than 24 million in 2014 [11], [12]. Shanghai is also with developed industrial and transportation hub in China which always suffers with heavy human-induced impacts and drastic increases of built-up areas. Baoshan area is the main industrial one with the developed iron and steel industries [13]–[15]. PAHs in Shanghai are considered be from various sources and expected to continue display by spatiotemporal differences to raise the potential impacts on the environment and human health [8], [13]. In this article, the main objective is as follows: ambient particle-bound PAHs with five size-range in Baoshan zone of Shanghai was collected in spring and early autumn periods during 2017. Thirteen species of PAHs as the typical organic contents were measured by GC-MS to measure the chemical characteristics, toxic evaluation and possible sources. And diagnostic ratio methods and air masses trajectories also were selected to evaluate the possible sources.

2 MATERIALS AND METHODS

2.1 Ambient particle sample collection in Shanghai sampling site

Baoshan sampling sites as shown in Fig. 1 were located in Shanghai University (Baoshan campus), Shanghai. They are near to the residential zone and with several main traffic roads. Steel Group is located about 15 km northeast of these sampling sites and there were also plenty of industrial plants near to campus. The spring sampling site (N 31.32°, E 121.40°) and the early autumn sampling site (N 31.32°, E 121.33°) were very near. The spring sampling phase was from 21 March to 11 April, 2017 and the early autumn sampling phase was from 14 September to 4 October, 2017. An Andersen high volume air sampler (AHV-1000R, Shibata Co., Tokyo, Japan) with 566 L/min flow rate was to collect the ambient particle samples with five size-segregated ($PM_{1.1}$, $PM_{1.1–2.0}$, $PM_{2.0–3.3}$, $PM_{3.3–7.0}$, coarse particles (>7.0 μm) in aerodynamic diameter). Each sampling case was last to 48 hours in principle. Meteorological factors were gathered from Tianqihoubao (http://www.tianqihoubao.com/) and air quality factors were gathered from Shanghai Environmental Monitoring Center.

Figure 1: Distribution of sampling sites in Shanghai, China.

2.2 Analysis of particle-bound PAHs by the gas chromatography mass spectrometry (GC/MS) and statistical analysis

The PAHs organic extraction method and the GC/MS analysis system were mainly referred from the previous study [8]. The standard chemicals of 16 species PAHs classified by United States Environmental Protection Agency (USEPA) were obtained from Sigma-Aldrich. The extract sample solution was enriched to 0.1 ml and analysed by GC-MS (Model Agilent 5973N, Agilent Technologies. Inc., USA). All the samplers were determined by three times and then made the suitable statistical analysis for the data processing by Microsoft Excel 2016. These particle-bound PAHs are: Dibenz[a,h]anthracene (DB[a,h]A), Benzo[a]pyrene (BaP), Benzo[a]anthracene (BaA), Benzo[b]fluoranthene (BbF), Benzo[k]fluoranthene (BkF), Indeno[1,2,3-c,d]pyrene (IcdP), Anthracene (ANT), Benzo[g,h,i]perylene (BghiP), Chrysene (CHR), Fluoranthene (FLA), Fluorene (FL), Pyrene (PYR), Phenanthrene (PHE), separately.

2.3 Carcinogenic risk and mutagenicity for particle-bound PAHs

Toxic equivalency factors (TEF) based on Benzo[a]pyrene (BaP) were chosen to obtain the assessment of the potential risk of exposure to a mixture PAHs and these methods also included the carcinogenic equivalency factors and the individual mutagenicity [8], [16], [17]. The carcinogenic equivalent (BaP-TEQ), potency equivalent (BaP-PEQ) and mutagenic equivalent (BaP-MEQ) for total PAHs were calculated to estimate human health risk with the respiratory system. BaPeq (TEQ), BaPeq (PEQ) and BaPeq (MEQ) from $PM_{1.1}$ and $PM_{1.1–2.0}$ samples were calculated by the published article [18].

3 RESULTS AND DISCUSSIONS

3.1 Size distribution characteristics of total and separate particle-bound PAHs patterns

Due to Nap, Ace and Ant as the main volatile PAHs [22], the other 13 species PAHs were considered as particle-bound PAHs research objective to analyse the size distribution and characterizations. Fig. 2 showed that total PAHs in ambient particles at each sampling event. The average TPAHs in TSP was 4.65 ng/m^3 (1.71–12.1 ng/m^3) in spring and 3.85 ng/m^3 (2.07–5.75 ng/m^3) in early autumn, respectively. The largest TPAHs occurred on 25–26 March (12.1 ng/m^3) and 14–16 September (5.75 ng/m^3), the lowest one occurred on 8–9 April and 3–4 October, about 47.9% (35.4%–56.2%) TPAHs in spring were mainly measured in $PM_{1.1}$, about 23.3% (20.0%–28.8%) were in $PM_{1.1–2.0}$. In early autumn, about 66.3% (57.0%–5.8%) TPAHs were in $PM_{1.1}$, 8.94% (6.75%–12.5%) were in $PM_{1.1–2.0}$. Also, TPAHs in TSP have a very strong correlation with these in $PM_{1.1}$ ($r = 0.99$), $PM_{1.1–2.0}$ ($r = 0.97$) of spring and $PM_{1.1}$ ($r = 0.95$), $PM_{1.1–2.0}$ ($r = 0.87$) of early autumn. It could be indicated that TPAHs in $PM_{1.1}$ played an important role to PAHs in ambient particles. Agudelo-Castañeda et al. studied 16 PAHs in $PM_{1.0}$ of Canoas in Brazil were 1.32 ng/m^3 in summer and 2.02 ng/m^3 in winter [18]. He et al. studied about 16 PAHs in $PM_{2.1}$ of Nanjing were about 58.2 ng/m^3 in 2010 with those PAHs in spring higher than those in summer [23]. Wang et al. also survived about 37 species PAHs of Shanghai in summer of 2015 were on average of 9.41 ng/m^3 in TSP and 4.97 ng/m^3 in $PM_{1.1}$ [8]. Ren et al. also measured 16 species of PAHs in Xi'an and Guangzhou of China in winter and summer of 2013, it was said that they were main in $PM_{1.1}$ and that in winter was higher than that in summer [24]. Compared with those PAHs from the reported articles, PAHs in this article were also dominant in $PM_{1.1}$ which seems very consistent with those reported. And all of these were showed the analogous distribution trend which PAHs were mainly in $PM_{1.1}$ and PAHs in summer were shown the relatively light concentrations than these in other seasons.

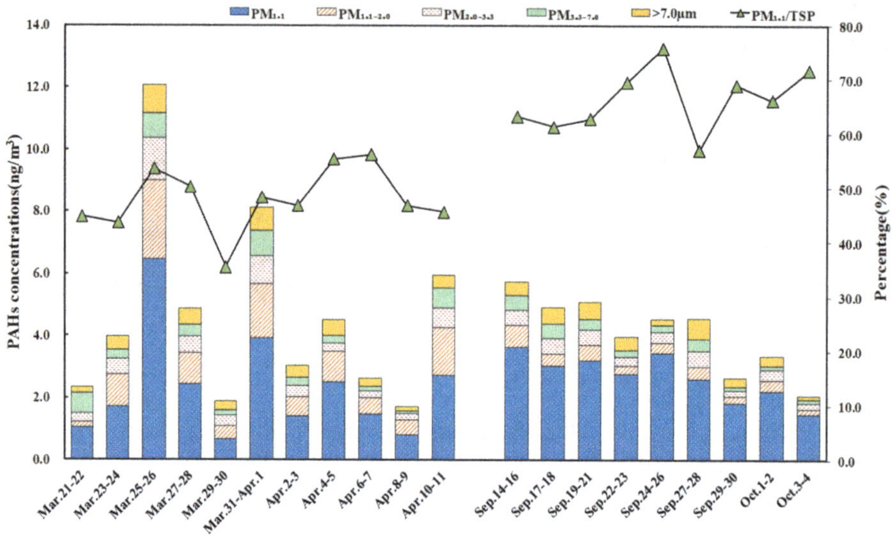

Figure 2: Size distribution of 13 PAHs in each sampling case during both spring and early autumn sampling periods.

3.2 Characteristics of total and separate PAHs patterns in $PM_{1.1}$

According to the molecular weight, PAHs could be divided into three groups, Flu, Phe and Ant with 2- and 3-ring PAHs as low molecular weight (LMW) PAHs, FLN, Pyr, Chy, BaA with 4-ring PAHs as middle molecular weight (MMW) PAHs, BbF, BkF, BaP, IndP, DB[a,h]A and BghiP which with 5- and 6-ring PAHs as high molecular weight (HMW) PAHs, separately [8], [19], [25]. As shown in Fig. 3, LMW PAHs groups, there were about 25.1% in $PM_{1.1}$, 18.9% in $PM_{1.1\text{–}2.0}$ in spring, and about 46.9% in $PM_{1.1}$, 11.9% in $PM_{1.1\text{–}2.0}$ in early autumn period. There are about 44.0% MMW PAHs in $PM_{1.1}$ and 22.2% in $PM_{1.1\text{–}2.0}$ in spring, about 67.5% in $PM_{1.1}$, 7.48% in $PM_{1.1\text{–}2.0}$ in early autumn. There are about 56.5% HMW PAHs in $PM_{1.1}$, 21.8% in $PM_{1.1\text{–}2.0}$ in spring, about 66.7% in $PM_{1.1}$, 9.92% in $PM_{1.1\text{–}2.0}$ in early autumn. It is easy to find that the over 50% HMW were located in $PM_{1.1}$. For LMW PAHs, this is the first time to find that LMW PAHs in early autumn were mostly distributed in $PM_{1.1}$. Fig. 3 also showed that all of the largest LMW, MMW and HMW in spring were on 25–26 March, largest LMW and HMW PAHs in early autumn were on 14–16 September while the largest MMW were on 19–21 September. In spring, average values of LMW:MMW:HMW in $PM_{1.1}$ were 1:8.39:15.50 and 1:5.64:7.94 in $PM_{1.1\text{–}2.0}$. In early autumn, the ratios of LMW:MMW:HMW in $PM_{1.1}$ were 1:7.44:10.2 and 1:3.34:6.22 in $PM_{1.1\text{–}2.0}$, Though TPAHs in spring were higher those in early autumn, the LMW, MMW, HMW PAHs in $PM_{1.1}$ were higher than those in spring which indicated that the PAHs might be distributed in finer particles of the early autumn which could be explained by temperature effects ($r = 0.60$). and HMW PAHs are expected to require a much longer time to partition to coarse particles than LMW PAHs and HMW PAHs only stay in fine particle with their non-volatile characteristic [8], [18], [26].

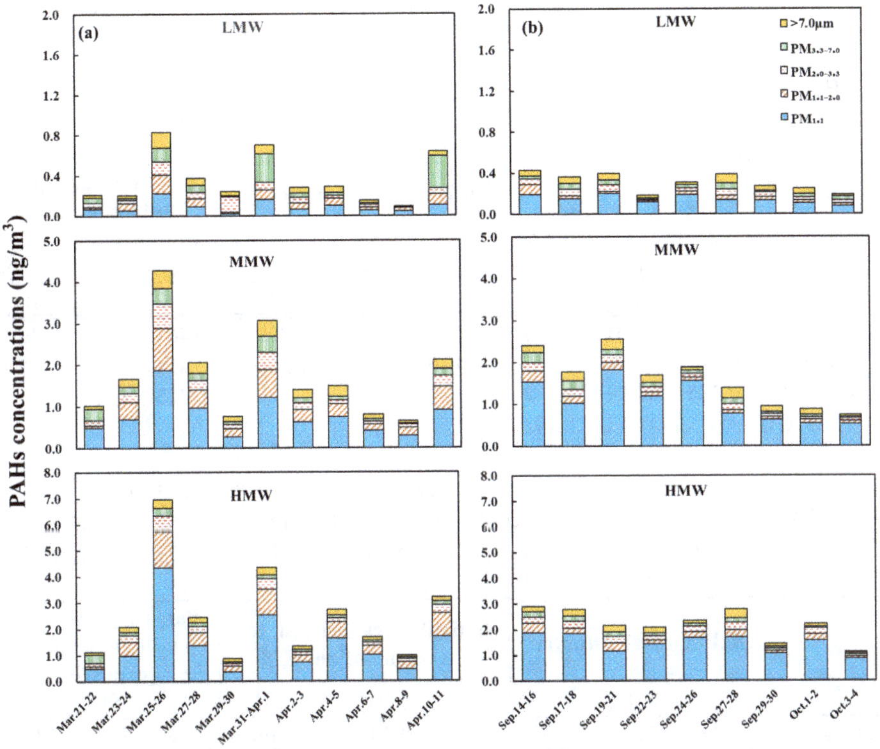

Figure 3: Variation of particle bounds LMW,MMW and HMW PAHs in each sampling case. (a), spring sampling period; and (b), early autumn period.

In addition, COMPAHs has been classified including Flu, Pyr, BaA, Chy, BbF, BkF, BaP, IndP and BghiP and COMPAHs/TPAHs is a characteristic diagnostic parameter for PAHs produced by stationary sources [23], [27]. The ratios of COMPAHs/TPAHs in $PM_{1.1}$ were 72.3%–92.7% in spring and 85.8%–92.7% in early autumn. The lower ratios of LMW/HMW and the higher ratios of COMPAHs/ TPAHs in $PM_{1.1}$ and $PM_{1.1-2.0}$ could be indicated that combustion sources might play an important role in fine particles. CANPAHs are to assess the carcinogenic potential of PAHs to the exposed people and usually including BaA, Chy, BkF, BaP, IndP and DB[a,h]A [22], [28]. Species CANPAHS are from COMPAHs group. The ratios of CANPAHs to TPAHs in TSP were 45.04% (39.9–48.6%) in spring and 49.26% (43.0%–59.5%) in early autumn. Moreover, over 70% of CANPAHS were a trend to distributed in $PM_{2.0}$ ($PM_{1.1}$+$PM_{1.1-2.0}$) in both spring and early autumn. Compared with those in spring, CANPAHS in early autumn were more trend to locate in $PM_{1.1}$. These PAHs are of interest due to their potential toxicity, carcinogenic, and mutagenicity.

3.3 Health risk evaluation of PAHs patterns in $PM_{1.1}$ and $PM_{2.0}$

Although atmospheric PAHs exist at relatively low concentrations, epidemiologic studies have linked long-term low-level PAH exposure with various health outcomes, including cancers among adult populations [16]. Carcinogenic risk and mutagenicity were calculated in combination with measured PAHs in fine particles. The BaP-equivalent comparison of

BaP-TEQ $_{(13PAHs)}$, BaP-TEQ $_{(6PAHs)}$, BaP-PEQ and BaP-MEQ of $PM_{1.1}$ and $PM_{2.0}$ in spring and early autumn were shown in Table 1.

Table 1: BaP-equivalent and individual cancer risk for inhalation for PAHs in $PM_{2.0}$ and $PM_{1.1}$ in Shanghai, during spring and early autumn sampling periods.

	BaP-equivalent(ng/m^3)			Individual cancer risk for inhalation			Potential cases per million people exposed		
	$PM_{2.0}$			$PM_{2.0}$			$PM_{2.0}$		
Spring	Ave.	Max.	Min.	Ave.	Max.	Min.	Ave.	Max.	Min.
(BAP-TEQ)13PAHs	0.424	1.261	0.118	4.67E-07	1.39E-06	1.30E-07	0.467	1.387	0.130
(BAP-TEQ)7PAHs	0.394	1.174	0.109	4.34E-07	1.29E-06	1.20E-07	0.434	1.292	0.120
(BAP-PEQ)6PAHs	0.385	1.194	0.108	4.23E-07	1.31E-06	1.19E-07	0.423	1.314	0.119
(BAP-MEQ)8PAHs	0.665	1.982	0.192						
Early autumn									
(BAP-TEQ)13PAHs	0.358	0.519	0.179	3.94E-07	5.71E-07	1.96E-07	0.394	0.571	0.196
(BAP-TEQ)7PAHs	0.328	0.480	0.158	3.61E-07	5.28E-07	1.74E-07	0.361	0.528	0.174
(BAP-PEQ)6PAHs	0.326	0.499	0.155	3.59E-07	5.48E-07	1.71E-07	0.359	0.548	0.171
(BAP-MEQ)8PAHs	0.554	0.790	0.329						
	BaP-equivalent(ng/m^3)			**Individual cancer risk for inhalation**			**Potential cases per million people exposed**		
	$PM_{1.1}$			$PM_{1.1}$			$PM_{1.1}$		
Spring	Ave.	Max.	Min.	Ave.	Max.	Min.	Ave.	Max.	Min.
(BAP-TEQ)13PAHs	0.299	0.953	0.070	3.28E-07	1.05E-06	7.66E-08	0.328	1.048	0.077
(BAP-TEQ)7PAHs	0.277	0.888	0.064	3.05E-07	9.77E-07	6.99E-08	0.305	0.977	0.070
(BAP-PEQ)6PAHs	0.274	0.914	0.064	3.01E-07	1.01E-06	7.01E-08	0.301	1.005	0.070
(BAP-MEQ)8PAHs	0.477	1.515	0.118						
Early autumn									
(BAP-TEQ)13PAHs	0.320	0.462	0.163	3.52E-07	5.08E-07	1.80E-07	0.352	0.508	0.180
(BAP-TEQ)7PAHs	0.294	0.430	0.144	3.23E-07	4.73E-07	1.59E-07	0.323	0.473	0.159
(BAP-PEQ)6PAHs	0.294	0.445	0.142	3.23E-07	4.90E-07	1.56E-07	0.323	0.490	0.156
(BAP-MEQ)8PAHs	0.490	0.687	0.300						

13 PAHs, Pyr, Phe,FLU, FLN, Chy, B[ghi]P, ANT, IndP, BkF, BbF, BaA, BaP, DB[a,h]A;
7 PAHs, IndP, BkF, BbF, BaA, BaP, DB[a,h]A, Chy;
6 PAHs, BkF, BbF, BaA, BaP, IndP, Chy;
8 PAHs, BkF, BbF, BaA, BaP, DB[a,h]A, Chy, IndP.

Average individual cancer risk for inhalation by BaP-TEQ $_{(13\ PAHs)}$, BaP-TEQ $_{(6\ PAHs)}$, BaP-PEQ both from $PM_{2.0}$ and $PM_{1.1}$ were under 10^{-6} which indicated that there are no cancer risk under these conditions. In spring, BaP-TEQ $_{(13\ PAHs)}$ in $PM_{2.0}$ were about 0.424 ng/m^3 BaP-equivalent indicated that potential cancer cases for per million people exposed with this condition was 0.467, and average BaP-TEQ $_{(13\ PAHs)}$ in $PM_{1.1}$ were 0.299 ng/m^3 with 70.5% contribution to that in $PM_{2.0}$. In early autumn, BaP-TEQ $_{(13\ PAHs)}$ in $PM_{2.0}$ was 0.358 ng/m^3 BaP-equivalent with the 0.394 potential cancer cases for per million people exposed under this condition and they are almost from $PM_{1.1}$. In $PM_{2.0}$, BaP-TEQ $_{(6\ PAHs)}$ in spring were 0.394 ng/m^3 and 0.328 ng/m^3 early autumn. According to these two different calculation methods of BaP-TEQ, max and average BaP-TEQ in spring were higher than in early autumn while the min BaP-TEQ were higher in early autumn. It was also amazing that the average BaP-TEQ $_{(13\ PAHs)}$ and BaP-TEQ $_{(6\ PAHs)}$ in $PM_{1.1}$ were higher in early autumn period while these in spring were higher in $PM_{2.0}$. Also the min BaP-TEQ in $PM_{1.1}$ of the early autumn period were higher than in spring. For the BaP-PEQ, the average in spring were 0.385 ng/m^3 in $PM_{2.0}$ with 0.423 potential cases for per million people and 0.274 ng/m^3 in $PM_{1.1}$ with 0.328 potential cases for per million people. In early autumn, the average of BaP-PEQ were 0.326 ng/m^3 in $PM_{2.0}$ and 0.294 ng/m^3 in $PM_{1.1}$. For the BaP-MEQ, the average values in spring were 0.665 ng/m^3 in $PM_{2.0}$ and 0.477 ng/m^3 in $PM_{1.1}$. In early autumn, the average of BaP-MEQ were 0.554 ng/m^3 in $PM_{2.0}$ and 0.490 ng/m^3 in $PM_{1.1}$. Both BaP-PEQ and BaP-MEQ could be considered that these values were directly affected from these relative PAHs from $PM_{1.1}$. In $PM_{1.1}$, average BaP-TEQ $_{(13\ PAHs)}$, BaP-TEQ $_{(6\ PAHs)}$, BaP-PEQ and BaP-MEQ in early autumn were higher than in spring. Over 50% BaP-equivalent of BaP-TEQ, BaP-PEQ and BaP-MEQ in $PM_{2.0}$ were from $PM_{1.1}$ and PAHs distributed in $PM_{1.1}$ could be considered to pose a stronger toxicity and do great harm to the health.

According to BaP-equivalent values, species dominant PAHs contribution rate to BaP-TEQ, BaP-PEQ and BaP-MEQ were calculated. In both $PM_{1.1}$ and $PM_{2.0}$ of spring, BaP, BbF and IndP could be considered as the main contributor to kinds of BaP-equivalent values, especially for BaP. In early autumn, BaP and BbF were shown to make the mostly contribution to BaP-TEQ, BaP-PEQ and BaP-MEQ in both $PM_{1.1}$ and $PM_{2.0.}$ It should be noticed that not only BaP but BbF and IndP also could be considered make the great contribution to kinds of health risks. It was obvious to find that BaP-TEQ$_{(13\ PAHs)}$, BaP-TEQ$_{(6\ PAHs)}$, BaP-PEQ and BaP-MEQ in $PM_{2.0}$ were main from these in $PM_{1.1}$. The average of them in $PM_{1.1}$ was larger in early autumn rather than spring. The main contributor PAHs to BaP-TEQ$_{(13\ PAHs)}$, BaP-TEQ$_{(6\ PAHs)}$, BaP-PEQ and BaP-MEQ in spring was obviously BaP. In $PM_{1.1}$ of early autumn, BaP and BbF played more important role to BaP-TEQ$_{(13\ PAHs)}$, BaP-TEQ$_{(6\ PAHs)}$ and BaP-PEQ while the BaP-MEQ. This result seems to consistent with the previous articles such as Agudelo-Castaneda et al. showed the BaP-TEQ (0.324 ng/m^3), BaP-PEQ (0.194 ng/m^3) and BaP-MEQ (0.489 ng/m^3) in $PM_{1.0}$ of metropolitan area of Porto Alegera area [16]. Wang et al. also studied to find that the toxicity of PAHs in $PM_{1.1}$ posed the significant health risks in Shanghai [8]. Ren et al. surveyed the BaP-TEQ and BaP-MEQ in atmosphere of two megacity named Xi'an and Guangzhou in China which showed PAHs in $PM_{1.1}$ were also showed more toxic in both Xi'an and Guangzhou [24].These PAHs with great health risks were belonging to HMW-PAHs and a group named COMPAHs. HMW PAHs and COMPAHs associated with $PM_{1.1}$ indicated the contribution of dense traffic, especially of heavy vehicles, as reported in previous studies. To know more about the sources of them is also one important study point.

3.4 Possible anthropogenic sources of atmospheric particles

PAHs are as one group of persistent pollutants which originate primarily from the incomplete combustion of fuel, especially from vehicular emissions and heating and so on [19]. In spring, LMW/HMW were 0.065 in $PM_{1.1}$ and 0.126 in $PM_{1.1\text{–}2.0}$, all COMB/TPAHs values in $PM_{1.1}$ and $PM_{1.1\text{–}2.0}$ were over 0.90 which indicated the main combustion sources in these research areas which Wang et al.'s study showed the similar sources for Shanghai [8]. Nowadays, diagnostic ratios have been widely calculated for characterizing the sources for atmospheric PAHs and several marker compounds are often used to indicate the prevalence of gasoline or diesel engines as well as the relative contribution of traffic, domestic heating, wood combustion and other sources [29], [30]. Fig. 4 showed typical PAHs diagnostic ratios in $PM_{1.1}$and $PM_{1.1\text{–}2.0}$ in both sampling periods.

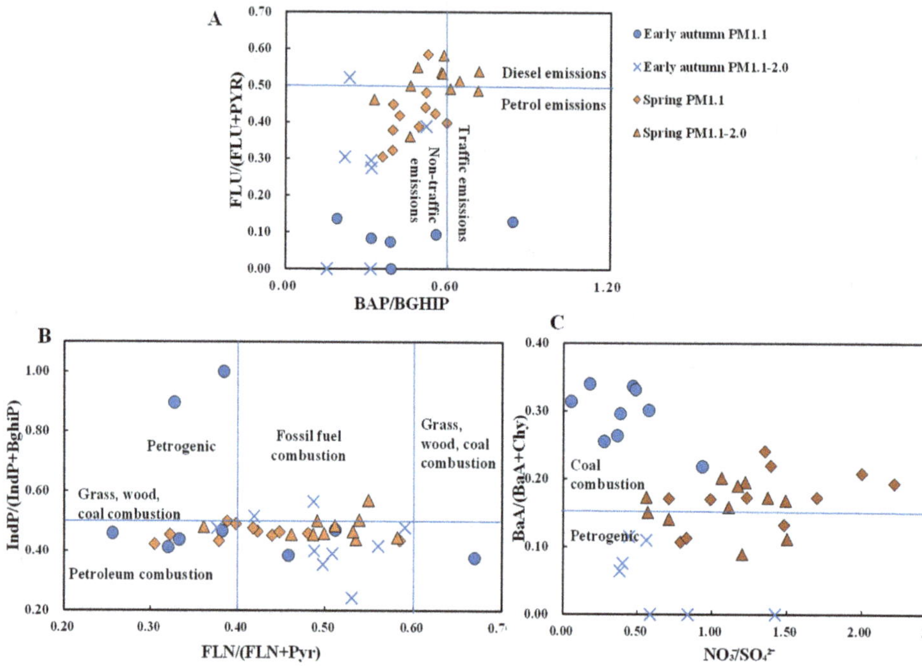

Figure 4: Correlation between PAHs diagnostic ratios and NO_3^-/SO_4^{2-} for source identification in $PM_{1.1}$ and $PM_{1.1-2.0}$ of spring and early autumn. (a) BaP/BghiP, Flu/(Flu+Pyr); (b) Fln/(Fln+Pyr), IndP/(IndP+BghiP); and (c) BaA/(BaA+Chy), NO_3^-/SO_4^{2-}.

In spring, the Flu/(Flu+Pyr) in $PM_{1.1}$ were almost under 0.50 to show the important sources of petrol emissions. Flu/(Flu+Pyr) in $PM_{1.1-2.0}$ indicated the sources of diesel emissions and petrol emissions [7]. Bap/Bghip in $PM_{1.1}$ were main got the efforts from non-traffic emissions, and $PM_{1.1-2.0}$ were sometimes got efforts from traffic emissions [7]. Fig. 4 showed values of IndP/(IndP+BghiP) in $PM_{1.1}$ and $PM_{1.1-2.0}$ were got the main efforts from petroleum combustions. The values of FLN/(FLN+Pyr) in $PM_{1.1}$ indicated two main sources such as petrogenic and fossil fuel combustion (e.g., coal combustion) while the $PM_{1.1-2.0}$ showed the main sources from combustion such as fossil fuel and grass, wood and coal combustion. The ratios of BaA/(BaA+Chy) in $PM_{1.1}$ showed that they might be got the efforts from petrogenic and coal combustion sources and mostly got the efforts from petrogenic sources in $PM_{1.1-2.0}$. Generally, it could be considered the traffic emissions and coal combustion are the important PAHs sources. PCA analysis also showed the main sources in $PM_{1.1}$ were from traffic emissions and coal and diesel combustion. In early autumn, Flu/(Flu+Pyr) in $PM_{1.1}$ and $PM_{1.1-2.0}$ showed the main sources of petrol emissions, BaP/BghiP in $PM_{1.1}$ occasionally were over 0.60 which indicated the traffic emissions to $PM_{1.1}$. FLN/(FLN+Pyr) in $PM_{1.1}$ were almost under 0.40 showed the important from petrogenic sources and these values in $PM_{1.1-2.0}$ were almost over 0.40 showed the main efforted from grass, wood, coal and fossil fuel combustion. IndP/(IndP+BghiP) in $PM_{1.1}$ and $PM_{1.1-2.0}$ were almost under 0.50 which showed the important sources of petroleum combustion. BaA/(BaA+Chy) in $PM_{1.1}$ were in the range 0.20–0.35 which showed the main sources from coal combustion. And the values in $PM_{1.1-2.0}$ were below 0.20 which showed the main effects from petrogenic sources. The mass ratios of NO_3^-/SO_4^{2-} in $PM_{1.1}$ and $PM_{1.1-2.0}$ were almost below 1.0 which showed the important sources of stationary sources. It was obviously shown that the important sources in $PM_{1.1}$ were petroleum combustion and coal combustion, and the sources in $PM_{1.1-2.0}$ might be fossil fuel combustion and petrogenic sources. PCA analysis also showed the main sources in $PM_{1.1}$ were from coal and diesel combustion. And it was interesting that the source in $PM_{1.1}$ and $PM_{1.1-2.0}$ not only in spring but in early autumn showed some obvious different source efforts.

Diagnostics ratios analysis showed petroleum and coal combustion which main from the industries and vehicle emissions. Air mass backward trajectories at four sampling events such as, S-MAX (25–26 March) with the max TPAHs in spring, S-MIN (8–9 April) with the light PAHs in spring and EA-MAX (14–16 September) which with the max PAHs in TSP in early autumn, EA-MIN (3–4 October) (light TPAHs in early autumn) were calculated on Hybrid Single-Particle Lagrangian Integrated Trajectory model (HYSPLIT) from NOAA Air Resources and shown in Fig. 5. Air masses at S-MAX sampling period across the northwest part of China to Shanghai, which is the main coal combustion area in China [8], [31]. At S-MIN, EA-MAX and EA-MIN sampling cases, the air masses were main from marine area with northeast direction and played cleaning role to the atmosphere. It could be considered that the atmospheric environment in spring was easy to get long-range transport air masses efforts with pollutants which could be as an important factor to explain the occurred of the relative severe air pollution events. Air masses at all sampling events were from the northeast direction where the steel group and port location indicated the important local source effect.

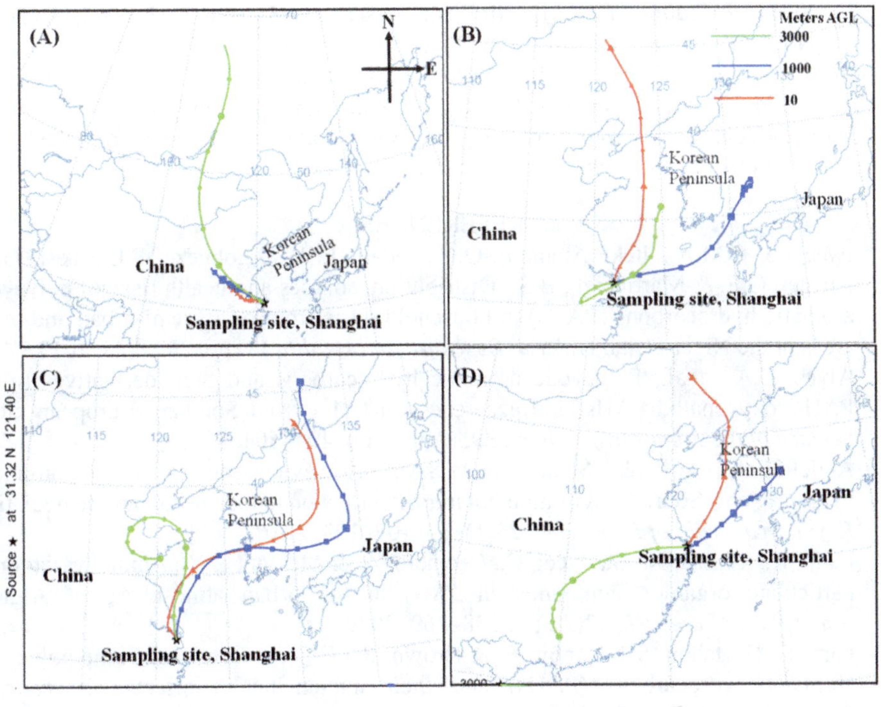

Figure 5: Air masses backward trajectories calculated by HYSPLIT model (GDAS meteorological data) with four sampling events. (a) S-MAX; (b) S-MIN; (c) EA-MAX; and (d) EA-MIN.

4 CONCLUSIONS

Particle-bound PAHs in ambient five size-ranges particle of Baoshan area of Shanghai were firstly collected to measure their chemical characterization, possible source and toxic evaluation during spring and early autumn. Chemical analysis indicated that PAHs were with higher accumulation efficiency in $PM_{1.1}$, especially in early autumn periods. We should do more study to evidence the conditions of PAHs accumulation efficiency, the influence of meteorological factors, high sunlight, along with the photochemical degradation of PAHs. Diagnostic ratios, PCA analysis and air mass backward trajectories showed that local stationary sources (e.g., diesel emission and industries source) are main PAHs sources which might be from steel industry and the possible relative transportation. Higher PAHs in $PM_{1.1}$ during early autumn could be explained by the effect from the steamship emissions on marine by air masses transport. BaP-equivalents of BaP-TEQ, BaP-PEQ and BaP-MEQ got great contribution from HMW groups, the most toxic one is BaP, which this result should be with more evidence by cytotoxicity experiments and simulation experiments and investigations. It was surprised that PAHs could be keep a relative high concentration with more finer particles though with good air quality. In addition to these PAHs, there are also several tens of components PAHs should be paid attention to, such as the nitro-PAHs and oxygen-PAHs, these ingredients also have serous effects on health and so on. Compared with particles' mass concentration, these PAH compounds and trace elements should also attract more attention to improve the environmental conditions.

ACKNOWLEDGEMENTS

Some works of this study were supported by the Special Funds for Innovative Area Research (No. 20120015, FY2008–FY2012) and Basic Research (B) (No. 24310005, FY2012–FY2014; No. 18H03384, FY2017–FY2020) of Grant-in-Aid for Scientific Research of Japanese Ministry of Education, Culture, Sports, Science and Technology (MEXT) and the Steel Foundation for Environmental Protection Technology of Japan (No. C-33, FY2015–FY2017).We would like to express great gratitude to several master students from the LU lab in Shanghai University who helped with our sampling work.

REFERENCES

[1] Iwegbue, C.M.A., Iteku-Atata, E.-O.C., Odali, E.W., Egobueze, F.E., Tesi, G.O., Nwajei, G.E. & Martincigh, B.S., Distribution, sources and health risks of polycyclic aromatic hydrocarbons (PAHs) in household dusts from rural, semi-urban and urban areas in the Niger Delta, Nigeria. *Exposure and Health*, **11**, pp. 209–225, 2019.

[2] Alves, C.A. et al., Polycyclic aromatic hydrocarbons and their derivatives (nitro-PAHs, oxygenated PAHs, and azaarenes) in $PM_{2.5}$ from Southern European cities. *Science of the Total Environment*, **595**(April), pp. 494–504, 2017.

[3] Abdel-Shafy, H.I. & Mansour, M.S.M., A review on polycyclic aromatic hydrocarbons: Source, environmental impact, effect on human health and remediation. *Egyptian Journal of Petroleum*, **25**(1), pp. 107–123, 2016.

[4] Ladji, R., Yassaa, N., Balducci, C., Cecinato, A. & Meklati, B.Y., Annual variation of particulate organic compounds in PM_{10} in the urban atmosphere of Algiers. *Atmospheric Research*, **92**(2). pp. 258–269, 2009.

[5] Kim, K.-H., Jahan, S.A., Kabir, E. & Brown, R.J.C., A review of airborne polycyclic aromatic hydrocarbons (PAHs) and their human health effects. *Environment International*, **60**, pp. 71–80, 2013.

[6] Wang, Z. et al., Gas/particle partitioning of polycyclic aromatic hydrocarbons in coastal atmosphere of the north Yellow Sea, China. *Environmental Science and Pollution Research*, **20**(8). pp. 5753–5763, 2013.

[7] Tobiszewski, M. & Namieśnik, J., PAH diagnostic ratios for the identification of pollution emission sources. *Environmental Pollution*, **162**, pp. 110–119, 2012.

[8] Wang, Q. et al., Studies on size distribution and health risk of 37 species of polycyclic aromatic hydrocarbons associated with fine particulate matter collected in the atmosphere of a suburban area of Shanghai city, China. *Environmental Pollution*, **214**, pp. 149–160, 2016.

[9] Khan, A., Ishaq, M. & Khan, M.A., Effect of vehicle exhaust on the quantity of polycyclic aromatic hydrocarbons (PAHs) in soil. *Environmental Monitoring and Assessment*, **137**(1–3), pp. 363–369, 2008.

[10] Zhang, Y. & Tao, S., Global atmospheric emission inventory of polycyclic aromatic hydrocarbons (PAHs) for 2004. *Atmospheric Environment*, 43(4), pp. 812–819, 2009.

[11] Ding, X.X. et al., Characteristics of size-resolved atmospheric inorganic and carbonaceous aerosols in urban Shanghai. *Atmospheric Environment*, **167**, pp. 625–641, 2017.

[12] Wu, M. et al., Analysis of perfluorinated compounds in human serum from the general population in Shanghai by liquid chromatography-tandem mass spectrometry (LC-MS/MS). *Chemosphere*, **168**, pp. 100–105, 2017.

[13] Huang, Y. et al., A multimedia fugacity model to estimate the fate and transport of polycyclic aromatic hydrocarbons (PAHs) in a largely urbanized area, Shanghai, China. *Chemosphere*, **217**, pp. 298–307, 2019.

[14] Feng, Y., Liu, Y. & Tong, X., Spatiotemporal variation of landscape patterns and their spatial determinants in Shanghai, China. *Ecological Indicators*, **87**, pp. 22–32, 2018.

[15] Chen, Y., Wang, J., Shi, G., Sun, X., Chen, Z. & Xu, S., Human health risk assessment of lead pollution in atmospheric deposition in Baoshan District, Shanghai. *Environmental Geochemistry and Health*, **33**(6), pp. 515–523, 2011.

[16] Agudelo-Castañeda, D., Teixeira, E., Schneider, I., Lara, S.R. & Silva, L.F.O., Exposure to polycyclic aromatic hydrocarbons in atmospheric $PM_{1.0}$ of urban environments: Carcinogenic and mutagenic respiratory health risk by age groups. *Environmental Pollution*, **224**. pp. 158–170, 2017.

[17] Jung, K.H. et al., Assessment of benzo(a)pyrene-equivalent carcinogenicity and mutagenicity of residential indoor versus outdoor polycyclic aromatic hydrocarbons exposing young children in New York city. *International Journal of Environmental Research and Public Health*, 7(5). pp. 1889–1900, 2010.

[18] Agudelo-Castañeda, D.M., Teixeira, E.C., Schneider, I.L., Rolim, S.B.A., Balzaretti, N. & e Silva, G.S., Comparison of emissivity, transmittance, and reflectance infrared spectra of polycyclic aromatic hydrocarbons with those of atmospheric particulates (PM_1). *Aerosol and Air Quality Research*, **15**(4), pp. 1627–1639, 2015.

[19] Shen, R., Wang, Y., Gao, W., Cong, X., Cheng, L. & Li, X., Size-segregated particulate matter bound polycyclic aromatic hydrocarbons (PAHs) over China: Size distribution, characteristics and health risk assessment. *Science of the Total Environment*, **685**, pp. 116–123, 2019.

[20] Dan, M., Zhuang, G., Li, X., Tao, H. & Zhuang, Y., The characteristics of carbonaceous species and their sources in $PM_{2.5}$ in Beijing. *Atmospheric Environment*, **38**(21). pp. 3443–3452, 2004.

[21] Traversi, D., Festa, E., Pignata, C. & Gilli, G., Aero-dispersed mutagenicity attributed to particulate and semi volatile phase in an urban environment. *Chemosphere*, **124**(1), pp. 163–169, 2015.

[22] del R. Sienra, M., Rosazza, N.G. & Préndez, M., Polycyclic aromatic hydrocarbons and their molecular diagnostic ratios in urban atmospheric respirable particulate matter. *Atmospheric Research*, **75**(4). pp. 267–281, 2005.

[23] He, J., Fan, S., Meng, Q., Sun, Y., Zhang, J. & Zu, F., Polycyclic aromatic hydrocarbons (PAHs) associated with fine particulate matters in Nanjing, China: Distributions, sources and meteorological influences. *Atmospheric Environment*, **89**, pp. 207–215, 2014.

[24] Ren, Y. et al., Composition and size distribution of airborne particulate PAHs and oxygenated PAHs in two Chinese megacities. *Atmospheric Research*, **183**, pp. 322–330, 2017.

[25] Wang, Y. et al., The ion chemistry, seasonal cycle, and sources of $PM_{2.5}$ and TSP aerosol in Shanghai. *Atmospheric Environment*, 40(16), pp. 2935–2952, 2006.

[26] Hassanien, M.A. & Abdel-Latif, N.M., Polycyclic aromatic hydrocarbons in road dust over Greater Cairo, Egypt. *Journal of Hazardous Materials*, **151**(1). pp. 247–254, 2008.

[27] Zhang, Y., Wang, X., Chen, H., Yang, X., Chen, J. & Allen, J.O., Source apportionment of lead-containing aerosol particles in Shanghai using single particle mass spectrometry. *Chemosphere*, **74**(4), pp. 501–507, 2009.

[28] Hong, H.S., Yin, H.L., Wang, X.H. & Ye, C.X., Seasonal variation of PM_{10}-bound PAHs in the atmosphere of Xiamen, China. *Atmospheric Research*, **85**(3–4). pp. 429–441, 2007.

[29] Niu, X. et al., Atmospheric levels and cytotoxicity of polycyclic aromatic hydrocarbons and oxygenated-PAHs in $PM_{2.5}$ in the Beijing-Tianjin-Hebei region. *Environmental Pollution*, **231**(1), pp. 1075–1084, 2017.

[30] Y. L. Zhang and F. Cao, Fine particulate matter ($PM_{2.5}$) in China at a city level, *Sci. Rep.*, vol. 5, 2015.

[31] Q. Wang and W. Wang, Size characteristics and health risks of inorganic species in $PM_{1.1}$ and $PM_{2.0}$ of Shanghai, China, in spring, 2017, *Environ. Sci. Pollut. Res.*, 2020.

SECTION 6
GLOBAL, REGIONAL AND LOCAL STUDIES

EL NIÑO SOUTHERN OSCILLATION (ENSO)-INDUCED PM_{10} VARIABILITY IN MINING AREAS WITHIN THE 2014–2017 PERIOD

HELI A. ARREGOCÉS[1,2], ROBERTO ROJANO[1] & GLORIA RESTREPO[2]
[1]Grupo de Investigación GISA, Facultad de Ingeniería, Universidad de La Guajira, Colombia
[2]Grupo Procesos Fisicoquímicos Aplicados, Facultad de Ingeniería, Universidad de Antioquia SIU/UdeA, Colombia

ABSTRACT

Given that El Niño Southern Oscillation (ENSO) phenomena had significant environmental impacts in 2015–2016, this study presents an exploratory analysis of PM_{10} variability in open-pit mining areas in northern Colombia. PM_{10} and meteorological variables were measured at three locations, and analyses were performed. The results obtained showed that ENSO plays an important role in the temporal distribution of daily precipitation, and its occurrence resulted in an 8.3% decrease in annual average precipitation within the period of February 2015–May 2016. However, it had no significant effect on temperature, relative humidity, and wind speed within the study period. Further, the annual mean PM_{10} concentration corresponding to all the stations considered in this study between 2014 and 2017 was 34 $\mu g/m^3$ (95% CI, 33–35 $\mu g/m^3$), and the standard deviation was ± 11.59 $\mu g/m^3$. The individual arithmetic means of PM_{10} concentrations at the different stations were 1.11–2.07 times the WHO air quality guidelines for PM_{10} but did not exceed the maximum level permissible specified by the Colombian norm. Furthermore, the average PM_{10} concentrations in all the samples corresponding to the El Niño classified Oceanic Niño Index (ONI) phase (37.14 $\mu g/m^3$; CI95% 36.06–38.22 $\mu g/m^3$) were numerically higher than those corresponding to the neutral classified ONI phase (32.54 $\mu g/m^3$; CI95% 31.82–33.27 $\mu g/m^3$), and the correlation coefficients between PM_{10} and meteorological parameters were found to be relatively low.
Keywords: PM_{10}, ENSO, ONI, variability, air quality.

1 INTRODUCTION

The El Niño Southern Oscillation (ENSO) is an atmosphere/ocean phenomenon of significance owing to its global effects and multiple interrelations. ENSO events are associated with economic losses due to different conditions resulting from the teleconnections that exist between them and other factors such as climate and public health in different areas. Additionally, ENSO is considered to be the most important climate cycle that is usually associated with the interannual variability climate, with the possibility of engendering extreme weather events, such as heavy rainfall, droughts, and storms [1]. Studies on the temporal and spatial behavior of $PM_{2.5}$ and PM_{10} have shown that their distribution strongly depends on meteorological variables [2], [3]. Consequently, the effects of meteorological variables on the formation and deposition of atmospheric particles are well established [4]–[6].

The ENSO phenomenon brings about dry conditions in the north and northeast of South America, and consequently opposite conditions in the southeast and southwest regions. This behavior is due to the anomalies that are associated with the Walker circulation and the Hadley cells on both sides of the equator [7]. These interannual climatic variations that cause precipitation anomalies have been reported in several studies [8]–[10].

Further, from a public health perspective, ENSO events are associated with changes in marine ecosystems and the spread of bacterial diseases, which are increasingly propagated owing to the local geographical conditions [11]. Similarly, some epidemiological studies on

WIT Transactions on Ecology and the Environment, Vol 252, © 2021 WIT Press
www.witpress.com, ISSN 1743-3541 (on-line)
doi:10.2495/AIR210191

air pollution have incorporated the correlation between meteorological variables and pollutant concentration levels [12], and several studies have been conducted to clarify how El Niño variability brings about climate responses owing to the patterns of the teleconnections that are associated with other factors. The results of these studies have provided evidence of the influence of ENSO on the variability of the concentration of particulate materials with sizes below 10 μm (PM_{10}) [13], [14]. Furthermore, several studies have been conducted in this regard, focusing on precipitation frequency levels, which significantly influence the transport and dispersion of PM_{10}. Specifically, Singh and Palazoglu [15], [16] studied the effects of ENSO events on particulate matter in six regions in the United States and revealed that such effects are detectable in terms of the seasonal and annual variability of air pollution episodes.

Therefore, given the occurrence of the El Niño phenomenon in northern South America, it can affect local-scale phenomena, particularly the concentrations of pollutants and particulate matter. However, its effects on PM_{10} concentrations in Colombia have not been previously investigated. Therefore, this study focuses on the El Cerrejón mine, which emits some of the highest levels of PM in northern Colombia, with primary aims to (1) statistically analyze the spatiotemporal behavior of PM_{10} concentrations and meteorological data in this mine area and (2) analyze the relationships between PM_{10} levels and ENSO. The remainder of this paper is organized as follows: Section 2 describes the data and methodology used, Section 3 provides the results and discussions considering existing literature, and Section 4 concludes the paper.

2 MATERIALS AND METHODS

The open-pit coal mine and the area affected by the related mining activities are located in northern Colombia (11° 5 N, 72° 40 W). East of this area is the Sierra Nevada de Santa Marta, and in the west are the Perijá Mountains, which lie along the Colombian border with Venezuela. This area includes three municipalities (Albania, Hatonuevo, and Barrancas), and has a total population of 59,089 inhabitants of which 34% are indigenous people. Fig. 1 shows the map of this mining region, with the PM_{10} sampling stations, Barrancas (BR), Las Casitas (LC), Patilla (PT), Provincial (PV), and Sol y sombra (SS) indicated. The figure also shows the locations of the mining emission sources, i.e., the La Puente, Tabaco, Patilla, Comunero, 100, and Oreganal pits.

2.1 Meteorological data

Meteorological measurements corresponding to the 2014–2017 period included hourly mean temperature (T, °C), wind speed (WS, m/s), wind direction (WD, degrees), relative humidity (RH, %), and precipitation (Pp, mm). The data were obtained from the stations with meteorological sensors, i.e., PV, BR, and LC, which are representative of the six main exploitation areas from north to south, i.e., La Puente Pit (LPUp), Tabaco pit (TABp), Patilla pit (PATp), Comunero pit (COMp), 100 pit (100p), and Oreganal pit (OREp). The daily averages of each variable were calculated from the hourly data based on the availability of valid measurements at least 75% each day. The daily averages of WD were calculated using the air quality tool of R-Studio software version 3.6.3, OpenAir [17].

2.2 Monitoring of PM_{10} data

PM_{10} data corresponding to the 2014–2017 period were collected. This period was chosen because it includes PM_{10} data corresponding to one year before and one year after the

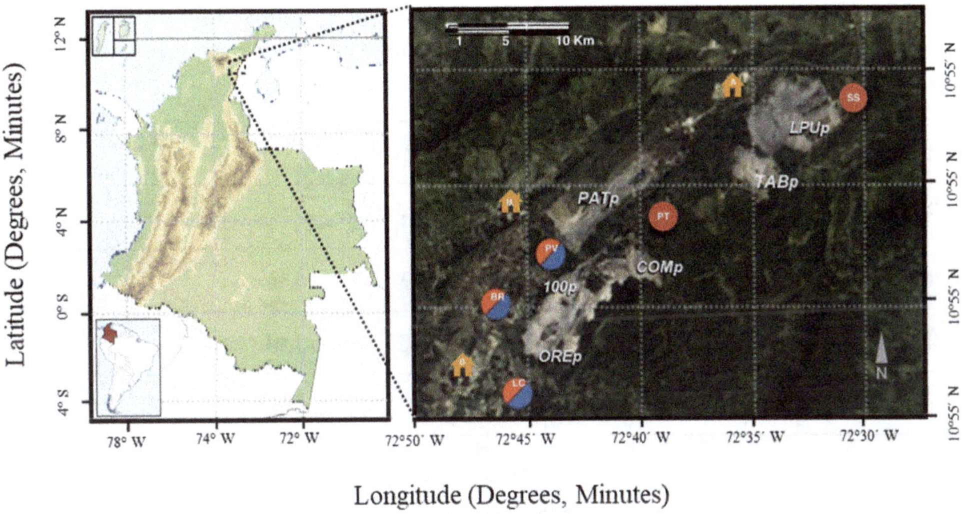

Figure 1: Geographical location of study area showing the PM_{10} sampling stations from north to south: SS (Sol y sombra), PT (Patilla), PV (Provincial), BR (Barrancas), and LC (Las Casitas); and the nearby municipalities, Albania (A), Hatonuevo (H), and Barrancas (B). The extraction pits are also shown. Left panel: Modified from Mapa de Colombia (orografía).svg Wikipedia, https://goo.gl/; Right panel: Google earth v7.1.5.1557, 2016–03–02, DigitalGlobe 2016. http://www.earth.google.com.

occurrence of the 2015–2016 El Niño event. Specifically, particulate matter samples were collected using a high-volume sampler (Hi-Vol TE- 6,070) obtained from Tisch Environmental Company (Cleves, OH, USA). Thereafter, PM_{10} samples were obtained using Whatman® quartz filters (8 × 10 in.), after which they were processed and analyzed using gravimetric techniques, according to the procedure established in the Code of Federal Regulations (Appendix J to Part 50, PM_{10}) [18]. The filters were conditioned at 35 ± 5% RH and 23 ± 1°C for at least 24 h before weighing using a Sartorius MC 5 Microbalance, packaging, and shipping to Columbia. The process of preparing and weighing the filters was implemented at the Center for Air Resources Engineering and Science, Clarkson University. Table 1 shows the geographical locations of the sampling stations and the variables measured at each station.

PM_{10} samples were obtained every 3 days for 24 h from 3 January 2014 to 28 December 2017. The readings obtained for BR, LC, PT, PV, and SS were 428,444, 322, 440, and 449, respectively. Quality control was performed on the analyzed samples.

2.3 Oceanic Niño Index (ONI) phases classification

El Niño occurrence was estimated using data from the ONI. The ONI data were obtained from the National Oceanic and Atmospheric Administration (NOAA). The values in red corresponded to El Niño, while those in blue corresponded to El Niña [19]. ONI values can be used to identify Pacific surface area anomalies over a moving average of three months

Table 1: Geographical coordinates of the designated sampling sites. The measured variables and the elevations of the sampling sites (above sea level, m) are also shown.

Site name	ID	Latitude (°)	Longitude (°)	Elevation (m)	Variables measured
Sol y Sombra	SS	11° 8′ 37.88″ N	72° 30′ 36.64″ W	117	PM_{10}
Patilla	PT	11° 30′ 00″ N	72° 40′ 15.60″ W	115	PM_{10}
Provincial	PV	11° 1′ 23.12″ N	72° 44′ 5.99″ W	156	PM_{10}, *T*, *Pp*, HR, ws, wd
Barrancas	BR	10° 57′ 34.38″ N	72° 46′ 45.52″ W	150	PM_{10}, *T*, Pp, HR, ws, wd
Las Casitas	LC	10° 57′ 1.69″ N	72° 44′ 28.23″ W	162	PM_{10}, *T*, *Pp*, HR; ws, wd

(Table 2). The warm (red) and cold (blue) periods are based on a threshold of ±0.5°C. If ONI values above +0.5°C are recorded within five consecutive months, the period is characterized as an El Niño period. Conversely, when the index values are below –0.5°C for the same number of months, the period is characterized as La Niña. In this study, the El Niño intensity was classified as weak, moderate, and strong when the ONI values were in the ranges 0.5–0.9, 1–1.4, and > 1.5, respectively.

Table 2: El Niño intensity within the 1990–2017 period based on Oceanic Niño Index (ONI) values and classifications from 1990 to 2017. (Data adapted from the NOAA.)

Year	DJF	JFM	FMA	MAM	AMJ	MJJ	JJA	JAS	ASO	SON	OND	NDJ
2014	–0.4	–0.5	–0.3	0	0.2	0.2	0	0.1	0.2	**0.5**	**0.6**	**0.7**
2015	**0.5**	**0.5**	**0.5**	**0.7**	**0.9**	**1.2**	**1.5**	**1.9**	**2.2**	**2.4**	**2.6**	**2.6**
2016	**2.5**	**2.1**	**1.6**	**0.9**	0.4	–0.1	–0.4	**–1**	**–0.6**	**–0.7**	**–0.7**	**–0.6**
2017	–0.3	–0.2	0.1	0.2	0.3	0.3	0.1	–0.1	–0.4	**–0.7**	**–0.8**	**–1**

3 RESULTS AND DISCUSSION

3.1 Meteorological conditions during ENSO

Fig. 2 shows the monthly variability of the meteorological parameters considered in this study. The annual average temperature was 29.5°C (percentile 25%, Q1 = 28.1°C; percentile 25%, Q3 = 30.7°C), and the annual average RH was 65% (percentile 25%, Q1 = 59%; percentile 25%, Q3 = 70%). Further, the average speed of the observed predominant ENE winds (frequencies ~ 30%) was 2.90 m/s (percentile 25%, Q1 = 2.00 m/s; percentile 25%, Q3 = 3.60 m/s).

ENSO played an important role in the temporal distribution of daily precipitation, bringing about a decrease in precipitation owing to the changes in temperature in the Pacific Ocean. This phenomenon resulted in low precipitation in the Colombian Caribbean zone, with the maximum and minimum precipitation values being 119.20 and 1.10 mm, respectively. The annual historical average of the registered precipitation in this area was 738 mm, with two rainy seasons, the first from February to May (short season) and the second from August to November (long season) [20].

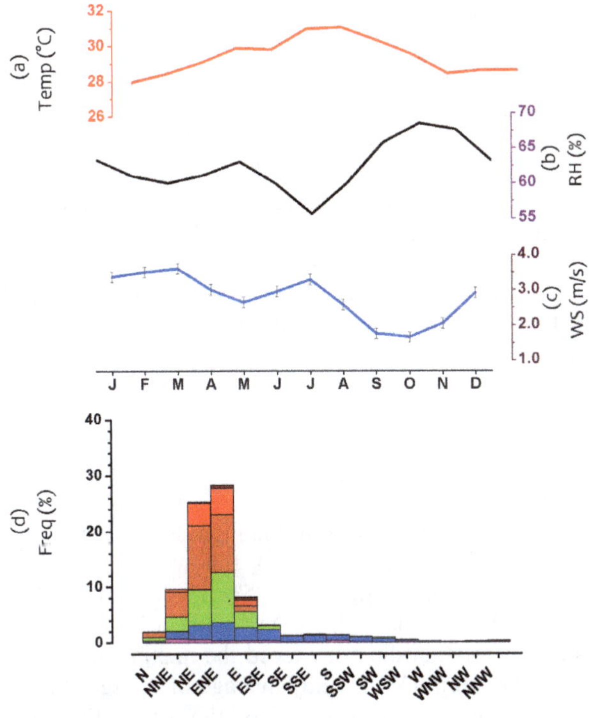

Figure 2: Monthly variability of meteorological parameters. (a) Temperature (°C); (b) Humidity relative (%); (c) Wind speed (ms^{-1}); and (d) Predominant wind direction (Freq, %).

Further, during the study period, an 8.3% decrease in annual average precipitation was observed owing to the occurrence of the ENSO phenomenon. This annual historical average precipitation diminished by 17% during the first rainy season (March–May), and a lower percentage decrease (4%) was observed during the second rainy season (August–November). These decreases are consistent with those reported by Córdoba-Machado et al. [21] and Poveda et al. [22].

To determine the origin of the differences between the stations in the extraction mine area (PV, BR, and LC) with respect to temperature, RH, precipitation, and wind, ANOVA was performed. In general, no significant differences between stations were observed with respect to these parameters ($p > 0.05$). The results showed that the meteorological conditions were uniform throughout the extraction area, including the El Cerrejón area. The temperature varied by 0.25–2.46% per year, with the LC station showing the highest variation level (2.46% per year). These temperature trends were not statistically significant ($p > 0.05$) at all the sites. Further, RH varied between –1.58 and 1.05% per year, with the LC station showing the highest variation level (–1.58% per year). The decreasing RH trend showed statistical significance ($p < 0.05$) at the LC station (–0.98°C/year). WS variation ranged between –5.75 and 2.17% per year, with the LC station showing the highest variation level (–5.75% per year). These WS variation trends were not statistically significant at all the stations, and at the BR station (–5.75% per year) and the PV station

(–1.08% per year), it was decreasing, while at the LC station (2.17% per year), it was increasing.

Pearson correlation analysis was performed to determine the level of correlation between ONI values and the meteorological parameters (T, WS, HR, and Pp) at the different study sites. The analysis showed positive correlations between the ONI values and temperature ($r^2 = 0.16$) and wind speed ($r^2 = 0.18$) at all the stations. Conversely, negative correlations were observed for humidity relative ($r^2 = -0.22$) and precipitation ($r^2 = -0.24$). Further, during the three seasons, the variations of the correlation coefficients between monthly precipitation and the ONI values were less significant during the El Niño period. Furthermore, at the three stations, the correlation coefficients varied in the ranges 0.006–0.012 and 0.158–0.491 in El Niño classified ONI phases (year = 2015) and neutral classified ONI phases, respectively.

3.2 PM_{10} concentrations

The annual mean concentration of PM_{10} for all the sampling stations in the study area between 2012 and 2016 was 34 µg/m^3 (95% CI 33–35 µg/m^3) and the standard deviation was ± 11.59 µg/m^3. In the analysis by season, the highest concentrations were observed at the LC station (41 µg/m^3; CI95% 40–42 µg/m^3), followed by the PV station (38 µg/m^3; CI95% 37–39 µg/m^3), BR station (35 µg/m^3; CI95% 33–36 µg/m^3), PT station (34 µg/m^3; CI95% 32–36 µg/m^3), and SS station (23 µg/m^3; CI95% 22–24 µg/m^3). The individual arithmetic averages at each station did not exceed the maximum permissible level (MLP) specified by the Colombian norm (MLP; daily average of 75 µg/m^3 and annual average of 50 µg/m^3 for PM_{10}). However, they were 1.11–2.07 times the value specified in the WHO air quality guidelines for PM_{10} (annual mean of 20 µg/m^3). Table 3 presents a summary of the statistics corresponding to the different monitoring sites throughout the study period. The site with the highest PM_{10} concentration was the LC station, which is located SW of the mining operation area (downwind), and that with the lowest PM_{10} concentration was the SS station, which is located NE of the mining operation area (upwind).

Table 3: Statistical data corresponding to all the stations within the 2014–2017 period.

Stations	Min	Max	Q_1	Median	Mean	Q_3	IC	Desv. Est.
PT	9.05	103.7	25.2	32.39	35.19	42.2	1.3	13.09
BR	6.65	96.22	24	33.1	34.35	42.1	1.3	14.23
LC	4.77	94.92	29.4	39.59	41.5	50	1.4	16.25
PV	5.95	87.39	28.6	35.99	38.46	46.6	1.3	14.08
SS	4.87	89.98	14.2	21.02	22.28	29.6	1.1	12.64
All	4.77	103.7	23.2	32.4	34.3	43.4	0.6	15.42

The Sen–Theil estimator, which uses the median of the slopes of all the lines connecting pairs of two-dimensional sample points, was used to estimate the time trends for PM_{10} sampling at each station within the 2012–2016 period. Thus, it was observed that PM_{10} showed increasing concentration trends at the PT, BR, and SS stations ($p < 0.01$), with the estimated slopes ranging between 6.2 and 7.7% per year; thus, the high significance observed for the BR and PT sites. Conversely, the LC and PV sites showed no significant trends ($p > 0.1$). Pearson correlation coefficients corresponding to the relationship between PM_{10} concentrations at the different sites were determined. All the values showed significance at 95% CI, with r ranging between 0.42 and 0.83, indicating a common PM_{10} emission source in the study area.

3.3 PM_{10} concentrations during ENSO

The average concentrations of PM_{10} in all the collected samples corresponding to the years in the El Niño classified ONI phase (37.14 μg/m^3; CI95% 36.06–38.22 μg/m^3) were numerically higher than those of the samples corresponding to the years in the neutral classified ONI phase (32.54 μg/m^3; CI95% 31.82–33.27 μg/m^3). A boxplot was used to illustrate the descriptive statistics of the PM_{10} concentrations in the neutral and El Niño phases (Fig. 3). The range of the values in the upper quartile (Q3) and lower quartile (Q1) for the neutral phase (22.88–43.70 μg/m^3) was lower than that corresponding to the El Niño classified ONI phase (26.32–47.60 μg/m^3). These results indicate that higher observed PM_{10} concentrations are related to the El Niño phenomenon.

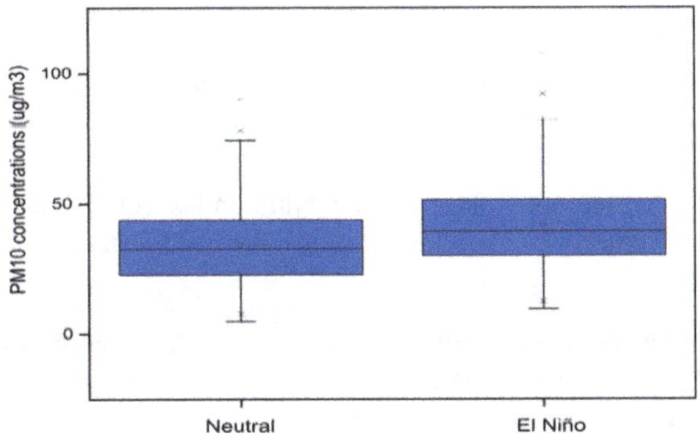

Figure 3: Box plots of daily PM_{10} concentrations for the neutral and El Niño classified ONI phases.

The results corresponding to different sampling sites were compared using ANOVA, which allowed the identification of statistically significant differences ($p < 0.05$) between PM_{10} concentrations in neutral and El Niño phases. The ANOVA results indicated that the difference between the monthly averaged PM_{10} concentrations corresponding to the neutral and El Niño phases were not statistically significant for BR ($p = 0.001$), PT ($p = 0.001$), and SS ($p = 0.036$) stations. However, they were significant for the LC ($p = 0.922$) and PV ($p = 0.330$) stations.

Further, Fig. 4 shows the monthly variation of ONI values between 2014 and 2017 as well as the variation of daily PM_{10} concentrations for all the sampling sites (PT, BR, LC, PV, and SS) for the same period. The highest PM_{10} concentrations were observed between August and December 2015 and January and March 2016. Similarly, a slight increase in PM_{10} concentration was observed between August and December 2015 and January and March 2016, compared with the other years considered in this study. This behavior can be explained considering the observed precipitation patterns. As mentioned in Section 3.1, for 2015, the annual historical average precipitation decreased by 17% during the first rainy season (March–May) and by 4% during the second rainy season (September–November).

Therefore, a higher level of precipitation could bring about a decrease in atmospheric PM_{10} concentrations owing to its scavenging effect.

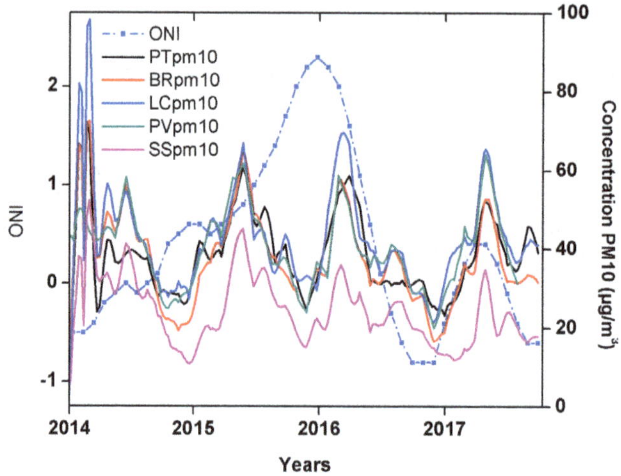

Figure 4: Monthly variation of the Oceanic Niño Index (ONI) and variation of daily PM_{10} concentrations at all the sampling sites between 2014 and 2017.

Pollutant concentration significantly depends on weather conditions, which influence the diffusion, dilution, and accumulation of pollutants [23], [24]. At the stations, where meteorological parameters were measured (i.e., BR, LC, and PV), PM_{10} showed a negative correlation with RH (r ranging from –0.15 to –0.34) and precipitation (r ranging from –0.14 to –0.25). However, at these stations, it showed a positive correlation with temperature (r ranging from 0.10 to 0.36) and wind speed (r ranging from 0.18 to 0.33). Although the correlation coefficients were relatively low, the influence of meteorological variables on the particulate matter concentration remained indisputable [25]. The low correlations may be explained by other factors that were not considered in this study: emissions rates in the mining area, regional biomass burning, and variability of the planetary boundary layer. Meteorological anomalies resulting from ENSO have an impact on PM_{10} concentrations. The most significant relationships can be attributed to precipitation anomalies, vertical temperature variability, and barometric pressure trends. Precipitation anomalies during ENSO fall below normal levels by up to 17%. El Niño effects on precipitation are associated with interactions between the ENSO seasonality, teleconnection patterns, and local climate [26]. Global model reanalysis data indicate that at different geopotential heights, highly significant temperature inversions occurred during ENSO (https://psl.noaa.gov/data/gridded/data.ncep.reanalysis.html). Vertical-scale temperature inversions generate higher levels of stability between atmospheric layers, which influences updrafts and provides less stability to surface air masses, thus creating poor circulation conditions [27]. This study does not consider the impact of emissions on pollution levels and the results do not allow the prediction of pollution events caused by increased local emissions. These values are shaped by the ability to relate the patterns of variability of local meteorological conditions leading to high PM_{10} concentrations, which helps to distinguish air quality conditions driven by anthropogenic activities from those induced by natural phenomena.

4 CONCLUSIONS

The results of this study showed that meteorological parameters, temperature, RH, and WS, did not vary significantly owing to the occurrence of the ENSO phenomenon. Conversely, precipitation varied significantly in this regard. Specifically, average precipitation diminished significantly during the first rainy season (March–May). However, the decrease was not significant during the second rainy season (August–November).

The individual arithmetic averages of PM_{10} concentrations at each station were higher than the value specified by the WHO air quality guidelines for PM_{10} but did not exceed the maximum level permissible specified in the Colombian norm. ENSO affected PM_{10} concentrations in the El Cerrejón mine, primarily owing to changes in precipitation; however, some critical factors cannot be ruled out. For example, contributions from communities located close to the stations. Further, a significant correlation was observed between the ENSO phenomenon and PM_{10} concentrations at stations located downwind of the extraction area, i.e., the LC and PV stations, which are located in towns that are close to the extraction site. However, at stations located in the center and upwind of the extraction area, i.e., the SS, PT, and BR stations, no correlations were observed in this regard.

ACKNOWLEDGEMENTS

The author would like to thank MinCiencias Colombia (Biannual Plan-Excellence Scholarship Program), the University of La Guajira and University of Antioquia.

REFERENCES

[1] Kovats, R.S., Bouma, M.J., Hajat, S., Worrall, E. & Haines, A., El Niño and health. *Lancet*, **362**(9394), pp. 1481–1489, 2003.

[2] Li, L., Wu, A.H., Cheng, I., Chen, J.-C. & Wu, J., Spatiotemporal estimation of historical $PM_{2.5}$ concentrations using PM_{10}, meteorological variables, and spatial effect. *Atmos. Environ.*, **166**, pp. 182–191, 2017.

[3] Tiwari, S. et al., Assessment of $PM_{2.5}$ and PM_{10} over Guwahati in Brahmaputra river valley: Temporal evolution, source apportionment and meteorological dependence. *Atmos. Pollut. Res.*, **8**(1), pp. 13–28, 2017.

[4] Alonso-Blanco, E. et al., Temporal and spatial variability of atmospheric particle number size distributions across Spain. *Atmos. Environ.*, **190**, pp. 146–160, 2018.

[5] André, F., Jonard, M. & Ponette, Q., Influence of meteorological factors and polluting environment on rain chemistry and wet deposition in a rural area near Chimay, Belgium. *Atmos. Environ.*, **41**(7), pp. 1426–1439, 2007.

[6] Zhang, Y. et al., The influence of relative humidity on the heterogeneous oxidation of sulfur dioxide by ozone on calcium carbonate particles. *Sci. Total Environ.,* **633**, pp. 1253–1262, 2018.

[7] Oort, A.H. & Yienger, J.J., Observed interannual variability in the Hadley circulation and its connection to ENSO. *J. Clim.*, pp. 2751–2767, 1996.

[8] Cazes Boezio, G., Talento, S. & Pisciottano Jalabert, G.J., Seasonal probability forecasts of Dec.-Jan.-Feb. precipitation in northern Uruguay and Rio Grande do Sul obtained with the coupled forecast system 2 of NOAA and statistical downscaling. *Rev. Bras. Meteorol.*, **27**(4), pp. 377–387, 2012.

[9] Grimm A.M. & Tedeschi, R.G., ENSO and extreme rainfall events in South America. *J. Clim.*, **22**(7), pp. 1589–1609, 2009.

[10] Bernal, G., Ruiz-Ochoa, M. & Beier, E., Variabilidad estacional e interanual océano-Atmósfera en la cuenca colombiana. *Rev. Cuad. del Caribe*, 2014.

[11] Timmermann, A. et al., El Niño: Southern oscillation complexity. *Nature*, **559**(7715), pp. 535–545, 2018.

[12] McMichael, A.J., Woodruff, R.E. & Hales, S., Climate change and human health: Present and future risks. *Lancet*, **367**(9513), pp. 859–869, 2006.

[13] Kim, J.-S., Zhou, W., Cheung, H.N. & Chow, C.H., Variability and risk analysis of Hong Kong air quality based on monsoon and El Niño conditions. *Adv. Atmos. Sci.*, **30**(2), pp. 280–290, 2013.

[14] Wie, J. & Moon, B.-K., ENSO-related PM_{10} variability on the Korean peninsula. *Atmos. Environ.*, **167**, pp. 426–433, 2017.

[15] Singh A. & Palazoglu, A., Climatic variability and its influence on ozone and PM pollution in six non-attainment regions in the United States. *Atmos. Environ.*, **51**, pp. 212–224, 2012.

[16] Singh A. & Palazoglu, A., A statistical framework to identify the influence of large-scale weather events on regional air pollution. *J. Appl. Meteorol. Climatol.*, **50**(12), pp. 2376–2393, 2011.

[17] Carslaw, D., The openair manual open-source tools for analysing air pollution data. *2015 David C. Carslaw*, London, p. 289, 2015.

[18] EPA, Method IO-3.5. Determination of metals in ambient particulate matter using inductively coupled plasma/mass spectrometry (ICP/MS). *Compendium of Methods for the Determination of Inorganic Compounds in Ambient Air*, 1st ed., Center for Environmental Research Information: Cincinnati, OH, pp. 1–35, 1999.

[19] C.P. Center, ENSO: Recent evolution, current status and predictions. 2015.

[20] Rojano, R.E., Manzano, C.A., Toro, R.A., Morales, R.G.E.S., Restrepo, G. & Leiva, M.A.G., Potential local and regional impacts of particulate matter emitted from one of the world's largest open-pit coal mines. *Air Qual. Atmos. Heal.*, **11**(5), pp. 601–610, 2018.

[21] Córdoba-Machado, S., Palomino-Lemus, R., Gámiz-Fortis, S.R., Castro-Diez, Y. & Esteban-Parra, M.J., Assessing the impact of El Niño modoki on seasonal precipitation in Colombia. *Glob. Planet. Change*, **124**, pp. 41–61, 2015.

[22] Poveda, G., Jaramillo, L. & Vallejo, L.F., Seasonal precipitation patterns along pathways of South American low-level jets and aerial rivers. *Water Resour. Res.*, **50**(1), pp. 98–118, 2014. DOI: 10.1002/2013WR014087.

[23] Tian, G., Qiao, Z. & Xu, X., Characteristics of particulate matter (PM_{10}) and its relationship with meteorological factors during 2001–2012 in Beijing. *Environ. Pollut.*, **192**, pp. 266–274, 2014.

[24] Unal, Y.S., Toros, H., Deniz, A. & Incecik, S., Influence of meteorological factors and emission sources on spatial and temporal variations of PM_{10} concentrations in Istanbul metropolitan area. *Atmos. Environ.*, **45**(31), pp. 5504–5513, 2011.

[25] Lee, J. & Kim, K.-Y., Analysis of source regions and meteorological factors for the variability of spring PM_{10} concentrations in Seoul, Korea. *Atmos. Environ.*, **175**, pp. 199–209, 2018.

[26] Boulanger, J.P., Leloup, J., Penalba, O., Rusticucci, M., Lafon, F. & Vargas, W., Observed precipitation in the Paraná-Plata hydrological basin: Long-term trends, extreme conditions and ENSO teleconnections. *Clim. Dyn.*, **24**(4), pp. 393–413, 2005.

[27] Power, S., Haylock, M., Colman, R. & Wang, X., The predictability of interdecadal changes in ENSO activity and ENSO teleconnections. *J. Clim.*, **19**(19), pp. 4755–4771, 2006.

HEALTH IMPACTS OF $PM_{2.5}$ AND NO_2 SHIP-RELATED AIR POLLUTION IN MATOSINHOS MUNICIPALITY, PORTUGAL

RAFAEL A. O. NUNES[1], MARIA C. M. ALVIM-FERRAZ[1], FERNANDO G. MARTINS[1], JUKKA-PEKKA JALKANEN[2], ELISA MAJAMÄKI[2] & SOFIA I. V. SOUSA[1]
[1]LEPABE – Laboratory for Process, Environment, Biotechnology and Energy, Faculty of Engineering, University of Porto (FEUP), Portugal
[2]Finnish Meteorological Institute, Finland

ABSTRACT

Air pollution is considered the most significant environmental risk to health, causing substantial economic consequences. International maritime traffic has been identified as a major air pollution source, emitting nitrogen oxides (NO_x), sulphur oxides (SO_x) and particulate matter (PM). Thus, this work aimed to study the local health impacts of $PM_{2.5}$ and NO_2 ship-related air pollution in Matosinhos municipality, where Leixões port is located (one of the major ports in Portugal), in 2014. Hourly shipping emissions were obtained using STEAM model with a resolution of 250 × 250 m for a 50 × 50 km domain. The contributions of $PM_{2.5}$ and NO_2 concentrations in Matosinhos were modelled using the AERMOD model. The annual mean background concentrations of $PM_{2.5}$ and NO_2 were acquired from the fixed monitoring stations of the QualAr network. Log-linear functions based on WHO-HRAPIE relative risks (RR) for $PM_{2.5}$ and NO_2 all-cause natural mortality were used to calculate the excess of premature deaths. The costs associated with the health impacts were estimated as the product of the additional premature deaths and the value of statistical life (VSL). It was estimated that shipping emissions caused 15 additional deaths (95% CI 19–21), corresponding to 29 million € (95% CI 17–40) for the Matosinhos municipality in 2014. These results show that shipping emissions are an environmental concern that affects the Matosinhos municipality population.

Keywords: local impacts, shipping emissions, atmospheric modelling, health effects, monetary valuation.

1 INTRODUCTION

Ship engines have been recognised as a significant source of pollutants, mainly nitrogen oxides (NO_x), sulphur oxides (SO_x) and particulate matter (PM) and associated with environmental, health, economic and climatic impacts [1]–[3]. In the last decades due to new and more restrictive climate and air pollution policies, the emissions of NO_x, SO_x and fine particulate matter of the main land-based sources (traffic, industrial processes and power generation) decreased considerably in Europe. [4], [5]. Oppositely, for shipping only in recent years the technical and legislative control measures became more restrictive and to be released more often. Among the most recent measures, it is important to highlight the CAP2020 sulphur limit, the creation of new Nitrogen Emission Control Areas (NECAs) and the open debate to control black carbon (BC) emissions, especially in the arctic regions [6], [7]. Moreover, the European Union, in cooperation with the International Maritime Organization (IMO), has been establishing with the shipping industry an ambitious objective of reducing greenhouse gases, sulphur oxides and nitrogen oxides emissions and trying a "zero-waste, zero-emission" approach for maritime transport [8]. Despite this, it remains essential to study the health effects of shipping emissions and measure these effects in monetary terms. In the last years, the associated health effects have been almost exclusively reported for regional and global scales [1]–[3], [9]. In these regional and global scale studies, authors concluded that the most affected areas were the coastal ones and those close to

WIT Transactions on Ecology and the Environment, Vol 252,
www.witpress.com, ISSN 1743-3541 (on-line)
doi:10.2495/AIR210201

maritime ports. Many case studies, where other anthropogenic emissions were considered, demonstrate the dominant influence of shipping emissions on the air quality of port cities. Using high-resolution local models is fundamental to analyse the air quality in more detail and make decisions at the local level to reduce the impacts caused by shipping emissions in port cities. To contribute to increase the study of the impacts of shipping emissions on human health in Portugal, because as far as known these studies are still scarce in Portugal, this study aimed to reduce this gap by estimating the local health impacts of $PM_{2.5}$ and NO_2 ship-related air pollution in Matosinhos municipality in 2014.

2 MATERIALS AND METHODS

2.1 Ship emissions inventory

The hourly shipping emissions for 2014 were obtained using the Ship Traffic Emission Assessment Model (STEAM) with a resolution of 250 × 250 m for a domain with 50 × 50 km. Shipping emissions were estimated based on the messages reported by the Automatic Identification System (AIS), both terrestrial and satellite-based, which allows to know the position of each ship and on the technical characteristics of each ship (ship type, size, machinery type, operative mode and emission factors). STEAM considers these parameters and estimates the emissions based on water resistance calculations, predicting the main and auxiliary engines power and loads, as well as the fuel consumption of each ship [10].

2.2 Dispersion model approach

Atmospheric simulations have been performed with the AERMOD Modelling System for a domain of 50 × 50 km. AERMOD is steady-state, multiple sources, Gaussian dispersion model designed for short-range (up to 50 km) dispersion of air pollutant emissions. The AERMAP pre-processor was used to determine source and receptor heights using Shuttle Radar Topography Mission Global Coverage (~90 m) terrain elevation data [11]. Onsite hourly meteorological data was generated using Weather Research and Forecasting (WRF) model for 2014 with a resolution of 1 × 1 km [12]. The NO_x emissions from the emission input data were converted into NO_2 concentrations using the Ambient Ratio Method 2 based on ambient NO_2 and NO_x monitoring data from EPA's Air Quality System for a 10 year period (2001–2010) [13].

2.3 Background concentrations

Hourly concentrations of PM_{10} and NO_2 for the Matosinhos municipality were obtained from the QualAr Monitoring Network operated by the Environmental Portuguese Agency [14]. Fig. 1 shows the location of the background $PM_{2.5}$ and NO_2 air quality stations considered in the present study. According to the available data, two stations were considered with data for PM_{10} and two for NO_2. As there were no available data for $PM_{2.5}$ (pollutant considered for the assessment of the health effects), a conversion factor of PM_{10} to $PM_{2.5}$ was applied. A country specific conversion factor for Portugal was estimated as the mean ratio of $PM_{2.5}$ to PM_{10} of the country all stations for 2014 and applied for Matosinhos municipality. This ratio was calculated based on the methodology used in the WHO Ambient air pollution database [15]. After the conversion, annual average concentrations of $PM_{2.5}$ and NO_2 were calculated and assigned to each parish, applying the inverse distance weighting method.

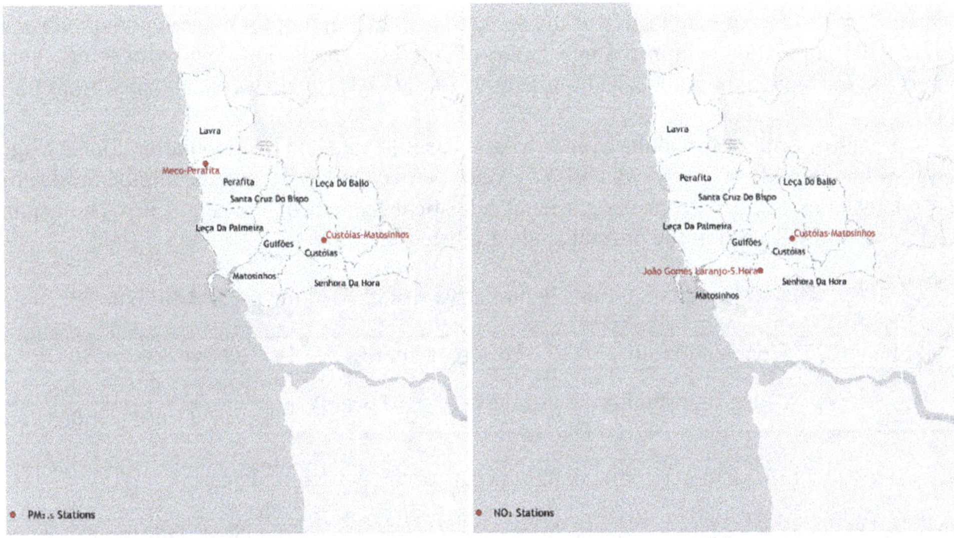

(a) (b)

Figure 1: Location of the background. (a) $PM_{2.5}$ air quality stations in the Matosinhos municipality (red dot); and (b) NO_2 air quality stations in the Matosinhos municipality (red dot).

2.4 Health impact assessment

The ship-related health impacts were evaluated based on two scenarios: (i) the baseline scenario (B-SCN) considering the background concentrations; and (ii) a non-shipping scenario (WTS-SCN) resulting of the difference between the background concentrations and the ship-related concentrations performed with AERMOD model, not considering the shipping emissions contributions.

The influence of shipping emissions on health was assessed comparing the differences between the B-SCN and WTS-SCN scenarios.

To assess the additional premature deaths attributable to $PM_{2.5}$ and NO_2 ship-related air pollution, relative risks (RR) were estimated using the log-linear functions derived from available epidemiologic studies, using the following expression [16]–[18]:

$$RR_{log-linear} = e^{\beta(C-C_0)}, \tag{1}$$

where C is the $PM_{2.5}$ or NO_2 concentrations (μg m^{-3}), C_0 is the endpoint-specific theoretical minimum-risk concentration, i.e. the concentration below which there are no additional health risks (No theoretical minimum-risk concentration was considered) and β-coefficient relates the change in the RR to an increase of the risk per 10 μg m^{-3}.

The attributable fractions (AF) were calculated following the attributable risk as:

$$AF = (RR - 1)/RR. \tag{2}$$

The excess burden of disease (EBD) was calculated as the increment in the number of deaths due to $PM_{2.5}$ and NO_2 ship-related air pollution in 2014, eqn (3):

$$\Delta EBDs = BI \times AF \times Pop, \tag{3}$$

www.witpress.com, ISSN 1743-3541 (on-line)

where BI is the baseline incidence of the selected health endpoint for a given population and Pop is the population within the age group of interest. Population data by age (one year intervals) for Matosinhos at local administrative level (LAU) 2 level were obtained from ESS [19].

All-cause mortality for adults over 30 years' baseline rate and the population for each age group with intervals of four years until 85+ years were calculated using the number of deaths for Matosinhos in 2014 from the National Statistical System of Portugal [20]. The health endpoints and RR used in the present study are listed in Table 1.

Table 1: Health endpoints and relative risk used in the present study.

Pollutant	Health endpoints	Group of interest	Relative risk (95% CI)
$PM_{2.5}$	All-cause mortality	Adults (age ≥ 30 years)	1.062 (1.040–1.083)
NO_2			1.0027 (1.0016–1.0038)

2.5 Assessment of socio-economic external costs of the burden of disease

External socio-economic costs of the burden of disease were calculated with the results of the additional deaths for the two scenarios previously mentioned.

Unit health costs (cost per death) were used to estimate the economic value of the burden of disease. The exposure cost of the additional deaths was calculated as the product of the exposure-response function (ERF) and its unit health cost according to eqn (4)

$$Exposure\ cost = (ERF) \times (Cost\ per\ death). \tag{4}$$

The value of statistical life (VSL) was used to calculate the cost of the additional premature deaths [21]. The specific VSL, which is how much society is willing to pay to avoid an anonymous death, for Portugal for 2015 were taken from the Health Economic Assessment Tools (HEAT), and following the benefit transfer approach, the value was adjusted for 2014 considering the income growth and price inflation as recommended in OECD [22], [23]. VSL was adjusted according to eqn (5)

$$VSL\ PT\ 2014 = VSL\ PT\ 2015 \times (1 + \Delta P + \Delta Y)^{\beta}, \tag{5}$$

where: $VSL\ PT$ 2015 is the VSL for Portugal in 2015 obtained from HEAT; β is the income elasticity of VSL (the value of 0.8 was used as recommended by OECD); ΔP is the increase in consumer price from the year 2015 to 2014; and ΔY is the real change in real GDP per capita growth from the year 2015 to 2014. All values previously described were obtained from OECD [24]. The VSL used in the present study was 1.87 million €/death.

3 RESULTS AND DISCUSSION

Figs 2 and 3 show the $PM_{2.5}$ and NO_2 concentrations respectively, for the B-SCN and WTS-SCN scenarios.

Table 2 shows the additional premature deaths associated with exposure to $PM_{2.5}$ and NO_2 ship-related air pollution for 2014 in Matosinhos municipality, as well as the external costs. It was estimated that shipping emissions caused 15 deaths (95% CI 19–21), which corresponded to 11% of premature deaths in Matosinhos municipality in 2014. Fig. 4 shows the spatial distribution of premature deaths associated with the exposure to $PM_{2.5}$ and NO_2 ship-related air pollution for 2014 in Matosinhos municipality. As can be seen, most mortality occurred in the parishes of Matosinhos (with the same name of the municipality) close to

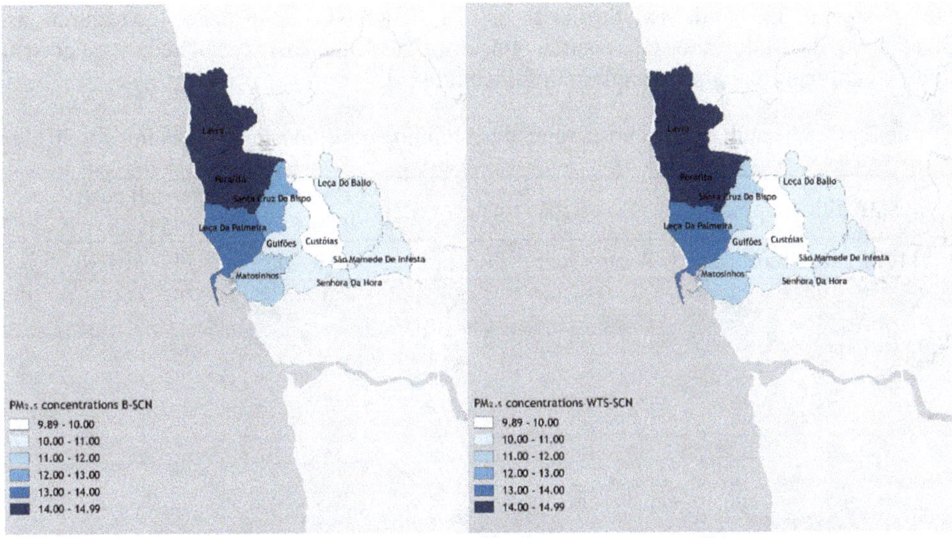

(a) (b)

Figure 2: $PM_{2.5}$ concentrations (at LAU2 level) for (a) the baseline scenario (B-SCN); and (b) the non-shipping scenario (WTS-SCN).

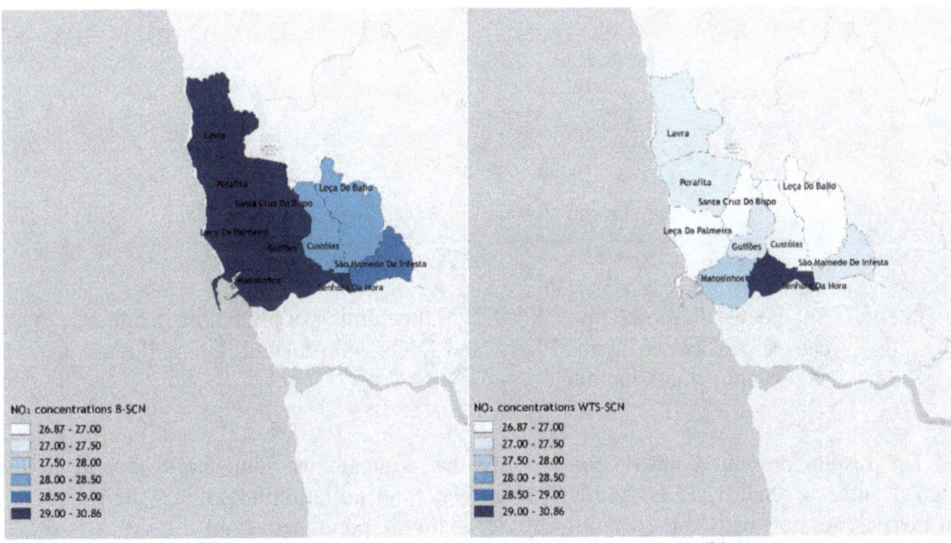

(a) (b)

Figure 3: NO_2 concentrations (at LAU2 level) for (a) the baseline scenario (B-SCN); and (b) the non-shipping scenario (WTS-SCN).

Leixões Port. These results appeared to be associated to a greater predisposition of this zone to shipping emissions (close to the port and coastline) and because these areas are densely populated. Regarding exposure costs, it was estimated a cost of 29 M€ (95% CI: 17 M€–40

www.witpress.com, ISSN 1743-3541 (on-line)

M€) related to the additional deaths from $PM_{2.5}$ and NO_2 ship-related concentrations. Considering the Matosinhos city council budget for 2014, the costs calculated in the present study would represent almost a quarter of the total [25].

Table 2: Deaths and external costs related to shipping emissions in Matosinhos for 2014.

Health endpoint	Risk group	Deaths (95% CI)	External costs (95% CI)
All-cause mortality	Adults (age ≥ 30 years)	15 (9–21)	29 (17–40) M€

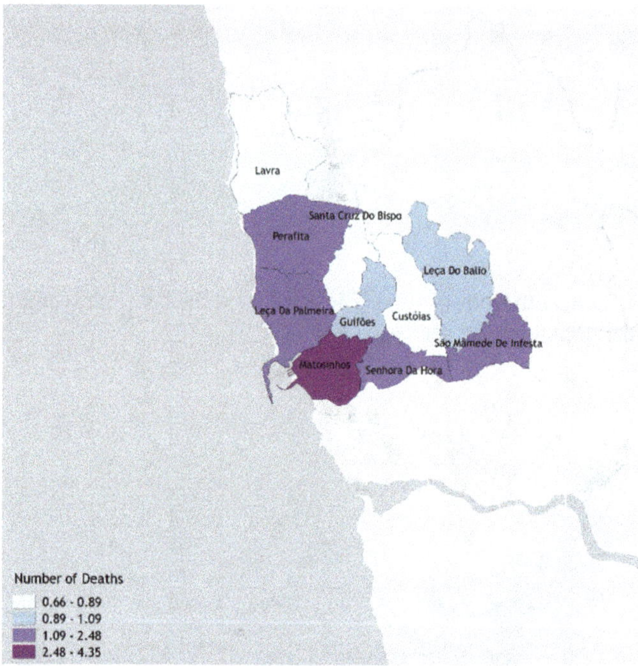

Figure 4: Spatial distribution (at parish level) of the number of premature deaths due to all causes associated with $PM_{2.5}$ and NO_2 ship-related air pollution in the Matosinhos municipality.

The results presented above demonstrate that Matosinhos municipality seems to be significantly affected by $PM_{2.5}$ and NO_2 ship-related air pollution, mainly by the proximity of the Leixões port with high shipping activity to highly populated areas.

4 CONCLUSIONS

In this study, local impacts of $PM_{2.5}$ and NO_2 ship-related air pollution in Matosinhos municipality for 2014 were assessed. Fifteen deaths (95% CI 19–21) related to shipping emissions were estimated, which corresponded to a cost of 29 M€ (95% CI: 17 M€–40 M€). Results show that the number of deaths reported in the present study represented 11% of premature deaths of the Matosinhos municipality and almost a quarter of the total Matosinhos city council budget for 2014. Most mortality occurred close to Leixões Port, which appeared to be associated to a greater predisposition of this zone to shipping emissions and residential densely populated areas close to the port. These findings contribute to understanding the local health impacts in the municipality of Matosinhos, highlighting the impact of the Leixões port presence close to highly populated areas. Moreover, the results could be used in the future to study mitigation measures to reduce shipping emissions contributions mainly close to the port area of Leixões. In the future the study of other major ports of Portugal and health endpoints should consider.

ACKNOWLEDGEMENTS

This work was financially supported by: Base Funding – UIDB/00511/2020 of the Laboratory for Process Engineering, Environment, Biotechnology and Energy – LEPABE – funded by national funds through the FCT/MCTES (PIDDAC) and project EMISSIP (POCI-01-0145-FEDER-032201), funded by FEDER funds through COMPETE2020 – Programa Operacional Competitividade e Internacionalização (POCI) and by national funds (PIDDAC) through FCT/MCTES. Rafael A.O. Nunes thanks the Portuguese Foundation for Science and Technology (FCT) for the individual research grant SFRH/BD/146159/2019. Sofia I.V. Sousa thanks the Portuguese Foundation for Science and Technology (FCT) for the financial support of her work contract through the Scientific Employment Stimulus – Individual Call – CEECIND/02477/2017. Jukka-Pekka Jalkanen would like to acknowledge the financial support from the European Union's Horizon2020 research and innovation programme under grant agreement #874990 (EMERGE project). This work reflects only the authors' view and INEA is not responsible for any use that may be made of the information it contains.

REFERENCES

[1] Sofiev, M. et al., Cleaner fuels for ships provide public health benefits with climate tradeoffs. *Nat. Commun.*, **9**(1), 2018.

[2] Corbett, J.J., Winebrake, J.J., Green, E.H., Kasibhatla, P., Eyring, V. & Lauer, A., Mortality from ship emissions: A global assessment. *Environ. Sci. Technol.*, **41**(24), pp. 8512–8518, 2007.

[3] Viana, M. et al., Estimated health impacts from maritime transport in the Mediterranean region and benefits from the use of cleaner fuels. *Environ. Int.*, 2020.

[4] Aksoyoglu, S., Baltensperger, U. & Prévôt, A.S.H., Contribution of ship emissions to the concentration and deposition of air pollutants in Europe. *Atmos. Chem. Phys.*, **16**(4), pp. 1895–1906, 2016.

[5] Chen, D. et al., Contribution of ship emissions to the concentration of $PM_{2.5}$: A comprehensive study using AIS data and WRF/Chem model in Bohai Rim Region, China. *Sci. Total Environ.*, 610–611, pp. 1476–1486, 2018.

[6] IMO, Prevention of air pollution from ships MARPOL Annex VI: Proposal to initiate a revision process, 2005.

[7] Klimont, Z. et al., Global anthropogenic emissions of particulate matter including black carbon. *Atmos. Chem. Phys.*, **17**(14), pp. 8681–8723, 2017.

[8] European Commission, Commission staff working document on the implementation of the EU Maritime Transport Strategy 2009–2018, Brussels, 2016.

[9] Nunes, R.A.O., et al., Estimating the health and economic burden of shipping related air pollution in the Iberian Peninsula. *Environ. Int.*, **156**, 106763, 2021.

[10] Jalkanen, J.-P., Brink, A., Kalli, J., Pettersson, H., Kukkonen, J. & Stipa, T., A modelling system for the exhaust emissions of marine traffic and its application in the Baltic Sea area. *Atmos. Chem. Phys. Discuss.*, **9**(4), pp. 15339–15373, 2009.

[11] EPA, Aermod Implementation Guide, *AERMOD Implement. Work. Off. Air*, 2009.

[12] Skamarock, W.C. et al., A description of the advanced research WRF version 3, NCAR Tech. Note, NCAR/TN-468+STR, *Natl. Cent. Atmos. Res. Boulder, Color.*, June, 2008.

[13] Podrez, M., An update to the ambient ratio method for 1-h NO_2 air quality standards dispersion modeling. *Atmos. Environ.*, **103**, pp. 163–170, 2015.

[14] APA, Qualar informação sobre a qualidade do ar, 2021. https://qualar.apambiente.pt/.

[15] World Health Organization (WHO), The Global Health Observatory: Explore a world of health data, 2021. https://www.who.int/data/gho/data/themes/topics/topic-details/GHO/ambient-air-pollution.

[16] WHO, AirQ+: Software tool for health risk assessment of air pollution, 2019. http://www.euro.who.int/en/health-topics/environment-and-health/air-quality/activities/airq-software-tool-for-health-risk-assessment-of-air-pollution. Accessed on: 25 Feb. 2019.

[17] Holland, M., Implementation of the HRAPIE recommendations for European Air Pollution CBA work, 2014.

[18] Héroux, M.E. et al., Quantifying the health impacts of ambient air pollutants: Recommendations of a WHO/Europe project. *Int. J. Public Health*, 2015.

[19] ESS, 2011 Census Hub, *2011 Census Database*, 2018. https://ec.europa.eu/CensusHub2/query.do?step=selectHyperCube&qhc=false.

[20] INE, Taxa de mortalidade pós-neonatal (‰) por Local de residência da mãe, 2018. https://www.ine.pt/xportal/xmain?xpid=INE&xpgid=ine_indicadores&indOcorrCod=0003291&contexto=bd&selTab=tab2.

[21] OECD, Valuing mortality risk reductions in regulatory analysis of environmental. Health and Transport Policies: Policy Implications, Paris, 2011.

[22] OECD, The cost of air pollution health impacts od road transport, 2014.

[23] OECD, Mortality risk valuation in environment, health and transport policies, 2012.

[24] OECD, OECD Data, 2019. https://data.oecd.org/.

[25] Câmara Municipal de Matosinhos, Orçamento e Grandes Opções do Plano 2014, 2013. https://www.cm-matosinhos.pt/servicos-municipais/comunicacao-e-imagem/noticias/noticia/orcamento-e-grandes-opcoes-do-plano-2014.

RELATIONSHIP BETWEEN POLLUTION LEVELS AND SOCIOECONOMIC VARIABLES: CONDITIONS IN NINE REGIONAL MUNICIPALITIES IN VALPARAÍSO, CHILE

PATRICIO HERRERA, JOSE ROJO & VALERIA SCAPINI
Centro de Investigación en Innovación, Desarrollo Económico y Políticas Sociales (CIDEP), Universidad de Valparaíso, Chile

ABSTRACT

Air pollution, which generates negative effects on people's health, is linked to 23% of deaths worldwide. Among the main air pollutants are particulate matter (PM_{10} and $PM_{2.5}$), which causes cardiovascular and respiratory diseases, affecting the whole community living in a sector exposed to emissions; and sulphur dioxide (SO_2), which can cause severe effects even with short exposures. Based on air pollution data obtained from National Air Quality Information System (SINCA) monitoring stations, meteorological information from the Meteorological Directorate of Chile, and socioeconomic information on income poverty level, multidimensional poverty level, unemployment, schooling, overcrowding and health, we constructed a data panel containing daily information at the community level, from 2016 to 2019. Using three ordinary least squares (OLS) regression models, we studied the relationship between socioeconomic variables and each of the pollutants. Our results show that, in general, there is a positive and significant relationship between the level of multidimensional poverty of a population and the totality of regional pollution levels; as well as a negative and significant relationship between unemployment and the different pollution levels. Finally, there is a negative and significant relationship between the level of citizen's education and pollution. The main objective of this study was to investigate the possible relationships between socio-economic variables and pollution, in order to generate evidence to aid implementation of environmental public policies.

Keywords: air pollution, particulate matter, sulphur dioxide, socioeconomic variables, poverty level, health risk, environmental monitoring.

1 INTRODUCTION

According to Chile's 2019 National Environmental Survey [1], carried out by the Ministry of the Environment, the main problem affecting the country's population is air pollution, as per 32.8% of those surveyed. The results indicated that 48% of those surveyed say they live in a municipality with fair to very bad air, and that in recent years, the country has stagnated (64%) or regressed (23.5%) with respect to its environmental situation. Also, respondents indicated that in their communities, the activities having a major impact on air pollution are: industries (energy production, pulp mills, copper smelters, cement plants) and land transportation (cars, buses and trucks), with 60% and 50% of the responses, respectively.

Although Chile has an environmental policy that has been implemented since the mid-1980s [2], by institutionalizing a governmental program through creation of the National Environmental Commission in 1994 (which in 2010 became the Ministry of the Environment [3]), progress tended to slow down because of the country's neo-extractivist economic growth model [4]–[6] and has intensified high-impact productive processes with a high environmental impact, as many industrial zones are located next to poor communities. Our object of study, the region of Valparaíso, is one of the territorial spaces with the greatest environmental conflicts.

The National Institute of Human Rights (INDH) [7] of Chile reports that there are currently 21 environmental conflicts in Chile, resulting from energy, mining and agro-industrial projects, where emissions and contamination of natural resources represent 60% of

WIT Transactions on Ecology and the Environment, Vol 252, © 2021 WIT Press
www.witpress.com, ISSN 1743-3541 (on-line)
doi:10.2495/AIR210211

the total number of conflicts [8], [9]. The territories of Concón, Nogales, El Melón, Puchancaví-Quintero and Petorca-La Ligua present worrying figures of multidimensional poverty, insufficient public health services and a lax environmental policy regarding the exploitation of natural resources and emission of polluting particles, to such an extent that some of these localities constitute "sacrifice zones", which increased their inequality indicators [6], [9].

The state has made unsuccessful attempts to reduce environmental damage to air, soil and biodiversity, in the Valparaíso region. Despite increased environmental regulation in the last decade (since the creation of the Ministry of the Environment), plus better investment in technology and monitoring instruments, both the state and companies continue to fail to resolve episodes of air pollution. From 2011 to now, sacrifice zones such as Puchuncaví-Quintero have had critical episodes of air saturation, as a result of sulphur particles, as well as other metals present in the productive process of the industrial park, creating high pollutant loads. The instruments and measurements have been inaccurate in clarifying toxicological damage to the population, in addition to other environmental disasters which were caused by oil spills, contaminating the flora and fauna in Quintero Bay. The paradox is that the episodes of contamination in localities within the Valparaíso region are the responsibility of state companies, which have produced long-term harmful effects on the region's population. Both the National Petroleum Company (ENAP) and the National Copper Corporation (CODELCO) play a major role in the emission of polluting particles, which complicates the State's role as a guarantor of protection for the right to live in pollution-free territories.

This study's main objective is to identify the relationships between socio-economic variables and pollution, in order to establish evidence for the implementation of environmental public policies in these territories. We tried to answer the following question: Is there a positive relationship between poverty and pollution? This research is divided into the following parts: Sections 2 and 3 contain the data and the methodology used, Section 4 presents the empirical results, and Section 5 is for the conclusions obtained from our research.

2 DATA

This paper uses different data sources. First, the pollution data panel obtained from the National Air Quality Information System (SINCA) website between January 2016 and June 2020. It was constructed with daily concentrations ($\mu g/m^3$) of PM_{10}, $PM_{2.5}$ and SO_2 from monitoring stations in the Valparaíso region that had information available for the period studied (totalling 25 monitoring stations from nine municipalities, corresponding to an area of 2369 km^2). Table 1 shows a list by municipality, with the names of the stations.

The location of the 25 monitoring stations is shown in Fig. 1.

Second, we used a meteorological data panel obtained from the Meteorological Directorate of Chile (MeteoChile). Daily data on temperature (°C), humidity (%), wind speed (degrees/knots) and rainfall (mm) were used. Thirdly, we obtained our socioeconomic data from the National Socio-economic Characterization Survey (CASEN). In particular, we used the population's years of schooling as an indicator of education; their unemployment rate as a labour indicator; for housing we constructed an indicator representing the percentage of people living with some degree of overcrowding in the community; and for health we constructed an indicator representing the percentage of people subscribing to a private health regime. Income poverty and multidimensional poverty correspond to the percentages of people living in these conditions.

We were able to respond to the set of research questions by merging the aforementioned databases, resulting in a panel of 8572 records for PM_{10}, as well as 9661 records for $PM_{2.5}$ and 16,579 records for SO_2. The descriptive statistics by pollutant are shown in Table 2.

Table 1: Monitoring stations with pollutants measured.

Municipalities	Stations	PM_{10}	$PM_{2.5}$	SO_2
Catemu	Catemu			X
	Romeral			X
	Santa Margarita			X
Concón	Colmo			X
	Concón	X	X	X
	Concón MMA		X	
	Junta de Vecinos			X
	Las Gaviotas			X
Panquehue	Lo Campo			X
Puchuncaví	La Greda	X	X	X
	Puchuncaví	X	X	X
	Los Maitenes	X	X	X
	Campiche			X
	Ventanas	X	X	X
Quillota	Cuerpo de Bomberos	X	X	X
	La Palma	X	X	X
	San Pedro	X	X	X
Quilpué	Quilpué	X	X	
Quintero	Quintero	X	X	X
	Centro Quintero	X	X	X
	Loncura	X	X	X
	Sur	X		X
	Valle Alegre	X		X
Valparaíso	Valparaíso		X	
Viña del Mar	Viña del Mar	X	X	

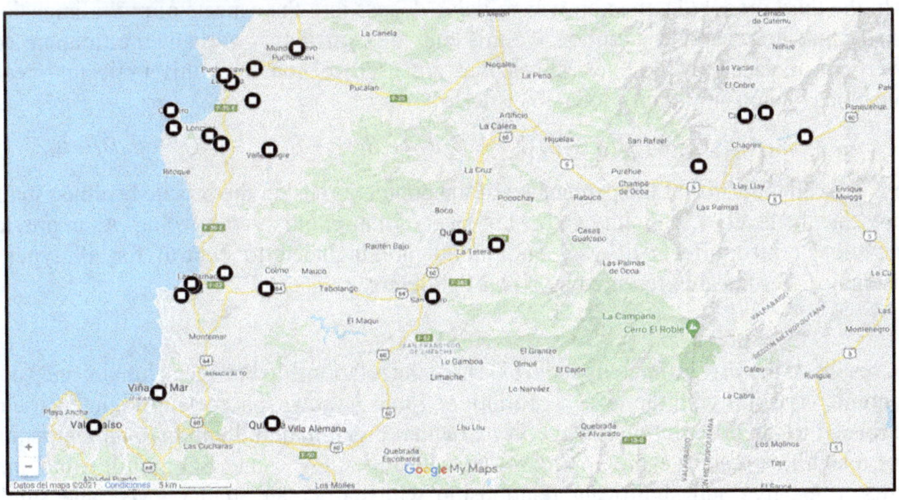

Figure 1: Location of the 25 environmental monitoring stations.

Table 2: Descriptive statistics by pollutant.

Variable	Observations	Mean	Standard deviation	Minimum	Maximum
PM_{10}	8,572	36.81	14.52	1	141
$PM_{2.5}$	9,661	15.18	7.06	1	84
SO_2	16,579	17.37	17.43	0.03	178.64

3 METHODOLOGY

Three ordinary least squares (OLS) regression models were designed to study the relationship between several air pollution (PM_{10}, $PM_{2.5}$ and SO_2) and socioeconomic variables. The first OLS model (1) seeks to explain the effect of socioeconomic variables on pollution more explicitly:

$$y_{it} = \beta_o + \beta_1 T_{it} + \beta_2 H_{it} + \beta_3 W_{it} + \beta_4 R_{it} + \beta_5 Z_i + \beta_6 D_i + \epsilon_{it}. \quad (1)$$

Here, y_{it} represents the daily concentrations of particulate matter (PM_{10}, $PM_{2.5}$) and sulphur dioxide (SO_2), measured in weight/volume ($\mu g/m^3$) expressed logarithmically, while the subscript *it* corresponds to both the municipality and the day, respectively. The explanatory variables considered are meteorological data and socio-economic variables. Among the meteorological data considered were: T representing temperature (°C), H representing humidity (%), W representing wind speed (degrees/knots), and R representing total rainfall (mm). Among the socio-economic variables considered, Z represents the income poverty level and D indicates the unemployment rate. Estimates include a fixed effect by year and month, thus controlling for all time-varying differences. Results are given by type of pollutant.

In the second model (2), income poverty's explanatory variable was replaced by the multidimensional poverty variable, leaving the model as follows:

$$y_{it} = \beta_o + \beta_1 T_{it} + \beta_2 H_{it} + \beta_3 W_{it} + \beta_4 R_{it} + \beta_5 M_i + \beta_6 D_i + \epsilon_{it}, \quad (2)$$

where M represents the multidimensional poverty indicator. Estimates include the temporal controls mentioned above. Results are presented by type of pollutant.

In the third model (3), the aim was to delve deeper into the aspects of multidimensional poverty that are related to pollution. To this end, the explanatory variable multidimensional poverty is replaced by three variables that are used to construct this indicator: health, education and housing; thus, model (3) is as follows:

$$Y_{it} = \beta_o + \beta_1 T_{it} + \beta_2 H_{it} + \beta_3 W_{it} + \beta_4 R_{it} + \beta_5 H + \beta_6 S_i + \beta_7 O_i + \beta_8 D_i + \epsilon_{it}. \quad (3)$$

Here, H is an indicator of the percentage of the population using the private health system, S represents the level of schooling and O the level of household overcrowding. As in previous estimations, fixed effects of year and month are included to control for all types of seasonality. Results are presented by type of pollutant.

4 RESULTS

The results of the first estimated model (1) show the relationship between climatic and socio-economic variables with the level of pollution: Table 3 shows that meteorological variables are related to the different pollutants. In particular, precipitation, humidity and temperature have a significant and negative relationship with PM_{10}, $PM_{2.5}$ and SO_2; while wind speed shows a negative and significant relationship with particulate matter $PM_{2.5}$ and sulphur dioxide (see columns 3, 4, 5 and 6); yet a positive and significant relationship with particulate matter PM_{10} in the model with controls (column 2).

Table 3: Summary eqn (1) – OLS.

	(1)	(2)	(3)	(4)	(5)	(6)
Variables	PM_{10}	PM_{10}	$PM_{2.5}$	$PM_{2.5}$	SO_2	SO_2
Water	–0.0271***	–0.0241***	–0.0177***	–0.023***	–0.0051***	–0.0069***
	(–0.214)	(–0.190)	(–0.139)	(–0.184)	(–0.0237)	(–0.0323)
Humidity	–0.0112***	–0.0136***	0.00102	–0.0012**	–0.011***	–0.0107***
	(–0.195)	(–0.237)	(0.0186)	(–0.0225)	(–0.163)	(–0.159)
Temperature	–0.0111***	–0.0467***	–0.0511***	–0.046***	–0.0179***	–0.00447
	(–0.0790)	(–0.334)	(–0.337)	(–0.301)	(–0.0857)	(–0.0214)
Wind speed	0.000578	0.0091***	–0.0208***	–0.005**	–0.058***	–0.049***
	(0.00303)	(0.0478)	(–0.101)	(–0.0242)	(–0.136)	(–0.116)
I. Poverty	0.726***	0.508**	–1.070***	–1.687***	2.627***	2.996***
	(0.0352)	(0.0246)	(–0.0590)	(–0.0930)	(0.169)	(0.193)
Unemployed	–9.385***	–10.15***	–0.357	4.136***	6.598***	8.622***
	(–0.154)	(–0.167)	(–0.00499)	(0.0579)	(0.115)	(0.150)
Observations	8572	8572	9661	9661	16,579	16,579
R-squared	0.110	0.175	0.164	0.259	0.071	0.096
Fixed eff. month	No	Yes	No	Yes	No	Yes
Fixed eff. year	No	Yes	No	Yes	No	Yes

Robust standard errors (SE) in parentheses below numerical values. *** $p<0.01$, ** $p<0.05$, * $p<0.1$.

In the case of socio-economic variables, our main focus of interest in this study, income poverty, showed a positive and significant relationship with PM_{10} and SO_2, while for $PM_{2.5}$ it is negative and significant. As for the unemployment variable, results show there is a negative and significant relationship for PM_{10} concentrations (columns 1 and 2). This could be because lower economic activity generates lower concentrations of pollutants. On the contrary, the relationship was positive and significant between unemployment and pollutants $PM_{2.5}$ and SO_2 (columns 4, 5 and 6). This relationship may be due to factors unrelated to economic activity.

Results of the second estimated model (2) are similar to those obtained in the previous section for the climate variables, as seen in Table 4. A negative and significant relationship is shown for most of the climate variables and the various pollutant emissions (see columns 1–6). For the socio-economic variables, the results indicate there is a positive and significant relationship between multidimensional poverty and the concentrations of all pollutants. On the contrary, the unemployment variable shows a negative and significant relationship with each pollutant, which we believe can be explained by the relationship between economic activity and pollution, as in the previous case.

Table 5 has results of the third estimated model (3), where it can be seen that the relationship between the meteorological variables and the different pollutants is similar to what was obtained previously. A negative and significant relationship is evident for all climate variables and the various pollutant emissions (columns 1–6).

As pertains to the elements in relation to multidimensional poverty, our results showed that schooling has a negative and significant relationship with all pollutants, except the uncontrolled model that explains PM_{10} (columns 2–6 in Table 5). Overcrowding also brings

Table 4: Summary eqn (2) – OLS.

	(1)	(2)	(3)	(4)	(5)	(6)
Variables	PM_{10}	PM_{10}	$PM_{2.5}$	$PM_{2.5}$	SO_2	SO_2
Water	–0.0256***	–0.0229***	–0.0167***	–0.0217***	–0.004***	–0.0068***
	(–0.202)	(–0.181)	(–0.131)	(–0.171)	(–0.0204)	(–0.0317)
Humidity	–0.0124***	–0.0146***	0.000107	–0.0029***	–0.011***	–0.0113***
	(–0.216)	(–0.255)	(0.00196)	(–0.0521)	(–0.165)	(–0.168)
Temperature	–0.0101***	–0.0459***	–0.0501***	–0.0438***	–0.017***	–0.00404
	(–0.0724)	(–0.328)	(–0.330)	(–0.289)	(–0.0803)	(–0.0193)
Wind_speed	–0.0075***	0.00237	–0.0256***	–0.0136***	–0.060***	–0.0487***
	(–0.0393)	(0.0124)	(–0.124)	(–0.0662)	(–0.142)	(–0.115)
M. Poverty	0.930***	0.818***	0.405***	0.811***	2.851***	3.602***
	(0.103)	(0.0910)	(0.0455)	(0.0911)	(0.180)	(0.228)
Unemployment	–12.20***	–12.20***	–3.473***	–1.441*	–3.451***	–3.102***
	(–0.200)	(–0.200)	(–0.0486)	(–0.0202)	(–0.0601)	(–0.0541)
Observations	8572	8572	9661	9661	16,579	16,579
R-squared	0.115	0.178	0.162	0.256	0.077	0.104
Fixed eff. month	No	Yes	No	Yes	No	Yes
Fixed eff. year	No	Yes	No	Yes	No	Yes

Robust standard errors (SE) in parentheses below numerical values. *** $p<0.01$, ** $p<0.05$, * $p<0.1$.

Table 5: Summary eqn (3) – OLS.

	(1)	(2)	(3)	(4)	(5)	(6)
Variables	PM_{10}	PM_{10}	$PM_{2.5}$	$PM_{2.5}$	SO_2	SO_2
Water	–0.025***	–0.0215***	–0.012***	–0.0163***	–0.006***	–0.008***
	(–0.193)	(–0.170)	(–0.0946)	(–0.129)	(–0.0262)	(–0.0380)
Humidity	–0.013***	–0.0154***	–0.004***	–0.0058***	–0.009***	–0.009***
	(–0.232)	(–0.270)	(–0.0649)	(–0.106)	(–0.136)	(–0.134)
Temperature	–0.010***	–0.0459***	–0.047***	–0.0449***	–0.018***	–0.00597*
	(–0.0688)	(–0.328)	(–0.311)	(–0.296)	(–0.0865)	(–0.0286)
Wind speed	–0.015***	–0.00601*	–0.065***	–0.0501***	–0.051***	–0.038***
	(–0.0807)	(–0.0314)	(–0.315)	(–0.244)	(–0.121)	(–0.0907)
Schooling	–0.0305	–0.162*	–0.494***	–0.514***	–0.160***	–0.235***
	(–0.0603)	(–0.320)	(–0.962)	(–1.002)	(–0.221)	(–0.323)
Overcrowding	–3.842***	–2.633***	–13.57***	–12.63***	–2.151***	–2.062***
	(–0.454)	(–0.311)	(–1.322)	(–1.229)	(–0.132)	(–0.127)
Health	–1.969***	–0.0636	–2.114***	–1.475***	–0.117	0.404
	(–0.407)	(–0.0131)	(–0.410)	(–0.286)	(–0.0146)	(0.0506)
Unemployment	–26.34***	–24.41***	–62.37***	–56.92***	1.408**	3.922***
	(–0.432)	(–0.401)	(–0.872)	(–0.796)	(0.0245)	(0.0683)
Observations	8572	8572	9661	9661	16,579	16,579
R-squared	0.118	0.181	0.240	0.316	0.086	0.114
Fixed eff. month	No	Yes	No	Yes	No	Yes
Fixed eff. year	No	Yes	No	Yes	No	Yes

Robust standard errors (SE) in parentheses below each value. *** $p<0.01$, ** $p<0.05$, * $p<0.1$.

a negative relationship that is significant for all pollutants (columns 1–6). Importantly, the health indicator showed there was a negative and significant relationship with $PM_{2.5}$. As for unemployment, the relationship takes on a negative and significant value for both pollutants, while for SO_2, the relationship is positive and significant.

5 CONCLUSIONS

Air pollution is one of the main problems affecting the human population worldwide. Although Chile has implemented an environmental policy since the mid-1980s, progress slowed down because of the economic growth model that has intensified production processes having high environmental impact, those that remain where many industrial areas have been, located next to poor communities. We focused on the Valparaiso region, a territorial space with the greatest environmental conflicts in Chile. We employed micro-level data about air pollution, meteorological indicators, and socio-economic variables to construct a panel of data with daily information at the municipality level, from 2017 to 2019. The main objective of this study is to identify the relationships between socio-economic variables and air pollution, in order to establish evidence for the implementation of improved environmental public policies in the Chilean territories.

Our results illustrate that the meteorological variables in all of the estimates made, relate in the same way to the different pollutants. In general, we show a negative and significant relationship between the climate variables and the various pollutant emissions.

In reference to socio-economic variables, we found that when we use income poverty, results showed a positive and significant relationship with PM_{10} and SO_2, while for $PM_{2.5}$ it is negative and significant. As for the unemployment variable, results showed there was a negative and significant relationship for PM_{10} concentrations (lower economic activity generates lower concentrations of pollutants),; while on the other hand, the relationship is positive and significant between unemployment and the pollutants $PM_{2.5}$ and SO_2 (perhaps due to factors unrelated to economic activity).

For multidimensional poverty, our results indicated a positive and significant relationship with the concentrations of all pollutants. On the contrary, the unemployment variable showed a negative and significant relationship with each pollutant, which can be explained by the relationship between economic activity and pollution, as in the previous case. In relating the elements to multidimensional poverty, in general the results demonstrated that schooling and overcrowding have a negative and significant relationship with all pollutants.

The health indicator showed us a negative and significant relationship with $PM_{2.5}$. As for unemployment, that relationship takes a negative and significant value for both pollutants; while for SO_2, the relationship is positive and significant.

Our study results are in line with our hypothesis, because we found there was a positive relationship between poverty and pollution. The relationship found between multidimensional poverty and the different pollutants is striking. This relationship is positive and significant at 1%. This will allow us to discuss the environmental and social inequalities that exist in the region productively. In considering future work, we recommend that this type of sampling and analysis be extended to other regions of the country, to see if similar results can be obtained. If so, investigation into the reasons for the negative relationship found between overcrowding and pollution is recommended.

REFERENCES

[1] Encuesta Nacional Ambiental (National Environmental Questionnaire), Ministerio del Medio Ambiente (Ministry of the Environment), Gobierno de Chile (Chilean Government).

[2] Ducoing, C. & Peres, J., *Natural Resources and Divergence. A Comparison of Andean and Nordic Trajectories*, Palgrave, 2021.

[3] Tecklin, D., Bauer, C. & Prieto, M., Making environmental law for the market: The emergence, character, and implications of Chile's environmental regime. *Environmental Politics*, **20**, pp. 879–898, 2011.

[4] Silva, E., Democracy, market economics, and environmental policy in Chile. *Journal of Interamerican Studies and World Affairs*, **38**, pp. 1–33, 1996.

[5] Taylor, D., *From Pinochet to the Third Way. Neoliberalism and Social Transformation in Chile*, Pluto Press, 2006.

[6] Astorga, R., *Gobernanza e institucionalidad ambiental en Chile: Cómo inciden en el crecimiento económico* (Governing and environmental institutionalization in Chile: How do they affect economic growth), CIEPLAN, 2019.

[7] INDH, Mapa de conflictos socioambientales (Map of socioenvironmental conflicts). https://mapaconflictos.indh.cl/#/.

[8] Fuenzalida, M. & Quiroz, R., La dimensión espacial de los conflictos ambientales en Chile (The spatial dimension of environmental conflicts in Chile). *Polis (Santiago)*, **11**, pp. 157–168, 2012.

[9] Allain, M., Conflictos y protestas socioambientales en Chile: Reflexiones metodológicas y resultados (Conflicts and environmental protests in Chile: Thoughts on methodology and results). *Revista de Sociología*, **34**, pp. 81–101, 2019.

[10] Bolados, P., Conflictos socio-ambientales/territoriales y el surgimiento de identidades post neoliberales (Valparaíso-Chile) (Social-environmental/territorial conflicts and the emergence of post-neo-liberal identities). *Izquierdas*, **31**, pp. 102–129, 2016.

ENHANCING THE SONIC URBAN ENVIRONMENT IN HISTORICAL SITES: EL-TAHRIR SQUARE IN EL-MANSHEYA, ALEXANDRIA, EGYPT

NIHAL ATIF, HASSAN ABDULSALAM & IBRAHIM MAROUF
Arab Academy for Science, Technology and Maritime, Faculty of Engineering,
Architecture Department, Alexandria University, Egypt

ABSTRACT

To improve the quality of life in cities, a variety of interventions are applied. Noise is a substantial and important sort of pollution that has a negative impact on the urban environment and human health. According to the complaint survey, it ranks second among environmental contamination complaints (conducted by EEAA (Egyptian Environmental Agency Affairs in 2019) [1]. The most significant source of noise in the city is traffic noise. In order to improve the urban sound environment, many physical techniques are being applied, especially in the heritage sites, which are considering the main attraction factors that ensure the economic and tourist life of the cities. Human sensory is affecting the marketing of the heritage site which aims at generating a strong place identity of the place that stems from its landscape. Not only do the physical characteristics enhance the place but also the relation between people and place is the main factor that improves the urban environment. Alexandria is Egypt's second-largest city after Cairo. It is located along the Mediterranean Sea, and El-Tahrir square is considered the oldest square in the city located in El-Mansheya. It is affected by a high level of noise due to the surrounding roads and the different social activities. The purpose of this paper is to clarify the design considerations for the most appropriate physical technique along with historical open spaces, with the goal of improving the quality of life (QOL) and the sonic urban environment specifically. The proposed methodology is focusing on how noise affects human perception and the urban environment. Then it delves into the various physical noise control approaches. The goal of this research is to evaluate the suitable design of a sonic urban environment that fulfils social, environmental, and economic perceptions while maintaining a balanced approach to the noise issue in order to improve QOL in historical sites.

Keywords: noise pollution, sonic urban environment, historical sites, traffic noise, acoustic sustainability, quality of life, noise reduction, acoustical level.

1 INTRODUCTION

The metropolis now accounts for more than half of global pollution [2]. The increased density of people in cities has resulted in an increase in numerous sorts of pollution. Continuation of a rapid population growth reaching 105% growth rate in 2020 that results in overcapacity and stress on the city's infrastructure. According to the latest world polls in 2017, Egypt is the sixth most populous country in the world, according to the WHO. Its population is predicted to be around 100,388 million people, resulting in a rapid increase in various forms of pollution [3]. According to the most recent report of Numbeo, Alexandria is ranked as the world's 10th most polluted city [4]. Noise is one of the major variables in the urban environment that has an impact on people's quality of life. Noise pollution is the second most influential pollutant on people's activities and perceptions, according to the WHO [2]. In Egypt, noise pollution is ranked second among environmental pollutants [1]. Historical sites consider as the main attraction point in the city that influences the site's economy. According to various research, each historical area has its sonic environment that should be preserved. Two types of sound impact the sonic environment, manmade and natural. Traffic noise is a manmade sound that comes from three different sources: roads, rail lines, and airports. The

WIT Transactions on Ecology and the Environment, Vol 252, © 2021 WIT Press
www.witpress.com, ISSN 1743-3541 (on-line)
doi:10.2495/AIR210221

highest source in the city of noise comes from the roadways. When the acoustic level reaches 70 decibels, it is considered desirable noise by the WHO. Roadways begin at 70 decibels, which is considered a source of noise [2].

2 SOUNDSCAPE

The research focuses on the study of the soundscape as a part of landscape environment design. The landscape is the art of modifying and arranging the different features in the open spaces, urban spaces, and before urban spaces. The soundscape is influenced by the acoustical level in the outdoor spaces. Sound within the urban environment may be generating by two sound sources, natural or manmade. Natural acoustics constitutes the sound that is produced by the natural sources in the natural soundscape, such as the sounds of any living organism [5]. Human-made acoustics are sounds generated from human activities. They include sounds generated from road traffic or sounds generated from human activities [6]. The high complexity of sound in the urban space can be attractive, enjoyable for some people, and not attractive for others, so an urban planner should respect those people who do not find the high complexity of sound attractive shown in Fig. 1 [7].

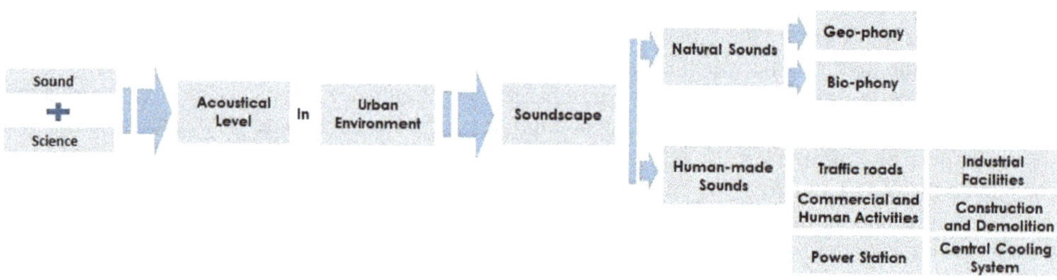

Figure 1: Soundscape sequence.

2.1 Noise and the urban environment

Sound is a means of communication, manufacturing tool, and noise [8]. Once sound starts to disturb humans and their responses, it has become an undesirable sound; therefore, it is considered a noise source [9]. There are different ways to redirect noise when we talk about controlling noise. It can be, in the path between the source and the listener, or the listener himself as shown in Fig. 2 [9]. Urban planners and designers always seek to control noise within the urban space in order to reach their aim at reducing the decibel level in the surrounding urban environment, preserve the identity of the sonic environment in each place and enhance the QOL [10]. Noise adversely affects development because of its direct and indirect harmful impact on society, life activities, public health, and human response. Thus, the development of a national plan for combating noise was mandatory, with the participation of all concerned ministries in the city, according to Cabinet directives. The plan comprised recognizing roles and duties of each ministry in light of the MSEA's (Ministry of State for Environmental Affairs, Egypt) strategy to mobilize efforts to face one of the most hazardous pollutants and identifying scientific solutions to combat it [11].

3 NOISE CONTROL PHYSICAL TECHNIQUES AND STRATEGY IN HISTORICAL SITES

There are four physical techniques to control noise pollution as shown in Fig. 2. The planning phase and the existing phase are the two relevant stages to divide these segments into. First;

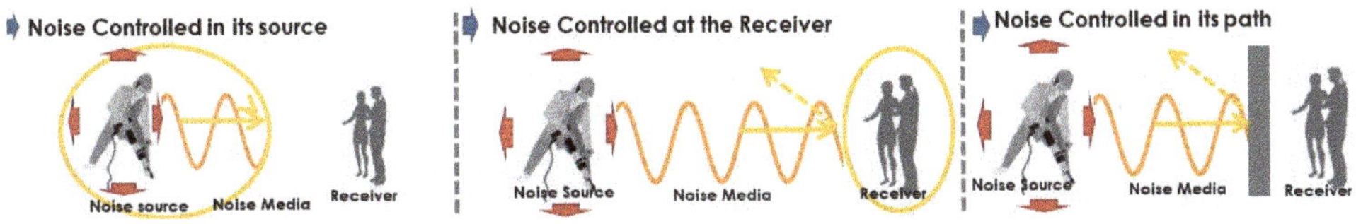

Figure 2: Noise control techniques. *(Source: Hill and Morrison [23], edited by the author.)*

acoustical site design is appropriate and starts in the planning phase of the site. That uses the arrangement of land use on a tract of land to minimize noise impacts. Acoustical site design starts with site planning. This stage control noise from its source.

Second, the acoustical architectural design is appropriate for the existing roadways in the architecture design phase. It incorporates noise reducing concepts in the details of individual buildings in the architecture phase. This stage controls the noise in the receiver. Third, in the construction phase, acoustical construction is ideal for existing streets. It entails the use of noise-reducing materials and procedures in the construction of buildings and sites. This stage regulates the receiver's noise. Fourth, between noise sources and noise-sensitive areas, noise barriers that are applied at the local level and the current situation can be erected. It manages the noise along its path [12]. Historical sites are divided into three types designed, Evolved and associative landscape. The designed landscape is clearly defined and created intentionally by man. The evolved landscape results from an initial social, economic, administrative and/or religious imperative, and has developed its present form by association with, and in response to, its natural environment. Associative landscape the inclusion of such landscapes on UNESCO's World Heritage List is justifiable by virtue of the powerful religious, artistic, or cultural associations of the natural element, rather than material cultural evidence, which may be insignificant or even absent [13]. According to the physical attributes of the historical sites, acoustical construction and noise barriers is suitable in the designed landscape, and Noise barriers are the only suitable in the evolved landscape all shown in Fig. 3. This paper is focusing on the designed history landscape.

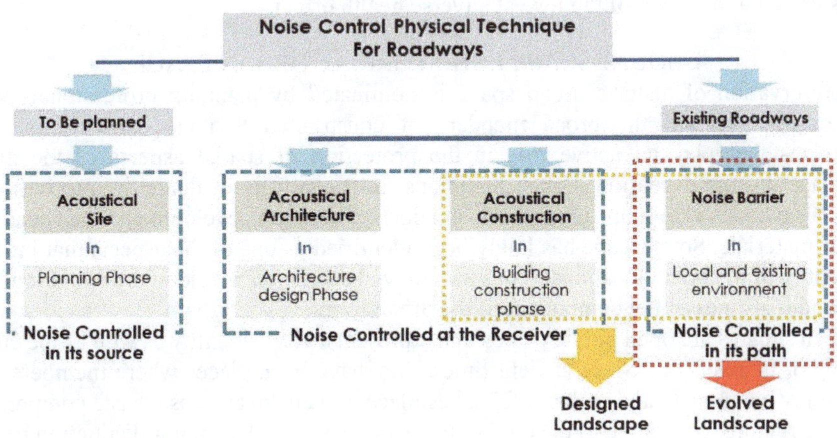

Figure 3: The noise control physical technique and stages. *(Source: [12], edited by the author.)*

The soundscape approach can be an effective way to create quiet zones that take into account not just the numerical limits of regulation, but also the acoustic comfort of the space. As indicated in Fig. 4, achieving sustainability is critical to improving one's quality of life. The question is how to prepare for noise reduction through sound design employing attention and/or acoustic masking in terms of the soundscape. The following methods can be used to reduce noise: (Planting: green walls, green roofs, green ground; Materials: using different acoustical materials; Water structure: The background of city noise is masked by the water structure; Noise barriers) [14], [15].

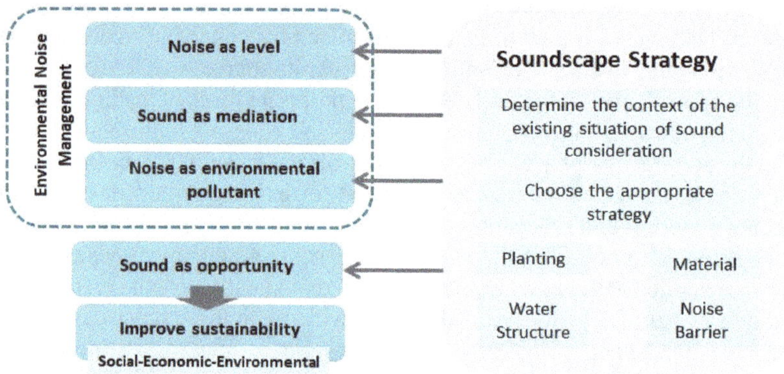

Figure 4: The soundscape design strategy [16].

In order to analyse the existing tool, there is different assessment tools should be considered along the planning process. The measurement, site survey, and questionnaire consider as the three main assessment tools that should be used in order to achieve a successful design. In conclusion, these guidelines, strategies, and assessment tools present a sound level which is a significant impact on human health and/or wellbeing is to be expected. When multiple adverse health effects are identified for a given environment, the guideline values are set at the level of the lowest adverse health effect.

4 SOUNDSCAPE IN HISTORICAL URBAN SPACE

The preservation of historic urban space is dominated by planning professionals with a predisposition toward the broad mandate of constructed heritage conservation; other professions that play an active role in the protection of spatial aspects of the historic townscape are the exception. Urban historians must carefully evaluate the motives of the numerous persons who shape urban form and document urban development when examining various materials. Soundscape has lately been identified as one of the aspects that influence the identity of historical locations. It has also played a significant role in the evolution of our cities and it is an essential component of the urban form.

Urban squares are open public places that represent a city's identity as well as the cultural heritage of its residents. Since ancient times, they have been places where members of the community congregate and "urban life" takes place. Urban squares, as a basic component of the city structure, contribute to the city's image and prestige. The major distinction between a public park and a public square, according to Levy [24], is that "on a square, citizens are not connected to manifestations of nature, but to the heart of urban culture, history, and memory" [17]. Due to changes in public space use over the previous few decades, many urban squares have lost their function and importance. As a result, creating urban squares as both public and open areas requires greater thought and effort.

In summary, the following are some basic design principles for urban squares [14], [15]:

- To create identity, sense of place and contribute to the overall city image;
- To promote public use and participation;
- To encourage social activities, communication, and social integration;
- To enhance the character of the environment;
- To create a public square that is legible, enjoyable, and long-lasting;
- To create both physically and socially accessible environments;
- To achieve environmental sustainability and low-cost maintenance through environmentally friendly design strategies;
- To promote art, cultural activities, and entertainment;
- To mitigate the impact of noise and enhance natural sounds.

People associate symbolic value with urban public squares, which are more than just physical areas. Sound is seen as a major phenomenon that shapes space and enhances a location's personality. This research shall investigate whether people's perception of the sounds in the respective heritage also reflects these differences. This is very important in order to understand the importance of sounds in preserving the authenticity of a historical place and to protect it from deterioration, similar to protecting a historical place from architectural or other changes. In total, the perception of a sound environment and can be understood as the auditory equivalence to the visual "landscape" and in an attractive way, it is rich the touristic vibes in the archaeological areas. The term soundscape denotes a sonic environment, formed within a context, as it is understood and perceived by people [18].

5 SOUNDSCAPE DESIGN DECISION MAKING

Any successful design decision making is created to achieve the goals for which it was envisaged. These goals may differ for similar design problems. As in historical sites when noise is extreme, the recession of the noise becomes a must in the design goal. Architectural and landscape architecture adapt the acoustical design, control, and analysis in the existing, new buildings and outdoor spaces. Architects may be involved in some acoustical services that can be categorized into one or more, as control of the noise and light from transportation systems, control of the noise and light related to building systems, Seismic, and vibration control, In addition, they are not involved in material and product testing, measurement, and reporting, enhancement of sound, light and electronic reinforcement [19].

The decision-making process of the landscape is divided into the planning processes and design considerations as shown in Fig. 5. These considerations should be well designed and attractive with great care and attention to detail, in order to reach a successful soundscape design [14]. Most of the design consideration along historical is an intrusion into the environment, blending neither with the landscape nor with the surrounding historical buildings for which they were built. Well typically a basic design, is squeezed, stretched, compressed, elongated, shortened and heightened, to fit the space and to realize the coveted shape and material of light and sound landscape elements. Once placed in the landscape it's viewed as a part of the environment. The landscape features and considerations such as this seem visually oppressive, out of place, and over dominant. Whereabouts in the design process, every significance should be adequate in the design to reach a high level of a suitable nightscape design in the ecology and the community in heritage sites [14], [20]. In all cases, the material, height, form, placing, colour, pattern, and visual impact are the design guides lines Architects should consider these issues in the preliminary design stages to fit the site importance.

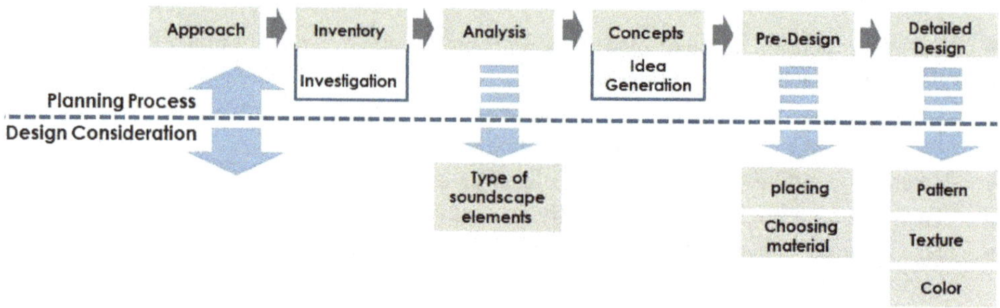

Figure 5: The sustainable planning process sequence. *(Source: [14] edited by the author.)*

In order to achieve a high level of successful soundscape in heritage sites, the approach stage combines the different assistance in the design [14]. Then it is followed by, the inventory step which is a visual analysis of the proposed site and the links that exist between the community, neighbourhood, and geographical area that are crucial steps in the process. The inventory stage can be divided into two attributes; physical and activity attributes.

The analysis stage is the part of the design process when the team's methodology can be used to review and analyse information gathered during the inventory phase. The goal is to distinguish between the community's and the local environment's design, planning, and environmental needs in connection to the proposed noise mitigation project. The concept stage is then dictated by the abstract ideas that emerge from the inventory and analysis of the current state [14], [19]. Furthermore, this step is specified by future needs and qualifications imposed by nearby human activities, site, and highway functions. After that, the concept and assessment development should be followed by the pre-design stage. It should also make use of the ideas and needs established during the creation of an appropriate preparatory design. The noise barrier must be structurally sound, motorist-friendly, long-lasting, acoustically effective, and visually appealing. Basic visual needs, location requirements, and transdisciplinary design criteria should all be considered in the design. At this point, the barrier's position, material, design configuration, and decisions should be decided. Finally, the detailed design is responsible for bringing the design concepts to life. Landscape architects and architects should offer materials, colours, textures, and grading features at this point in the design process. The goal of an appropriate noise barrier design is to use visual design concepts to lessen the visual impact. [14], [19], [20].

6 CASE STUDY (EL-TAHRIR SQUARE, ELMANSHEYA, ALEXANDRIA, EGYPT)

Alexandria is Egypt's second-largest city. It is a seaside city on the Mediterranean Sea that stretches 32 km along the coast in the country's north central direction. It is a major economic centre in Egypt that is known for its high levels of traffic noise [1]. The city of Alexandria has a population of about 4 million people, which, along with the rapid development of annual traffic and the flaws in new buildings, results in increased noise pollution [2]. After garbage disposal, water pollution, unclean and messy pollution, noise pollution is the fourth source of pollution in Alexandria [4]. El-Mansheya Square (El-Tahrir Square) consider as one of the oldest historical squares in Alexandria, which is surrounded by up to 14 historical listed buildings. The square is affected by the adjacent traffic follow and different commercial activities through the day and night which affect the QOL with sound pollution along the square.

6.1 Inventory

As shown in Fig. 6, Alexandria has three main squares, which are all located within a historical context. El-Shohada Square, El-Tahrir Square, and Saad Zagloul Square are all surrounded by historic buildings. These squares are considered as one of the most appealing destinations for locals and visitors alike. People believe walking and sitting in the squares to be one of the most motivating ways to relax [4]. El-Mansheya square was chosen as a case study zone as El-Shohad square is under renovation and Saad Zagloulis more acoustic pleasant than El-Mansheya.

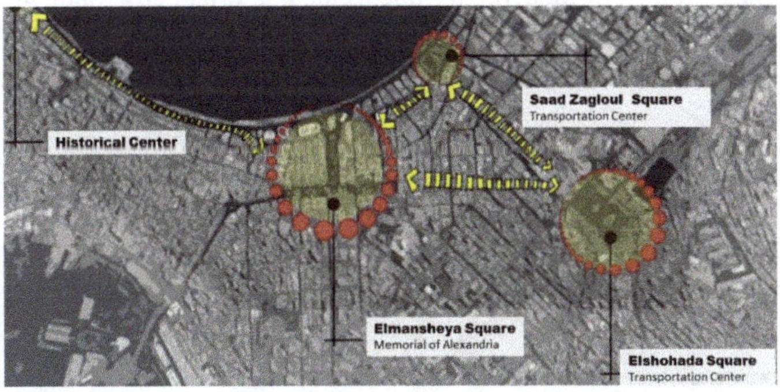

Figure 6: Alexandria squares. *(Source: [21] edited by the author.)*

The inventory step will be analysed according to site surveying of the physical attributes and activity attributes that influence the site.

6.2 Physical attributes

Different Aspects influence the site which is the width of the sidewalk where the different social activities as restaurants, shopping and street vendors, the width of the street where the source of traffic noise, the square width were different social activities, sitting area, picnics, etc. and the surrounding buildings feature shown in Fig. 7.

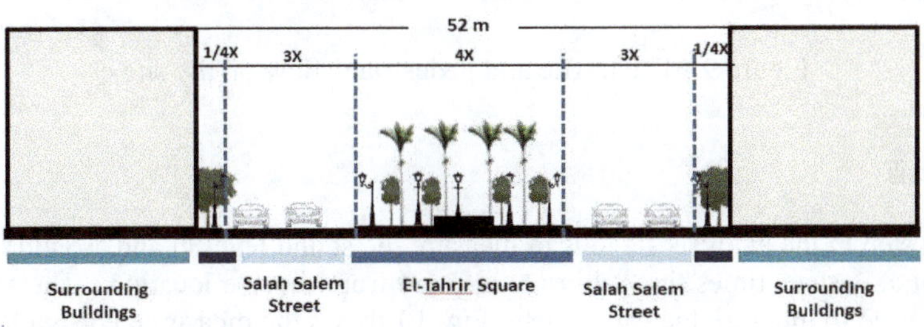

Figure 7: Physical attributes surrounding the site.

6.3 Social attributes

The site is surrounded by different activities day and night. The site is the CBD (Central Business District) and the government interest of Alexandria. As illustrated in Fig. 8, two forms of acoustics dominate the sound quality of the site: human-made acoustics and natural acoustics. Human-made acoustics are divided into two categories. The first source of traffic noise is congestion along the road, which is created by cars horns, tire friction with the tarmac, car motors, and the deflection of cars' speed on the buildings. Second, there are social activities such as social gatherings, sitting in cafeterias, and social contact that takes place around the plaza. Approximately 70% of the users are pedestrians and 30% vehicles there are two main congestion points at the intersection between Ahmed Orabi Street and Salah Salam Street. As well the intersection is between El-Sabaa Banat Street and Salah Salam Street is shown in Fig. 9. Natural acoustics are divided into two categories: wind sound and animal sound. There is also a lack of biodiversity due to the bad urban environment. Human made acoustics are always considered as unwanted sound at the variance of the natural acoustics that has a positive effect on humans psychologically.

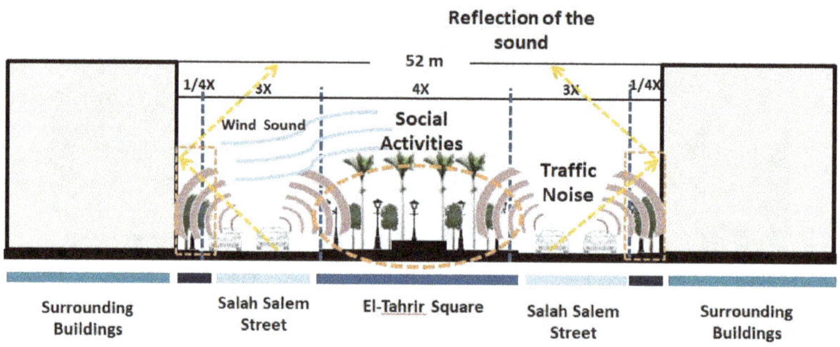

Figure 8: Activity attributes: Acoustics around the promenade.

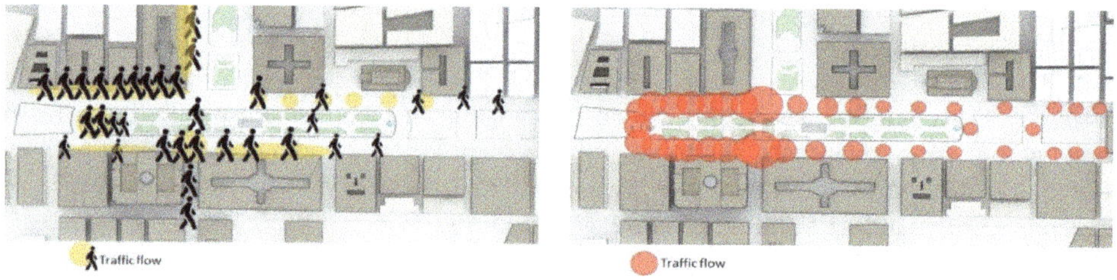

Figure 9: The traffic and pedestrians flow on the site.

6.4 Analysis

The initial step in the analysis stage is to measure the sound level in and around the square. It was shot at various times, in various zones, and from diverse locations. The sound level meter was used to measure the experiment. Fig. 10 shows the measurements gathered at ten different spots. The measurements were taken on weekdays and weekends, at various times throughout the day, as shown in Tables 1 and 2.

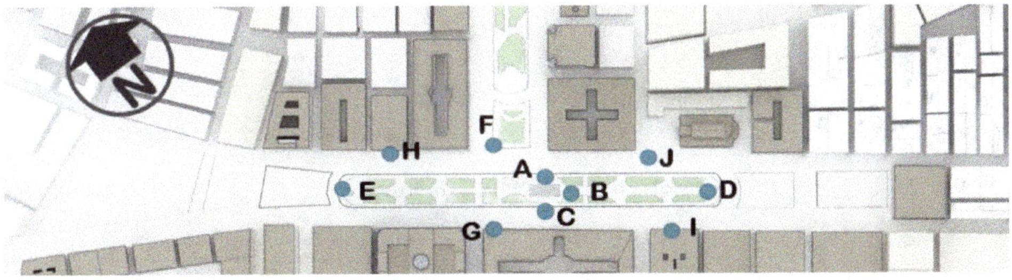

Figure 10: The measurements points.

The average sound level in and around the square on weekdays ranged from 84 to 71 decibels, and on weekends from 81 to 67 which is higher than the acceptable sound level and needs to be improved in order to improve the sonic urban environment [1]. The traffic noise map depicts the impact of traffic noise on the square as shown in Fig. 11.

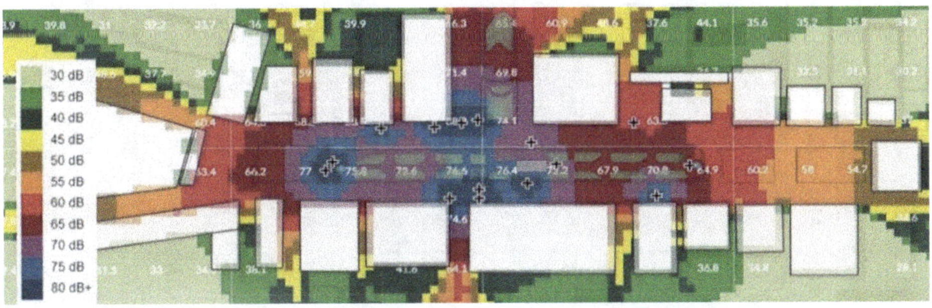

Figure 11: Noise mapping of sporting promenade. *(Source: [22], edited by the author.)*

The previous analysis clarifies that El-Tahrir square is affected by a high level of noise that disturbs the users throughout the site and the history of the site. In order to develop the surrounding sonic urban environment in the site different design considerations should be mentioned:

- Soundscape strategy should be applied in the renovation methodology
- The soundscape construction and noise barrier has been identified as the most appropriate physical technique as the site is considered a designed historical landscape
- Soundscape elements should be applied with appropriate design considerations that integrate with the site context and existing features to support its ability to be a part of the historical context without compromising any of the site's constraints,
- To reduce noise, soundscape elements and noise barriers with appropriately sized apertures for individuals arriving at the site should be installed surrounding the edges, especially in traffic jam points. It should be designed in a Roman style to match Alexandria and El-Mansheya's historical architectural style.
- It is recommended to do a questionnaire assessment in order to measure the user sonic satisfaction along with the site during the analysis step.
- To improve the natural, integrate the soundscape elements and noise barrier with various landscape components and plants,

Table 1: Measurement of sound level in El-Tahrir square in Weekdays in March 2021.

Time		A		B		C		D		E		F		G		H		I		J	
		Max	Min	Max	Min	Max	Min	Max	Min	Max	Min	Max	Min	Max	Min	Max	Min	Max	Min	Max	Min
Morning	8:00	71	64	66	60	68	62	65	60	69	64	74	71	75	71	74	71	69	64	69	64
	10:00	72	67	68	63	70	65	67	63	70	66	75	72	77	73	76	73	71	66	70	67
	12:00	81	72	73	64	75	71	71	66	81	73	81	75	83	75	83	75	81	72	80	72
	14:00	82	74	74	66	78	73	73	68	80	75	81	76	84	76	83	75	80	73	81	73
	16:00	81	74	74	65	76	72	72	67	80	74	80	75	83	75	84	75	80	73	81	74
Evening	18:00	80	72	76	66	78	71	76	65	83	74	85	75	84	75	85	77	78	67	79	71
	20:00	84	76	77	67	79	71	78	67	84	75	87	77	85	77	86	75	75	66	80	74
Night	22:00	84	76	75	69	83	74	73	66	82	78	88	82	86	80	85	81	78	70	83	77
	00:00	72	66	70	60	71	64	70	59	73	65	74	66	75	67	74	68	70	65	74	67

Table 2: Measurement of sound level in El-Tahrir square in Weekends in March 2021.

Time		A		B		C		D		E		F		G		H		I		J	
		Max	Min	Max	Min	Max	Min	Max	Min	Max	Min	Max	Min	Max	Min	Max	Min	Max	Min	Max	Min
Morning	8:00	67	62	66	60	68	62	65	60	67	62	67	71	69	61	72	61	64	60	64	59
	10:00	68	65	68	63	70	65	67	63	68	65	69	72	71	62	73	64	67	61	66	60
	12:00	72	65	70	64	75	71	71	66	74	70	80	75	71	65	76	70	72	65	70	64
	14:00	75	68	73	66	76	73	73	68	76	72	82	76	84	76	78	72	78	72	76	69
	16:00	76	70	74	65	76	72	72	67	79	74	83	75	83	75	81	75	78	72	77	69
Evening	18:00	80	72	75	66	77	71	76	65	83	74	85	75	84	75	85	77	78	67	79	71
	20:00	82	74	76	67	78	71	77	67	84	75	86	77	85	77	86	75	75	66	80	74
Night	22:00	83	75	76	67	80	74	73	66	82	78	87	82	86	80	85	81	78	70	83	77
	00:00	74	66	70	60	71	64	70	59	73	65	74	66	75	67	74	68	70	65	74	67

- Many retail shops for trading produce a lot of noise, the choice of durable, self-sufficient, and low-maintenance material is recommended.,
- Street vendors should be controlled on the site as they are a source of noise
- The adoption of higher, denser evergreen vegetation that lures birds towards the city centre should help to lessen noise. As well as a substantial wall overrun with climbers, there is a lack of biodiversity along with the site. On the front, this should be required. At the end of the analysis, it goes to some recommendations in order to improve the 3 pillars of sustainability according to the framework.

7 CONCLUSION

The impact of noise pollution on quality of life is discussed in this paper. Many scientists and planners agree that soundscape should be included as part of the planning process in order to incorporate the positive aspects of sound into the urban environment. According to the planning process and design consideration technique, the research examines the various factors that influence the sonic urban environment. The study's purpose is to create a sonic urban environment in historical sites that is compatible with existing roadways. Acoustics, sociological qualities of diverse noises, social demographic profiles of users, environmental conditions, and economic aspects are all factors to consider. It suggests that soundscape design will play an important role in the planning of cities that are both sustainable and pleasant to live in. People may develop a sonic urban environment by prioritizing safety and comfort, which implies they require a walkable city and liveability in their neighbourhood. There are certain guidelines for creating a liveable and walkable environment that lowers the usage of personal automobiles and allows for easy access with no barriers along the promenade.

The ongoing deterioration of El-Tahrir square in El-Mansheya urban environment necessitates development. The study was reviewed by pointing out the current state of the El-Tahrir square, which required to be improved in order to improve QOL in general and the sonic urban environment in particular. This applied study intends to reveal the issues in terms of the three pillars of sustainability, leading to design considerations particular to the local area. Its goal was to investigate and assess the acoustic performance of squares in Alexandria. The case-study method was semi-qualitative because the information needed for the social pillar focused around diverse users' perceptions, whereas the environmental pillar was concerned with sound measurements and the economic pillar with the specification of possible materials and processes. According to the case study, El-Tahrir Square is afflicted by a high level of sound that exceeds 80 dB during day and night, which has an impact on QOL along, despite the fact that noise levels should not exceed 65 dB according to the EEAA. Different forms of noise barriers could be optimized to enhance the human experience based on their benefits and drawbacks. Finally, it is recommended that noise management should be incorporated into the design process in order to achieve the goal of improving the soundscape, which is a significant aspect of the quality of life along the promenade. This study found that soundscape is an important feature of city planning that should be incorporated in the design to reduce the harmful effects of noise.

REFERENCES

[1] EEAA, Egyptian Environmental Law. Cairo, Egypt: Egyptian Ministry of State for Environmental Affair, 2015.

[2] WHO, World Health Organization. GHO: Interactive chart: Public health and environment (PHE): ambient air pollution, 2016. http://gamapserver.who.int/gho/interactive_charts/phe/oap_exposure/atlas.html.

[3] DataCommons, Egypt population, 2019. https://datacommons.org/ranking/Count_Person/Country/asia?h=country%2FEGY. Accessed on: 4 Aug. 2021.
[4] Numbeo, 2019. http://www.numbeo.com/pollution/rankings_current.jsp.
[5] Drever, J., *Soundscapes and Acoustic Ecology: Origins and Prospects.* The World Forum for Acoustic Ecology: London, 2005.
[6] Bendtsen, H., Noise barrier design: Danish and some European design, 2010. California: UCPRC-RP-2010-04.
[7] Elmqvist, T., Designing the urban soundscape, 2013. http://www.thenatureofcities.com/2013/08/25/designing-the-urban-soundscape/. Accessed on: 26 Feb. 2019.
[8] Hedfors, P., Site soundscapes, landscape architecture in the light of sound. Doctoral thesis, Swedish University of Agricultural Sciences, Sweden, 2003.
[9] Hällgren, N., Urban sound design: Can we talk about it? *An Interdisciplinary Journal of Sound and Sound Experience*, pp. 36–50, 2012.
[10] Beatley, T., Celebrating the natural soundscapes of cities, USA, 2013. http://sustainablecitiescollective.com/nature-cities/111251/celebrating-natural soundscapes-cities. Accessed on: 10 Aug. 2014.
[11] Irvine, K.N., Green space, Soundscape and urban sustainability. *Local Environment,* **14**(2), pp. 155–172, 2009.
[12] FHWA, *Design of Wood Highway Sound Barriers*, USDAm: United States, 1994.
[13] Hällgren, N., Urban sound design: Can we talk about it? *An Interdisciplinary Journal of Sound and Sound Experience*, pp. 36–50, 2012.
[14] Wang, K., *The Aesthetic Principles of Soundscape in Architectural Design and Built Environment.* Texas A&M University: USA, 2013.
[15] Stevens, M., *Quietening Open Spaces towards Sustainable Soundscapes for the City of London*, The City of London Corporation: London, 2010.
[16] Desarnaulds, V., Costanzo, E., Carvalho, A. & Arlaud, B., Sustainability of acoustic materials and acoustic characterization of sustainable materials. *Twelfth International Congress on Sound and Vibration,* Lisbon, pp. 1–8, 2005.
[17] Egan, M.D. *Architectural Acoustics*, McGraw-Hill: New York, 1988.
[18] VIC, *A Guide to the Reduction of Traffic Noise For Use by Builders, Designers and Residents,* Frontiers of Architectural Research, 1994.
[19] Schomer, P.D., *The Acoustical Society of America Standards Program: Organization and Scope*, The Acoustical Society of America, 2012.
[20] Kotzen, B., Plants and Environmental Noise Barriers. School of Architecture and Construction, University of Greenwich: London, pp. 265–275, 2004.
[21] Google Maps. https://www.google.com/maps/@31.2005972,29.8930479,1. Accessed on: 25 Jun. 2021.
[22] Noise tool. https://noisetools.net/dbmap/?config=1,0,20,70,95,2,4,3,0,1,1,0&grid=(10,10,70,40),1&view=40,25,16. Accessed on: 23 Jun. 2021.
[23] Hill, M. & Morrison, J., The relationship of selected community college faculty attitudes, socialization experiences, and reference group identities. *Political Science,* **1**(1), pp. 25–50, 1976. DOI: 10.1080/03616975.1976.10588061.
[24] Levy, B., Urban square as the place of history, memory, identity. *Memory of the City, Kulturklammer, Belgrade*, eds D. Drazic, S. Radisic & M. Simu, pp. 156–173, 2012. http://cdn.intechopen.com/pdfs-wm/45404.pdf.

Author index

WIT*PRESS* ...for scientists by scientists

Water Pollution XV

Edited by: ***S. MAMBRETTI****, Polytechnic of Milan, Italy and* ***J. S. PALENCIA JIMENEZ****, Polytechnic University of Valencia, Spain*

Environmental problems caused by the increase of pollutant loads discharged into natural water bodies requires the formation of a framework for regulation and control. This framework needs to be based on scientific results that relate pollutant discharge with changes in water quality. The results of these studies allow the industry to apply more efficient methods of controlling and treating waste loads, and water authorities to enforce appropriate regulations regarding this matter.

Water pollution problems are essentially interdisciplinary. Engineers and scientists working in this field must be familiar with a wide range of issues including the physical processes of mixing and dilution, chemical and biological processes, mathematical modelling, data acquisition and measurement, to name but a few. In view of the scarcity of available data, it is important that experiences are shared on an international basis. Thus, a continuous exchange of information between scientists from different countries is essential.

Papers presented at Water Pollution 2020, the 15th International Conference in the series of Monitoring, Modelling and Management of Water Pollution, are contained in this volume and highlight research works from scientists, managers and academics from different areas of water contamination.

WIT Transactions on Ecology and the Environment, vol. 242

ISBN: 978-1-78466-379-7 **eISBN: 978-1-78466-380-3**
ISSN (print): 1746-448X **ISSN (online): 1743-3541**

Published 2020 / 190pp

www.ingramcontent.com/pod-product-compliance
Lightning Source LLC
LaVergne TN
LVHW062228200926
848484LV00006B/1497